Modern Learning Theory:
Foundations and Fundamental Issues

1985

Modern Learning Theory: Foundations and Fundamental Issues

Thomas J. Tighe

New York   Oxford

OXFORD UNIVERSITY PRESS

1982

Library of Congress Cataloging in Publication Data

Tighe, Thomas J.
  Modern learning theory.

  Bibliography: p.
  Includes index.
  1. Learning, Psychology of.   I. Title.
BF318.T53   153.1'5   81-11185
ISBN 0-19-503027-3   AACR2

Printing (last digit): 987654321

Printed in the United States of America

The following sources have given permission to use quotations from the publications indicated: The American Association for the Advancement of Science, excerpt from *Science*, 1958, Vol. 128, pp. 969–977; the *Brown Alumni Monthly*, excerpt from February 1963 issue, pp. 8–14; Harper & Row, Publishers, Inc., excerpts from *The Psychology of Learning*, 1935 and revised edition 1952 by E. R. Guthrie, copyright 1935, 1952 by Harper & Row, Publishers, Inc., and from *The Psychology of Human Conflict* by E. R. Guthrie, copyright 1938 by Harper & Row, Publishers, Inc.; Holt, Rinehart and Winston, excerpts from *Behavior: An Introduction to Comparative Psychology* by J. B. Watson, copyright 1914 by Holt, Rinehart and Winston; F. S. Keller and W. N. Schoenfeld, excerpts from *Principles of Psychology: A Systematic Text in the Science of Behavior*, New York: Appleton-Century-Crofts, 1950; Alfred A. Knopf, Inc., excerpt from *Beyond Freedom and Dignity*, by B. F. Skinner, copyright © 1971 by B. F. Skinner, reprinted by permission of Alfred A. Knopf, Inc.; Lippincott/Harper & Row, excerpts from *Psychology from the Standpoint of a Behaviorist* by J. B. Watson, copyright 1919 by Lippincott/Harper & Row; Prentice-Hall, Inc., excerpts from *General Psychology in Terms of Behavior* by S. Smith and E. R. Guthrie, copyright 1921, reprinted by permission of Prentice-Hall, Inc., Englewood Cliffs, N.J., and from *Verbal Behavior* by B. F. Skinner, copyright © 1957, reprinted by permission of Prentice-Hall, Inc., Englewood Cliffs, N.J.

To the memory of my parents,
Thomas and Anna Vaughan Tighe

And for Lisa and Mark

Preface

This book provides an introduction to learning theory as it has developed through the study of animal learning. The premise of the book is that developments in learning theory must be treated in historical context if they are to be properly understood and evaluated. Theoretical controversies and "new developments" abound in the field of learning, and a historical perspective is the best way to bring into focus the enduring issues and the major continuities and contrasts in explanations of the learning process. Such a perspective, too, best conveys a sense of the main achievements of the field.

The history of learning theory shows two broad types of theoretical activity. The period extending roughly from 1925 to 1955 was characterized by the construction of global theories of learning, that is, theories intended to account for the learning of all forms of behavior in all species. The period from 1955 to the present has been characterized by the cultivation of more restricted theories intended to account for particular phenomena or subareas of learning. The separation of these two types of theoretical effort is approximate. Some important limited theories coexisted with global theory, and the past decade has seen a renewal of interest in the broad theoretical questions of learning. Nevertheless, global theory clearly has historical, and therefore, conceptual primacy.

The present volume deals with global theories of learning, including the historical forces that gave rise to global theories, the controversial issues and major research associated with them, and their relations to contemporary developments. The field of learning is properly introduced from the perspective of global theory because the philosophy, concepts, and data of global theory constitute the foundations of modern learning theory and because the fundamental questions which occupied the global theorists are with us today. While this book is primarily intended to acquaint the beginning student with these foundations and fundamental issues, it also seeks to foster in all who study learning a greater awareness of our common roots and concerns.

Many people should be acknowledged for their contribution to the writing of this book. I will always be grateful to Richard D. Walk who early encouraged me in the study of learning and who provided a model of balanced and dedicated scholarship. Colleagues at Dartmouth—Rogers Elliott, Carol Fowler, George Potts, James Rose, and George Wolford—were valuable resources and gave helpful advice on various aspects of the manuscript. The manuscript also benefited from the work of several reviewers, and in this regard I am grateful for the suggestions and comments of Richard Shull and Peter Holland. I am also very appreciative of the support provided by the editorial staff of the Oxford University Press.

Special thanks should be given two people. My colleague Robert Leaton was an unfailing source of encouragement and sound suggestions throughout the project. His knowledge of the field and cogent theoretical analyses were invaluable in resolving many a writing impasse, and I am most grateful for his generous assistance. Louise Tighe contributed in two vital ways. First, she served as a sounding board for organizational ideas and read much of the manuscript in draft form. Her contribution in these regards is probably best measured by what did *not* survive her judgment; the quality of the final product has been enhanced throughout because of her continual review. Second, she carried a considerably heavier share of our family responsibilities, thereby allowing me more time to write. In large measure, this book is hers as well as mine.

Finally, I thank my children, Lisa and Mark, for their understanding acceptance of the reduction in our time together occasioned by the writing of this book.

Hanover, New Hampshire   *T. J. T.*
August, 1981

Contents

Modern Learning Theory:
Foundations and Fundamental Issues

1
Introduction

Psychologists define learning as any permanent change in behavior resulting from experience, and the psychology, or science, of learning seeks to determine the conditions and principles which govern such changes. To put this more broadly, the psychology of learning is concerned with the ways in which we are shaped by experience. The significance of the field lies in the fact that virtually all psychological activity, whether cognitive, emotional, motivational, or motor, *is* shaped by experience, often definitively so. To a large extent, we are what we have learned to be.

Consider yourself, for example. A moment's reflection will produce numerous self-evident instances of learning in activities ranging from the most mundane to those vital to your well-being and happiness. You have learned to tie your shoelaces, to drive a car, to dance, to avoid certain foods, to distinguish and label an enormous number of features of your environment, to feel guilty about some actions, to inhibit emotional reactions, to cooperate with others in certain situations and to compete in others, to value success, to fear failure, and so on and on. In these and countless other such instances some enduring change in your makeup was brought about as a consequence of interaction with your environment.

Consider, too, the role of learning in your future development. Whatever your aspirations and goals, their achievement depends upon the elaboration and modification of your existing knowledge, skills, and habitual ways of behaving—the continual fashioning of yourself, so to speak, through appropriate experiences to meet particular demands imposed throughout the full range of your abilities.

Consider, finally, our hopes for the development of human society as a whole. Efforts in this direction arise only because we believe that the psychological makeup of human beings is not fixed so as to produce a constant ratio of good and ill in society. Rather, we feel that we are capable of achieving a more harmonious and beneficial society if we but more fully understand and better control the conditions which direct our development.

The contribution of experience is thus manifest at every level of our individual and social being. Of course, the effects of experience are always conditioned by the organism's physical (genetic) makeup. But it remains true that the full potentialities of any organism are realizable only through interaction with its environment. Experience is not only the best teacher, it is the only teacher.

These reflections suggest the vast scope, complexity, and significance of the topic of learning, and they indicate as well the difficulty of coming to grips with the problem of learning. How does one attack the question of how organisms change through experience when such change can occur at so many levels of psychological activity and can be studied in such varied environmental and behavioral contexts? One answer is to begin by studying the learning process in lower organisms. The argument for such an approach is that the fundamental principles of learning might be detected more readily in simpler life forms, particularly in view of the precise control which can be exercised over the life history and environment of such subjects. Basic principles uncovered in this fashion might then be applied in analysis of more complex phenomena of learning. The study of learning in lower animals is, in fact, one of the major approaches taken by psychologists, and until quite recently it could be fairly said that our views of human learning were largely based upon information derived from the study of learning in a relatively small number of infrahuman species.

The beginning student of the psychology of learning is likely to be disturbed by the extent of the psychologist's reliance upon lower animals as a source of information about human behavior. The student's interest is to better understand human behavior, and he or she is likely to be disappointed when the subject must be approached from the seemingly remote point of animal behavior. But the student's reaction in this regard is no different from that of most psychologists who themselves were likely to have undertaken the study of learning in order to better understand their own behavior or that of others, and who only later accepted the study of animal behavior as an effective means toward that goal.

Strictly speaking, human beings are members of the animal world, but the study of learning in infrahumans has become familiarly known as the field of "animal learning." Animal learning can, of course, be viewed as a subject of considerable interest in its own right, but it is clear that within psychology the study of animal learning has always been viewed as an important route to understanding human behavior. In the following chapter we shall discuss a particular combination of historical circumstances which gave impetus to the study of animal behavior toward this end and which assured it a prominent place in the psychology of learning. But at this point it may be useful to advance two considerations bearing on the question of the relevance of research with animal subjects.

First, if one accepts the basic assumption of psychology that humans are part of the natural order, then there is every reason to expect that the psychological processes of humans and those of lower organisms might exhibit some important common features. In this regard, it is interesting to note that we appear to be considerably less reluctant to accept the assumption of a continuity of neurophysiological processes between humans and lower animals than to accept the assumption of a continuity of psychological processes. Studies of neurophysiological processes in animals are generally more likely to be viewed as applicable to humans than are studies of parallel psychological processes. For example, we are not disturbed when we learn that conceptions of the nature of neural transmission in man are based largely on research on infrahuman nervous systems, yet studies of learning are likely to be viewed as specific to the species involved, even though learning appears to have its biological basis in the phenomena of neural transmission. Logically speaking, one form of extrapolation is as sound as the other. And indeed, continuity of neurophysiological processes implies a continuity of psychological processes.

The second consideration is that an approach through animal learning should be judged primarily by how well it works rather than by intuitive impressions of its appropriateness. The verisimilitude of a scientific model is irrelevant to evaluation of its usefulness. Has the study of animal learning brought into focus truly significant and general conditions of learning? Are explanatory concepts derived from animal research helpful in the analysis and understanding of complex human behavior? Are we, as a consequence of this approach, better able to predict and control our own behavior and that of others? These are the kind of questions that bear most pertinently on this issue. Students will form their own answers during the course of this book. At this point we simply urge an open mind on the issue, confident that affirmative answers will then be forthcoming.

The psychology of learning has of course proceeded along lines other than the study of animal behavior. In fact, techniques permitting objective, experimental study of rote verbal learning and memory were introduced around the turn of the century within a few years of the beginnings of experimental analysis of animal learning. During the first half of this century, these two areas, animal learning and verbal learning, formed the major concerns of learning psychologists. However, the study of verbal learning was a relatively self-contained endeavor in the sense that the data and concepts from this line of inquiry did not lend themselves to broad analysis of learning. The study of animal learning, on the other hand, has been associated from the outset with efforts to develop comprehensive theories and principles of learning applicable to all forms of behavior change. Throughout this period, then, broad theoretical and applied analyses of human learning proceeded not from the research most directly and uniquely concerned with human learning, but from research on animal learning.

The past decade has provided clear signs of a change in this conceptual state of affairs. There has been tremendous increase in research with human subjects, this increase reflecting in large part the application of techniques and principles developed in the study of animal learning. And recent developments arising within the tradition of research on verbal learning and memory are now exerting a major influence on general conceptions of the learning process. Nevertheless, the study of animal learning remains a major source of knowledge about the learning process, and familiarity with the field of animal learning and with the body of theory arising from that field is indispensable to understanding how psychology has answered the question of how we are shaped by experience.

Our coverage of theory focuses upon efforts to construct global, or all-encompassing, theories of learning. But before turning to the theories themselves we must consider the historical conditions that gave rise to global theory, since the rationale, aims, and even the content of the theories were strongly determined by those conditions.

2
Historical Background
of Modern Learning Theory

Our starting point, like any starting point in the history of science, is a somewhat arbitrary one. It is the nature of science to obscure the beginnings of its major ideas and movements. Even when significant conceptual developments are widely accepted as the work of one or but a few individuals, close examination is likely to reveal the contribution, direct or indirect, of many earlier and perhaps unheralded investigators. Broad movements or programs within a field tend to develop continuously out of the successes and failures of previous approaches, rather than as the result of a single formulation. In short, science proceeds by collective and cumulative effort and thus significant advances in theory and research are apt to be fed by many sources near and remote in time. So it is with the beginnings of modern learning theory.

By 1925, the techniques of study which were to provide the empirical basis of modern learning theory had long been available, and in fact earlier application of these techniques had established principles of learning which were to be central to later theories. Moreover, the basic rationale of modern learning theory, its broad conception of the subject matter, and its descriptive vocabulary had all been formulated within the psychology of the early 1900's. And animal psychology itself had been an active and respected enterprise since the turn of the century. But it was not until the period extending roughly from 1925 to 1950 that these methodological and conceptual ingredients became fully fused into a new approach to the study of learning, an approach which commanded the allegiance and enthusiastic efforts of a large number of psychologists. For the first time, the science of learning became something other than the insights and research programs of individuals or small groups working in relative independence; it became instead a broadly mounted effort to understand the learning process by application of commonly accepted analytical methods.

What was the nature of this approach which found such widespread acceptance among psychologists of the 1920's to 1950's? To answer this question requires consideration of three earlier developments in psychology which, in their later combination, define the essential features of the approach. These developments were the discovery of conditioning, the rise of behaviorism, and the emergence of animal psychology as a scientific discipline.

CONDITIONING

The term *conditioning* refers to two procedures for producing changes in behavior and in the laboratory. One procedure, designated *classical conditioning*, was introduced and studied extensively by the Russian physiologist Pavlov. The other procedure stems primarily from the work of the American psychologist Thorndike and is termed *instrumental conditioning*. Both conditioning procedures were introduced around the turn of this century.

In classical conditioning the behavior change is brought about by pairing a stimulus which reliably elicits a particular response with a stimulus which is neutral with respect to that behavior. The standard illustrative experiment is conditioning of the salivary response in dogs. The placement of meat powder in a dog's mouth unconditionally elicits salivation, but if, say, a tone is regularly paired with the placement of meat powder, then it, too, comes to elicit salivary flow. Pairing of the tone and meat powder has brought about a change in the dog's response to the tone.

In instrumental conditioning the behavior change is produced by making a particular stimulus event contingent upon occurrence of a given behavior. The standard illustrative experiment is conditioning of bar pressing in rats. A hungry rat is placed in a small, plain box containing a movable bar or lever which the rat is likely to accidentally depress in the course of moving about. If it is arranged that each depression of the bar results in delivery of a bit of food to a point inside the box where the animal is likely to find and eat it, then bar pressing soon comes to be emitted at a high rate. There has been an increase in the behavior instrumental to receipt of food.

The seeming simplicity of these procedures and of the behaviors to

which they are usually applied, coupled with their intuitive obvious-
ness as ways of changing behavior, makes it difficult to appreciate
their great value in the study of learning. In the course of the past 70
years, each of these procedures has been applied to a variety of behaviors in many species and in relation to an enormous number of
variations in particular conditions of training. This work forms the
core of the experimental psychology of animal learning and has had
a profound influence on all of psychology. Why have these procedures merited such attention and effort? Why is it that psychologists,
and particularly psychologists of the 1920's to 1950's, have tended to
see conditioning as the key to understanding learning?

First and foremost, the conditioning procedures give learning psychologists a means of reliably producing in an efficient, objective,
and standard way an instance of the phenomenon they seek to understand. Conditioned responses have been observed to remain virtually intact in animal subjects despite an interval of several years
between conditioning and retesting. Conditioning meets the criteria
defining *learning*, then, in that the behavior changes are long-term
and result soley from practice or experience. These persistent changes
are produced in a relatively easy and rapid fashion, and the outcomes
of conditioning experiments can be verified by any investigator.

Procedural Generality of Conditioning

Procedures for producing learning in the laboratory, no matter how
advantageous from the viewpoint of method, would be of little interest if they were of limited applicability, but conditioning has proved
to have great generality. Classical and instrumental conditioning have
been successfully applied to organisms ranging from the simple earthworm to man, and few psychologists would quarrel with the assertion
that *any* organism within this phylogenetic range can be conditioned
by either procedure. The variety of responses and stimuli which have
been employed in conditioning experiments is so large as to make an
exhausitve listing infeasible, but consideration of several samples of
each type of experiment should help make clear the broad behavioral
and situational generality of conditioning. Let us first consider some
instances of classical conditioning taken from published experiments.

A two-second presentation of a bright light unconditionally elicits a rearing and withdrawal movement in the anterior segment
of an earthworm. A six-second presentation of a vibratory stim-

ulus (mild vibration of the surface on which the worm rests) is initiated four seconds prior to light onset. One hundred such pairings given at 50-second intervals result in a significant and linear increase in the rearing-withdrawal reaction upon onset of the vibratory stimulus *only*.

A rat is placed in a restraining device and its tail fitted with electrodes to allow delivery of a moderately intense one-second shock. This stimulus reliably elicits a change in the rat's heart rate of the order of 100 beats per minute. Ten seconds prior to the onset of shock a light is turned on immediately in front of the subject and remains on for 11 seconds, terminating with the shock offset. After only a few such pairings of the light and shock at two-minute intervals, presentation of the light alone is found to produce a drop in heart rate of about 70 beats per minute.

Insertion of a pacifier into the mouth of a three-day-old human infant reliably elicits sucking. A 16-second tone is initiated one second in advance of the pacifier presentation, and tone and pacifier are withdrawn at the same instant. After 20 such pairings separated by one-minute intervals, presentation of the tone alone is found to elicit sucking responses.

College students are exposed to pairings of a nonsense syllable, such as *yof*, with many different meaningful words having a common positive evaluative meaning, e.g., *pretty, sweet, healthy,* and *love*. Words having a common negative evaluative meaning component, e.g., *thief, bitter, ugly, sick*, are paired with another nonsense syllable, *xeh*. Following this experience, the subjects are asked to rate the nonsense syllables on a scale of "pleasantness-unpleasantness." *Yof* is now found to be rated as significantly more pleasant than *xeh*.

Despite the vast range in organismic and behavioral complexity in these examples, the learning in each case is effected by the same basic procedure—the pairing of a stimulus which unconditionally elicits a behavior with a stimulus which is initially indifferent with respect to that behavior. In each case the alteration in behavior takes the same *general form*, namely, the initially neutral stimulus acquires the power to evoke a reaction resembling that evoked by the unconditioned stimulus. This is not to say, of course, that learning proceeds in identical fashion in each of these cases, or that the stimulus pairing pro-

cedure necessarily involves the operation of precisely the same processes in each case. But these illustrations do make clear that the
procedures of classical conditioning can be used to produce and study
changes in both simple and complex activities over a wide range of
subject populations.

A similar case can be made with respect to instrumental conditioning. Consider the following examples, which are also taken from
actual experiments.

A small T-shaped maze was fashioned from clear glass, its stem
approximately 7 inches long and each arm about 3 inches long.
A land snail was placed in the maze at the base of the stem and
immediately beneath a bright light which drove the negatively
phototropic (light-aversive) subject toward the choice point.
Whenever the snail turned into the left arm, it was punished by
a slight rise in floor temperature brought about by current flow
through an embedded coil. If the snail turned into the right arm,
it reached a dark box and was rewarded by being left in the dark
for a short time. Under this procedure, the initial "runs" through
the maze from the start point to the dark box were highly circuitous and included many errors. But several months' training
at the rate of two trials per day resulted in sustained errorless
performance with a fourfold reduction in "running" speed.

A rat is placed in an enclosed, 3-foot-long alley with a hurdle at
its midpoint. The rat is simply left to its own devices over a two-
hour test period, and no effort is made to deliver reward or punishment. The rat is observed to immediately jump the hurdle
into the side opposite its initial placement and then, several seconds later, to recross the hurdle. Hurdle-jumping from side to
side continues in this manner at a relatively constant rate of about
once every ten seconds and persists, despite the effortfulness of
this response, throughout the entire test period. What accounts
for such compulsive, "neurotic" behavior? The behavior was acquired as a result of several previous training sessions in which
grid floors in each compartment were wired so that every hurdle
jump postponed onset of foot shock for 20 seconds.

An experimenter leans over the crib of a three-month-old infant
and looks at the baby with an expressionless face for three minutes. Another experimenter, hidden from the infant's view, records the number of discrete voice sounds the baby makes in this

period. A number of such observations are made over a two-day period to establish a baseline rate of vocalizing. On the following two days, the observations are repeated, except that now whenever the baby vocalizes the experimenter smiles at the baby, emits clucking sounds, and lightly touches the infant's abdomen. This "social reward" lasts about one second, and at other times the experimenter assumes an expressionless face. On the following two days, the observations of vocalization are continued but once again under the condition of the first two days, i.e., without social reward for vocalization. When these procedures were applied to a group of 21 infants, the mean daily rate of vocalization under reward increased 86 percent over that observed under the baseline treatment. When reward was withdrawn during the final two days, the rate of vocalization declined to the baseline level. Clearly, social vocalization of infants, and more generally their social responsiveness, may be modified by responses adults make to them.

College students were given a series of ten-minute interviews in a quasi-therapy situation during which they were asked to describe spontaneously their personality characteristics and traits without questions or comments from the interviewer. For one group, the interviewer said "mm-hm" and nodded his or her head whenever the subjects evaluated themselves in a negative way, while for another group this feedback was omitted. The interviews were tape-recorded and later scored for number of negative self-references. Over the series of interviews, the "mm-hm" group showed a significant increase in frequency of negative self-references, while the control group showed no change. Moreover, subsequent questioning of the "mm-hm" subjects indicated that the manipulation of their verbal behavior was accomplished without their awareness.

Again, we see the operation of a common procedural principle in widely differing instances. In each case learning is produced simply by manipulating the consequences of a given behavior. More specifically, an increase in a given behavior was brought about by imposing a contingency between that behavior and either the presentation of a positively valued stimulus or the withdrawal or postponement of a negatively valued stimulus. Instrumental conditioning, then, like classical conditioning, is a means for experimentally modifying an enormous variety of behaviors.

General Characteristics of Conditioned Behavior

The generality of conditioning is more than a matter of widespread applicability of procedures. In addition, conditioned behavior itself exhibits a number of important general properties. Many of these properties were delineated by Pavlov in the course of his experiments on conditioning of the salivary response and were later found to characterize not only other classically conditioned responses but instrumentally conditioned behavior as well.

The most important characteristics of conditioned responses center around manipulation of the *reinforcing stimulus* within each type of conditioning. By reinforcing stimulus is meant the stimulus which is primarily identified with strengthening of the new stimulus-response relation. Thus, in classical conditioning the unconditional stimulus (UCS) is the reinforcing stimulus in that its presentation following a neutral stimulus is observed to strengthen the relation between that neutral stimulus and the behavior elicited by the UCS. The reinforcing stimulus in instrumental conditioning is the event that is made contingent upon the desired behavior and observed to bring about a change in that behavior. In general, the reinforcing stimulus in instrumental conditioning may be any stimulus which the animal ordinarily approaches or avoids.

Withdrawal of the reinforcing stimulus from either the classical or the instrumental conditioning sequence results in a weakening of the conditioned response. If the reinforcing stimulus is consistently withheld, the conditioned response eventually disappears and is said to be *extinguished*. The degree to which a conditioned response resists extinction is affected by a number of training variables, but none is more important than the conditions of reinforcement during training (e.g., the nature of the reinforcing stimulus and the frequency of its application).

Although extinction of conditioned responses is generally accomplished quite readily, two characteristics of extinguished responses point to long-term effects of conditioning. First, if an animal is reintroduced to the conditioning situation some time after extinction of the conditioned response, the response is likely to immediately reappear in considerable strength, a phenomenon known as *spontaneous recovery*. However, such recovery is short-lived, and unless the experimenter renews the reinforcing operation the conditioned response soon disappears again. Second, the *reconditioning* of an extinguished conditioned response is generally accomplished more rapidly than the original conditioning. A related fact is that a second extinction of a conditioned response is also likely to proceed more rapidly than the first extinction. In general, successive reconditionings and re-extinctions tend to occur with increasing rapidity, testifying to cumulative and permanent effects of conditioning experiences.

Another important characteristic of conditioned behavior is that it may occur in the presence of stimuli other than those prevailing during conditioning. Such *generalization* of conditioned behavior is proportional to the degree of similarity between the stimulus settings. Thus, for example, if a drop in heart rate has been conditioned to a 1,000-cps (cycle per second) tone, then a similar but smaller drop in heart rate would be likely upon presentation of an 800-cps tone, a still smaller drop upon presentation of a 400-cps tone, etc. Similarly, a conditioned bar press response would be likely to persist, although in diminished form, despite considerable alteration in specific aspects of the conditioning situation, e.g., in size or color of the apparatus. Generalization, then, greatly broadens the effects of conditioning experiences since through generalization conditioning may influence behavior in any subsequently encountered situation having stimulus properties in common with the conditioning situation.

Conditioned behavior can be made stimulus-specific despite the phenomenon of generalization. This is accomplished by withholding the reinforcing stimulus in the presence of those stimulus complexes which occasion the generalized behavior while at the same time continuing to present the reinforcing stimulus in the presence of those stimuli which define the conditioning situation. Thus, in the heart rate example noted above, the generalized responses to tones other than the 1,000-cps tone could be eliminated by alternating nonreinforced presentations of these tones with reinforced presentations of the 1,000-cps tone. Under such procedures, conditioned responses can be brought under very precise stimulus control, and the organism is said to have formed a *stimulus discrimination*.

There are two important points to bear in mind about the foregoing features of conditioning. The first is that they characterize *all* classical and instrumental conditioned responses, regardless of the species or behavior concerned. The second is that they can be shown to have important adaptive significance. Consider generalization and discrimination. Some degree of generalization is essential if learned behavior is to be useful to the organism, since it is most unlikely that any learning situation recurs in exactly the same form. Generalization frees learning from a rigid dependence upon the specific stimuli prevailing during the learning experience and makes likely the continued occurrence of the learned behavior in the face of normal fluctuations in the environment. On the other hand, persistent generalization of learned responses to similar but nonreinforced situations would clearly be maladaptive to the organism. Discrimination provides the mechanism whereby such inappropriate generalized responses are eventually eliminated from the organism's repertoire. The functional significance of these and other aspects of conditioning were early perceived and very persuasively argued by Hull (1929).

A number of additional common features of conditioned behaviors could be cited here, but those listed should be sufficient to establish that conditioning exhibits regularities of a fundamental, lawful, and adaptive nature. As such, conditioning is clearly more than simply a reliable means of producing behavior changes; it provides a model suitable to study of the major aspects of learned behavior in all species.

The psychologists of the 1920's and 1930's were well aware that conditioning might prove a powerful tool in the study of learning and were therefore concerned to assess the reliability and generality of conditioning phenomena. This attitude is nicely illustrated in an exchange of papers centering on some early observations of generalization by Anrep (1923). Working in Pavlov's laboratory, Anrep conditioned dogs to salivate to a tactile stimulus applied to a specific point on the skin and then tested generalization of the conditioned response by applying the tactile stimulus to other points on the skin. Figure 2-1, reproduced from Anrep's report, shows the magnitude of one subject's salivary response to stimulation of the point used in conditioning (point zero) and to stimulation of points progressively more distant from the locus of conditioning (points II, VI, etc.). Response magnitude is expressed as the percentage of the salivary flow to the conditioning stimulus (point zero). The two curves depict the outcome of generalization tests conducted early (upper curve) as opposed to late (lower curve) in the testing sequence. Although there was a general decline in response magnitude from the early to the later testing, the important point is that both tests provide clear evidence of generalization of the conditioned salivary response, with the magnitude of the generalized responses decreasing with increasing distance from the locus of conditioning. These data and similar data from several other subjects were interpreted by Anrep as establishing that tactile conditioned responses show a spatial gradient of generalization.

However, on the basis of statistical evaluation of these and related data emanating from Pavlov's laboratory, Loucks (1933) argued that convincing evidence for the generalization of such responses had not, in fact, been provided. In particular, Loucks's analysis seriously ques-

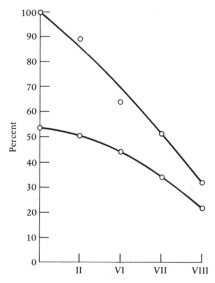

Figure 2-1. Generalization of a tactile conditioned salivary response in an individual subject.
(After Anrep, 1923, p. 412. By permission.)

tioned the reliability of Anrep's data, which had stood as the clearest demonstration of generalization. Concerned about the doubt which Loucks's analysis cast upon the conditioning model, Bass and Hull (1934) undertook a careful experimental check on Anrep's results. In introducing their study, Bass and Hull noted,

> Even if only apparent, such a disagreement, left too long in a state of uncertainty, might easily have an unfortunate influence on the development of the branch of science particularly involved. It was partly with a view to assisting in the clearing up of this doubt and partly to secure evidence bearing on the comparative psychology of the irradiation or generalization of tactile conditioned reflexes, that the present investigation was undertaken. (p. 47)

In Bass and Hull's experiment, college students were tested for generalization of a conditioned galvanic skin response (GSR). The GSR is a change in electrical resistance of the skin resulting from activity of the sweat glands. Like the salivary response, the GSR is mediated by the autonomic nervous system and is an involuntary response. Using a brief shock to the wrist as the UCS, the GSR was conditioned

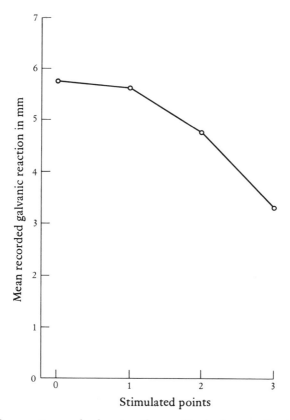

Figure 2-2. Composite graph showing the gradient of generalization of a tactile conditioned galvanic skin reaction in a human subject, based on eight subjects. Zero is the point conditioned. Each point represents the mean from 256 measurements.
(From Bass and Hull, 1934, p. 60. Copyright 1934 by the American Psychological Association. Reprinted by permission.)

to a tactile-vibratory stimulus applied to a specific point on the body, e.g., the shoulder. Then, following Anrep's procedure, tests for generalization were given by applying the tactile stimulus to three new points, e.g., the back, thigh, and calf. Figure 2-2 shows the average magnitude of the GSR as the point of stimulation was moved progressively farther from the locus of conditioning (point zero). The data of Figure 2-2 are clearly in agreement with those of Figure 2-1, and the thoroughness of Bass and Hull's observations left no doubt as to the statistical reliability of their gradient of generalization. Bass and Hull were at pains to underscore the striking similarity between their data and those of Anrep and emphasized that "the present results are in complete accordance with the findings of Anrep, despite the wide difference between the two investigations in the subjects used, the stimuli employed, and the reactions measured" (p. 65). It was such observations of the generality of conditioning processes that convinced psychologists they were on the right track in approaching learning through the study of conditioning.

Conditioning and Complex Behavior

Still another, and major, attraction of conditioning as a means to study learning is that conditioning has a good deal of face validity for the description of human learning in natural life situations. That is, many instances of our everyday learning can easily be thought of in terms of conditioning processes. For example, consistent with the basic principle of classical conditioning, we know that we may come to fear the sound of the dentist's drill, that we may salivate copiously at the sight or even the mere thought of food, that we are likely to respond in anticipation of the starter's gun, that words like *radical* and *gay* evoke in many individuals the emotional reaction elicited by the behaviors they denote. Consistent with the basic principle of instrumental conditioning, we know that we are likely to repeat activities which yield immediate gratification even when they imply more remote aversive consequences, that we may habitually avoid physical or social situations which have been associated with discomfort in the past, that at almost every turn our activities are subject to rewards and punishments, both subtle and obvious, designed to encourage socially approved forms of behavior.

Everyday learning also shows parallels with particular properties of conditioned responses. Thus, for example, consistent with the phenomenon of generalization of conditioned responses, we commonly observe that a child who has been bitten by a dog will now fear other animals as well. Similarly, a child who has learned to label a cat "kitty" may also call any other four-legged, furry creature "kitty." We know, too, that an effective way to eliminate an acquired fear of dogs would be to expose the child to these animals for several sessions while ensuring that the original fear-evoking event, i.e., the UCS, does not recur; and we know that generalized inappropriate verbalizations such as the "kitty" response will gradually be eliminated as a result of feedback from parents and teachers, i.e., as a result of discrimination training.

Specific conditions affecting the strength of conditioned responses also find analogs in natural life situations. A particularly striking example is concerned with the pattern of reinforcement during conditioning. It is known that conditioned instrumental behaviors become extremely resistant to extinction when the reinforcing stimulus is applied in such a manner as to bring about a high but variable ratio of behavior to reinforcement. Such a schedule closely resembles the schedule of payoffs found in "addictive" gambling games, such as slot machine playing.

Such analogs of conditioning phenomena abound, and the foregoing illustrations could be extended indefinitely. To pursue these analogies is to run the risks of exaggerating the comparability of conditioning and learning in the natural environment and of oversimplifying their relationship. But there is also a distinct value in drawing parallels between conditioning and everyday learning situations, particularly when the parallels are not merely isolated points of agreement but make up a coherent set of relationships. In such cases, the analogies provide more than just a comforting reassurance that conditioning has some general relevance; rather, they provide an immensely useful guide to the study and control of human learning.

An early illustration of this usage of conditioning is provided by Symonds (1927), who sought to call the attention of educators to some important implications of conditioning for classroom learning. Drawing only on the data of Pavlovian conditioning, Symonds listed 23 laws of conditioning and showed that many of these dictated some clear course of action in the classroom. For example, one law of conditioning is that conditioned responses develop when the UCS follows the neutral stimulus but not when the UCS precedes the neutral stimulus. One implication which Symonds drew from this fact for educational procedure is that in learning an association the unfamiliar element should be presented first. For instance, in learning words in a new language, the word in the foreign language should be presented before its equivalent in the vernacular. In such fashion, Symonds was able to generate a fairly comprehensive list of useful suggestions for effective teaching procedures.

The value of analyses of this type is not a question of whether one has arrived at the best possible set of training procedures for the tasks in question. Their primary value is that they enable a systematic approach to the enormous complexities of learning in natural life situations. In particular, they provide a means of zeroing in on conditions which are likely to prove significant for the learning involved, and they provide a rational basis for decisions between alternative courses of training. Ultimately, of course, the usefulness of "analysis by analogy" is a question of whether one is in fact able to deal more effectively, both experimentally and in the field, with the learning problem in question. By this criterion, there is no question that the data of conditioning have provided, and continue to provide, a useful framework for dealing with many practical problems of human learning, and this assertion will be documented on numerous occasions in the course of this text.

To recapitulate, we have pointed out features of conditioning which made this approach particularly attractive to learning psychologists. Conditioning gave the investigator standard procedures for producing and studying learning under full experimental control. These procedures were applicable across a wide range of species, developmental levels, and behaviors. All conditioned behaviors exhibited common properties which appeared to be of fundamental biological or adaptive significance. Finally, the principles of conditioning not only were applicable to simpler forms of learning but promised to aid understanding of complex human learning as well.

These features of conditioning are embodied in a famous experiment reported by Watson and Raynor in 1920. Basically, Watson and Raynor asked whether conditioning could be used to bring about significant forms of human emotional learning. To answer this question, they attempted to classically condition a fear reaction in a young child. The child, Albert, was selected for testing because he was a particularly well-developed, healthy, stable, and unemotional child; the experimenters felt they could do him little harm by their relatively innocuous procedures. At the age of nine months, Albert was tested for reaction to a variety of objects suddenly presented to him one at a time. These included such stimuli as a white rat, a rabbit, a dog, a monkey, cotton wool, and masks. Albert's usual reaction was to immediately reach for the objects and play with them. At no time did he show any sign of fear to these stimuli. At this age, a test was also made to determine if a fear reaction could be elicited by a loud sound, a stimulus which the experimenters had found effective in eliciting fear reactions in other young children. The sound was made by striking a hammer upon a steel bar suspended behind the child's back. Upon first hearing the unexpected loud sound, Albert "started violently, his breathing was checked and the arms were raised in a characteristic manner. On the second stimulation the same thing occurred, and in addition the lips began to pucker and tremble. On the third stimulation the child broke into a sudden crying fit" (p. 2). This was the first time an emotional situation in the laboratory had produced any fear or even crying in Albert. Having now determined a stimulus which unconditionally elicited a marked fear reaction and having also a variety of objects known to evoke positive reactions, Watson and Raynor were in a position to attempt to condition fear to one of the latter stimuli.

Conditioning was initiated at the age of 11 months and three days. The white rat was presented, and Albert predictably reached out to touch and play with it. Just as his hand touched the animal the bar was struck behind his back. After two such experiences Albert was whimpering, and in order not to disturb the child too seriously no further tests were given for one week. Upon presentation of the rat one week later, the effect of the earlier stimulations were apparent in that Albert initiated a reaching movement but uncharacteristically withdrew his hand before it touched the rat. There followed seven systematic conditioning trials in which the rat and loud noise were presented together. When the rat was subsequently presented *alone*, Albert immediately began to cry, turned sharply, fell over, and began to crawl rapidly away, a reaction which Watson and Raynor termed "as convincing a case of a completely conditioned fear response as could have been theoretically pictured" (p. 5).

About one week later, tests for generalization of the conditioned fear response were undertaken by presenting a variety of objects which had previously been shown to evoke approach responses or other positive reactions. In these tests, a rabbit was now found to elicit virtually the same fear reaction as that given to the rat. A dog also now frightened Albert, and a fur coat made him cry. He avoided a cotton wool ball, and he was even "pronouncedly negative" when shown a Santa Claus mask. In contrast to these reactions, Albert continued to play happily with blocks whenever these were presented in the test situation. Apparently, the fear response generalized only to stimuli having a furry quality.

Following these tests, a few reconditioning trials were given and Albert then had no further exposure to any of the experimental stimuli throughout the following month. Tests at the end of this period showed that the conditioned emotional reactions persisted, although with some loss in intensity.

In discussing their findings, Watson and Raynor pointed out that their data clearly suggest that many of the phobias in psychopathology are in fact conditioned emotional reactions of either the direct or the generalized type.

By present-day standards, the Watson and Raynor experiment is lacking in methodological precision, and as we shall see in later chapters, current conceptions of the nature of conditioning differ considerably from the simple response-transfer notion held by Watson and Raynor. Nevertheless, the Watson and Raynor experiment stood as a bold and dramatic demonstration of the generality of conditioning procedures and phenomena and of the relevance of conditioning to significant problems in human behavior. In view of such observations, it is understandable that conditioning came to be seen by many learning psychologists as the key to understanding learning.

BEHAVIORISM

Behaviorism refers to a particular set of beliefs about how psychology ought to go about its business. The term has come to have fairly distinct meanings regarding methodology, but these derive directly from behaviorism as formulated by its founder and chief promoter, John B. Watson. The importance of Watson's contribution can scarcely be exaggerated. Indeed, modern psychology can be said to begin with Watson. A full appreciation of Watson's role would require an historical treatment of psychology as a whole that would take us too far from our present concerns. Our aims here will simply be to review the major features of Watson's behaviorism and to make clear its formative influence on modern learning theory.

In many respects, behaviorism defined itself more clearly by what it stood in opposition to than by what it stood for. Its particular opponent was structuralism, the dominant psychological system of the day, and some acquaintance with structuralism is required if we are to fully understand the rationale of behaviorism and the attraction it held for psychologists.

Structuralism can be viewed as the acme of psychology's attempts to fashion a workable science of the *mind*. It was largely the creation of Edward B. Titchener, who held that psychology's task was the analysis of conscious experience by means of self-observation, or introspection. Titchener was well aware of the risks involved in basing a systematic psychology upon observations of private experience, but he argued that a scientific analysis of mental experience was possible if sufficient care was exercised in observational procedures. If the stimulus and instructional conditions accompanying a self-observation of conscious experience were standardized and accurately specified, then the observation had the status of an experiment, subject to verification by another observer. Accordingly, introspection was to be conducted under strict rules governing stimulus conditions, observer attitude (or "frame of mind"), and report procedures. Self-observation of the mind as practiced by the structuralists, then, was anything but an off-hand or common sense review and reporting of the flow of one's thoughts and feelings. It was a constrained and rigorous procedure to be practiced only by those who had undergone special preparation and training, and its data were to be evaluated by the usual scientific criteria of reproducibility and inter-observer agreement.

The structuralists believed that all conscious mental processes are made up of a number of simple processes blended together and that these elemental processes could be defined through introspection. Titchener distinguished three classes of elements—sensations, or the elements of perceptions; images, or the elements of ideas; and affections, or the elements of feelings. These elements in turn could be described in terms of basic properties, such as quality, intensity, vividness, and duration. The elements and their properties were the basic stuff, or structure, of mental experience, and the business of introspection was to discover how different groups and arrangements of elements gave rise to the different complex processes that constitute human consciousness—thought, memory, imagination, emotion, perception, and so on.

From the outset there were disagreements about the elemental composition of many mental phenomena, but Titchener was confident that persistent introspective effort would resolve these uncertainties. And in truth, structuralism made some progress, particularly in extending our knowledge of sensation and perception. However, despite many years of labor little headway was made in understanding other complex mental processes, and in almost all areas the disagreements of the introspectionists did not disappear with time but instead became increasingly frequent, sharp, and irreconcilable.

One particularly notable controversy occurred in the study of thought processes, and it raised the critical question of whether thinking is in fact accessible to the introspective technique. One group of prominent introspectionists concluded that thinking could go on without any observable conscious accompaniments. Another group, under Titchener, argued to the contrary that thinking was always analyzable into particular sensations and images if only one introspected properly. A dispute of this type is intolerable to any discipline professing itself a science. Here was an important problem under investigation by two groups using the same methodology, making their observations under the same conditions—and claiming diametrically opposed findings. Nor was there any objective criterion for resolving the difference. In the "imageless thought controversy" and other such disputes, the fatal flaw of structuralism stood clearly revealed. It was against the background of such impasses that behaviorism arose with its promise of a truly new and workable approach to understanding human activity.

Another aspect of structuralism which fostered a receptive attitude toward behaviorism was its highly restrictive nature. Since introspection could be carried out in a meaningful way only by well-trained human adults, the structuralists had little interest in other populations, such as children, the abnormal, or animal subjects. Even for the normal adult, introspection had little bearing on understanding the mental activities of everyday life. It was one of the cardinal points of structuralism that successful introspection required ridding oneself of common sense orientations and purging mental processes of any tinge of practical involvement or self-interest. The focus was exclusively upon understanding the generalized, normal human mind through observation of strictly delimited conscious experiences under essentially artificial environments. It was a matter of Titchenerian dogma that a concern with either individual differences in mental functioning or practical psychological problems would prove detrimental to a proper science of the mind.

It well may be that any psychology based on the direct analysis of mental experience must of necessity be a highly constrained matter if it is to have any chance of success. But many psychologists, particularly American psychologists, grew impatient with such a determinedly pure and narrow conception of their science and with its failure to speak to practical problems of society. Behaviorism, in contrast, was to offer a methodology applicable to *any* organism and was to make practical application of its findings not only a matter of legitimate concern but a test of its validity.

Some signs of change were apparent prior to Watson's formulation of behaviorism. There were vigorous although as yet small and isolated programs in areas clearly outside the pale of structuralism, areas such as mental testing, educational psychology, and developmental psychology. Of more general significance was the rise of functionalism, a movement which can be regarded as transitional between the systematic approaches of structuralism and behaviorism. As the name suggests, functionalism held that psychology ought to be primarily concerned with the *uses* of the mind rather than with its structure or content. In advocating this change in emphasis, the functionalists were not seeking the rejection of structuralism but rather were attempting to bring psychology more into line with what they perceived to be the realities of 20th century biological science. Essentially, the functionalists argued that the concepts of Darwinian biology required that the mind be seen as an instrument of evolution. Psychological processes, like physical processes, could not be properly understood, they claimed, unless they were viewed as activities which enabled the species to better compete in the struggle for existence. Given such an orientation, it follows that mental processes are best studied in relation to their ends in the natural environment and in the context of meaningful problems and activities. In such a program the techniques of introspection were of limited use, and while the functionalists did not reject introspection, they were led to place greater emphasis on observation of mental processes from the outside, so to speak. That is, mental functions were to be understood by inference from observation of total organismic activities. These important ideas found widespread acceptance among psychologists, with the result that during the early part of the century the study of mental processes underwent a significant and healthy broadening in objectives and methodology.

It was within this same period that Watson developed his behavioristic psychology. Watson had received his psychological training in the leading functionalist school, and he was in full agreement with the emphasis upon function and practical application in psychology. But to Watson's mind, functionalism provided a mere beginning to the revisions he felt to be necessary if psychology was ever to achieve its goal of providing an objective account of human activity. What Watson felt was required was nothing less than the abandonment of all previous programs toward that goal and a fresh start on the basis of totally new premises. At that time psychology had a 50-odd year history as a science, that is, as a deliberate, self-conscious effort to apply scientific method to the study of man. Underlying its program was the assumption that man was to be understood by study of the mind, and structuralism was the paramount expression and fruit of that premise. But in the foundering and impasses of structuralism Watson saw the inevitable fate of any psychology which took the mind as its subject matter. Hence, what was required was not merely the modification of existing procedures, but an altogether new way of attempting to understand human beings.

Watson elaborated this view in a series of papers and books, the most important of which appeared between 1913 and 1925. These writings are still remarkable for their vigorous, down-to-earth quality and for the sense of conviction and urgency they convey. One suspects that no small part of Watson's appeal to his contemporaries was the style of the man—direct, lively, and forceful. The essential points of his attack on traditional psychology are best expressed in his own words.

"States of consciousness," like the so-called phenomena of spiritualism, are not objectively verifiable and for that reason can never become data for science.

In all other sciences the facts of observation are objective, verifiable and can be reproduced and controlled by all trained observers. . . . Psychology, on the other hand, as a science of "consciousness," has no such community of data. It cannot share them, nor can other sciences use them. Not only can psychologist A not share his data with physicist A, but also he cannot share them with his brother psychologist B. Even if they existed, they would exist as isolated, unusable "mental" curiosities.

The psychologists' use of "introspection" as its principal method has been another very serious bar to progress. . . . All that introspective psychology has been able to contribute is the assertion that mental states are made up of several thousand irreducible units; for example, the thousands of sensation units like redness, greenness, coldness, warmth, and the like, and their ghosts called images, and the affective irreducibles, pleasantness and unpleasantness. . . .

But the truth or falsity of this assertion is inconsequential, since no other human being can make an introspective observation upon anyone but himself. Whether there are ten irreducible sensations or a hundred thousand (even granting their existence), whether there are two affective tones or fifty, matters not one whit to that organized body of worldwide data we call science. (Watson, 1924a, pp. 1–3)

"Consciousness" is neither a definable nor a usable concept . . . it is merely another word for the "soul" of more ancient times. (Watson, 1924b, p. 3)

The time seems to have come when psychology must discard all references to consciousness; when it need no longer delude itself into thinking that it is making mental states the object of observation. We have become so enmeshed in speculative questions concerning the elements of mind, the nature of conscious content (e.g., imageless thought . . .) that experimental students are beginning to feel that something is wrong with the premises and the types of problem which develop from them. There is no longer any guarantee that all mean the same thing when the terms now current in psychology are used. . . . One must believe that two hundred years from now, unless the introspection method is discarded, psychology will still be divided on the question as to whether auditory sensations have the quality of "extension," whether intensity is an attribute which can be applied to color, whether there is a difference in "texture" between image and sensation; and upon many hundreds of others of like character.

Our psychological quarrel is not with the systematic and structural psychologist alone. The last fifteen years have seen the growth of what is called functional psychology. This type of psychology decries the use of elements in the static sense of the structuralists. It is stated in words which seem to throw emphasis upon the biological significance of conscious processes rather than upon the analysis of conscious states into introspectively isolable elements. . . . Surely if . . . [mental] concepts are elusive when looked at from a content standpoint, they are still more deceptive when viewed from the angle of function. (Watson, 1914, pp. 7–9)

Such attacks were timely and did much to crystallize the growing conviction that psychology's first approach to creating a science of man had gone seriously awry. It had become bogged down in speculative questions which could not be subjected to clear experimental test, and it had become divorced from humanity's vital interests. The remedy Watson proposed was direct, drastic, and courageous—abandon utterly the study of mental experience and construct a new science based solely on the study of behavior, on what organisms do or say. By limiting itself to the behavior of organisms, psychology could clearly meet the cardinal requirement of natural science, that of public verifiability of its observations. But a call to substitute behavior for mind would scarcely have been heeded on that basis alone. It was heeded largely because Watson made a convincing case that a comprehensive psychology *could* be based on behavioral data alone.

Basically, Watson argued that an effective behavioral psychology was already in existence in the field of animal psychology and that the methods of that field need only be extended. Animal psychologists could observe nothing but their subject's behavior, yet had made considerable progress in the comparative study of such psychological processes as perception, learning, memory, motivation, and problem solving. Both classical and instrumental conditioning, as we have seen, originated with animal subjects, and they provided persuasive examples of how learning could be understood solely in terms of observed behavior and environmental events. In such a framework it was pointless to speculate about the subject's conscious experience, for example, about whether or how the animal experienced the formation of an association. The important thing was that the psychologist was able to tell which conditions led to behavior change and which did not. If the animal psychologists had been able to build up reliable and useful information about their subjects' learning processes, sensory capacities, instinctive reactions, problem-solving performance, and the like simply by measuring the subjects' behavior in appropriate settings, then why, asked Watson, should we not expect similar benefit from application of this basic method to the human animal?

Such a human psychology would perforce be objective, but it could be objected that it would be restricted in the scope and significance of its subject matter. Watson countered this objection in two ways. First, he argued vigorously that any psychological phenomenon which could not be defined in terms of measurable behavior simply was not appropriate subject matter for a science of man. Secondly, he showed that the most significant topics within traditional psychology could in fact be given meaningful behavioral definition. Consider, for example, his treatment of thinking, seemingly the most subtle and "internal" of psychological processes. By logical argument supported by general observations and some direct experimental evidence, Watson showed that it was plausible to regard thinking as subvocal talking, that is, as minimal movements of the musculature involved in speech production. Conceived of in this way, thinking consists of chains of responses differing only in magnitude from overt behaviors, and with appropriate instrumentation a person's thought processes could therefore be studied as objectively as any publicly observable behavior. Similarly, Watson defined emotions, or feelings, in terms of response patterns, largely visceral in nature, linked to specific settings and traceable to either an instinctive or a conditioned origin. Perception, traditionally conceived of as the study of the relations between mental experience and physical stimulation, became for Watson the study of the relations between differential reaction, either verbal or motor, and changes in physical stimulation.

Watson's translation of mental phenomena into behavior was begun in a series of short papers and culminated with the publication in 1919 of a general text (*Psychology from the Standpoint of a Behaviorist*) which provided a comprehensive treatment of human psychology without resort to consideration of mental experience in any guise.

The thesis of this text is that a full explanation of human activity can be achieved solely by examining behavior, publicly observable reactions, in relation to present and past environmental stimulation. Watson concluded that such a program not only was adequate to the traditional concerns of psychology but would in fact greatly enlarge the scope of the field, since any activity was open to analysis, no matter how complex or practical in nature, so long as it could be reliably measured. Moreover, Watson emphasized that by attempting to understand behavior as a function of environmental variables, psychologists would be gathering the only really *useful* information about organisms, namely, knowledge of what an organism is likely to do under given conditions.

There is little need to elaborate further on the general nature of Watson's influence upon psychology because the student who has passed through an introductory course in the field already has experienced the full flavor of the behavioral approach, so well has the core of Watson's teaching been incorporated into psychology. But it will be useful to underscore here several aspects of Watson's behaviorism which were to be of particular importance in shaping the thinking of learning psychologists for many years to come. These aspects are the development of the stimulus-response paradigm, Watson's reliance upon learning and conditioning as general explanatory concepts, and the strong peripheral response bias in theory construction.[1]

Stimulus-Response Psychology

As we have seen, Watson sought to understand psychological phenomena in terms of direct links between behavior and environment. To carry out this program in any particular case requires three steps. First, the phenomenon of interest must be defined in terms of a specific response or set of responses. While Watson emphasized that any response must be reducible to muscular or glandular activity, he recognized that most activities of psychological interest occur at a level of complexity which renders impracticable the direct measurement of the underlying muscular or glandular action. In practice, then, a *response* is generally specified at a more molar level and can refer to any aspect of behavior which can be reliably measured, regardless of its degree of complexity. Second, the psychologist must define those aspects of the environment, or stimuli, deemed likely to control the response. A stimulus, like a response, can be specified at many levels of complexity, provided the specification permits the stimulus observation to be made by others. Third, evidence is sought on the nature of the relationship between the stimulus and the response, preferably by experimental test, i.e., by varying the stimulus under controlled conditions in order to directly observe its effect on the response. By progressive isolation and testing of stimulus-response relations, one can presumably isolate all, or at least the most significant, determinants of the behavior in question and in this way arrive at a full explanation of that activity. This approach, the stimulus-response, or S-R, paradigm, was to become the standard paradigm not only for the study of learning but for much of modern psychology.

There is a strong air of practicality about stimulus-response psychology as formulated by Watson. While psychologists might not agree on the best definition of a given psychological phenomenon, they could certainly agree on one or more reasonable response definitions and thus could get about the business of producing data directly relevant to understanding the phenomenon in question. Moreover, stimulus-response psychology promised a fairly immediate test of its success or failure. If the activities of humans and animals are largely a function of enviornmental stimulation as claimed by the model, then on the basis of experimentally determined S-R relations, one ought to be able to predict behavior in the natural world from a knowledge of its antecedent stimulus conditions. And if such prediction proved possible, then the way was open to control human behavior by arranging the environment. These aspects of Watson's S-R program were a major factor in its appeal, particularly in the case of American psychologists.

General Behavior Theory, Learning, and Conditioning

Watson also had new things to say about the contribution of heredity and environment to human nature. His contemporaries were prone to find a hereditary basis for many psychological phenomena. William James, the foremost functional psychologist, attributed to humans no less than 24 basic instincts, or unlearned action tendencies, which, taken together, accounted for a large part of the motivational and emotional behavior of humans. James's list included such complex processes as imitation, rivalry, pugnacity, anger, sympathy, fear, acquisitiveness, constructiveness, curiosity, sociability, shyness, cleanliness, and shame. It was also widely assumed that hereditary mechanisms played a key role in shaping an individual's intelligence, temperament, talents, and interests.

Watson's approach to the heredity-environment issue was characteristically direct and fresh. He argued that from the behavioral viewpoint, the problem reduces to determining what behaviors are unlearned and what behaviors are learned and that the logical way to go about this determination is to find out what behavior the newborn child arrives with and how that behavior becomes elaborated into the complex activities of the adult. Accordingly, Watson set about systematically observing infant behavior, attempting to discern and classify the earliest, and presumably unlearned, behavior patterns. In this approach, he was in marked contrast to his contemporaries who based their views on the heredity-environment issue solely upon observation of adult behavior.

Observation of several hundred babies over the first days and months of life convinced Watson that the infant came equipped with relatively little in the way of organized behaviors. He did find that the infant exhibited at birth or shortly thereafter a large repertoire of reflexes of a directly adaptive nature (e.g., crying, sucking, grasping, orienting) as well as a large variety of uncoordinated manual responses, such as movements of the arms, legs, and hands. He also observed three types of emotional reaction—fear, rage, and love— each elicited by a restricted number of stimuli. The infant's fear reaction was elicited only by loud sounds and loss of support, the rage response only by restriction of movement, and the love reaction (defined in terms of smiling, gurgling, and the like) by tactual stimulation of various body zones. These relatively simple adaptive and emotional reactions appeared to make up the bulk, if not the entirety, of the infant's unlearned behavior.

Equally important, Watson noted that opportunities for the modification of these behaviors through experience were present from the moment of birth and increased rapidly with age. Watson found that when he attempted to chart the development of behavior over the first weeks of life, it soon became impossible to partial out the contribution of learning to any behavioral alteration or to any emerging behavior. For example, he noted that expressive reactions, such as crying, were strongly influenced even in the first days of life by the type and amount of attention these behaviors received from the caretaker; that is, infants quickly learned to cry if crying was rewarded by the caretaker's attention. As he pursued the genesis of behavior over the first year of life, the picture of development which emerged for

Watson was one in which the unlearned component of behavior, never very large or complex, was soon lost sight of as a consequence of the early and ever-increasing influence of learning. He was thus led to deny categorically the existence in humans of any instinctively determined activities of the type that James had postulated and to deny as well the inheritance of psychological characteristics in any form. Instead, he posed the view that all complex behavior develops through training out of the restricted unlearned response repertoire of the infant and that differences in adult behavior derive essentially from differences in training. The strength of Watson's convictions in this regard is seen in his often quoted statement:

> Give me a dozen healthy infants, well-formed, and my own specified world to bring them up in and I'll guarantee to take any one at random and train him to become any type of specialist I might select—doctor, lawyer, artist, merchant-chief and, yes, even beggar-man and thief, regardless of his talents, penchants, tendencies, abilities, vocations, and race of his ancestors. (1924b, p. 82)[2]

Watson arrived at these strong views about the time that classical conditioning was becoming widely known to American psychologists, and it is not surprising that he seized on conditioning as the process whereby the environment molded the developing individual. Classical conditioning not only met the demands of behaviorist methodology but also appeared well-suited to Watson's view that complex behavior develops from the unlearned stimulus-response associations of the infant. Thus, as we have seen, Watson himself had demonstrated experimentally that the infant's unlearned fear of loud sounds could be used to condition a host of new stimulus-response associations. In like manner, he argued, all of the adult's psychological processes develop through chains of naturally occurring conditioning experiences, all having their origin in the reflexes, manual responses, and emotional reactions of the infant.

Watson's radical environmentalism, buttressed as it was by the newly discovered phenomenon of conditioning, held a strong appeal to many psychologists. Some of this appeal reflected dissatisfaction with the structuralists' conception of the mind as a static and uniform product of nature. But doubtless much of the appeal lay in the fact that Watson had brought to the fore and made preeminent as an explanatory device what is in many ways the most significant and attractive capacity of humans—their great capacity for change and growth through experience. Of course, Watson's views were not without strong opposition from many quarters, and later research was to show that Pavlovian conditioning could not carry the explanatory burden which Watson had imposed upon it. Nevertheless, largely as a result of Watson's influence, the study of learning was elevated to a key role in psychology. Regardless of the validity of Watson's particular views about the ways in which people are shaped by experience, his research and writings established the idea that learning is the major determinant of almost any significant aspect of human activity and consequently that the study of learning provides the best approach to a comprehensive understanding of human nature. This viewpoint was to inspire the study of learning over the next several decades.

Response Bias in Theory Construction

Apart from the rational basis of behaviorism, there was also a certain emotional aspect to the appeal which behaviorism had for many psychologists. It may seem strange to see the word *emotional* applied to a scientific system, but as Heidbreder pointed out (1933), a system is not only the basis of procedure but a basis of morale as well. Watsonian behaviorism gave its adherents a strong sense of fruitful participation in a program of superior scientific merit. The key to this sense was the behaviorist's uncompromising commitment to a behaviorally based objectivity. As stressed earlier, the constraints and shortcomings of structuralism had primed receptivity to the tenets of behaviorism, and many psychologists came to believe with Watson that any form of subjectivism was strictly antithetical to science. A thorough-going behavioral objectivity promised a safeguard against the pitfalls of traditional subjective psychologies and, more broadly, promised to free psychology fully from the influence of religious and philosophical views of the nature of man. In this light, behaviorism was clearly more than an alternative methodology; it took on the character of a "crusade against the enemies of science" (Heidbreder, 1933, p. 259).

It is important to understand this aspect of behaviorism because the zeal for objectivity extended beyond the issues of the nature of acceptable data and subject matter. It influenced as well the form which explanations of learning were to take. Ideally, the behaviorist preferred explanations expressible solely in terms of associations between directly observable stimuli and responses. But even within the confines of animal learning, such a simple input-output model was soon challenged by a number of observations suggesting the importance in learning of processes arising within the organism. For example, it was early observed that only some portion of a given stimulus input might enter into association with a conditioned response. Thus, an animal conditioned to approach, say, a striated circle might later be found to approach a striated patch of any shape but not a bare circle, or vice versa. Such observations pointed to the operation of a stimulus selection mechanism within the subject and appeared to violate the concept of the organism as a passive input-output stimulus-response machine. Faced with such an observation, the behaviorist could respond by postulating a central (brain) selection process, defined strictly in terms of its behavioral manifestations (as in the foregoing illustration), and then set about a stimulus-response analysis of that selection process. Such an approach would be entirely within the bounds of Watsonian objectivity, since at every step the central process would be defined by observed stimulus and response events. But instead, the typical reaction to such observations was to seek some *peripheral response* mechanism which might salvage a direct input-output model of learning. In the foregoing case, for example, it has been suggested that the animal's visual scanning responses may not be distributed equally over the stimulus complex and thus the receptor system may pick up primarily, say, the striated aspect of the stimulus. Such an explanation puts the selection process in peripheral (muscle) activity, however minute, rather than in central (brain) activity. Since efforts were seldom made to measure the postulated scanning responses, there was little pragmatic difference between these theoretical alternatives. But to the behaviorist, the conceptual difference was a very real one. A hypothetical process based on muscle action was clearly thought to be more "real" and accessible than a postulated central process, and it was less suspect as a possible entranceway for "immaterial agents."

The foregoing instance is but one of many similar episodes in the study of learning. In fact, a number of the major issues in the history of learning theory center on challenges to the peripheral input-output interpretation of learning, and time and again, as we shall see, learning theorists betrayed a strong peripheral response bias in handling these challenges. That is, apparent departures from a stimulus-response model of learning were accounted for not in terms of hypothesized perceptual or cognitive mechanisms but in terms of hypothesized muscular and glandular processes, however minimal or implicit, however tentatively specified, however remote from measurement technology. Not all behaviorists shared the aversion to centrally based explanatory concepts, but this was certainly the majority feeling, and it accounts for the character of much of modern learning theory. To understand the reluctance of many learning theorists to employ centrally based explanatory concepts, one must bear in mind the origins and intensity of the behaviorists' distaste for anything which smacked of the subjective, the mysterious, or the immaterial.

Present-Day Status of Behaviorism

It would be misleading if our brief sketch of behaviorism were to terminate leaving the impression that orthodox Watsonian behaviorism still dominates learning theory. In fact, the essential message of behaviorism has been so fully incorporated into psychology that Watson's behaviorism is no longer recognizable as a distinct system. It is now simply a given that objectivity is the first prerequisite for sound psychological analysis, but objectivity as conceived by contemporary psychologists exhibits none of the virulent anti-mentalism of Watson's objectivity. It has long been consensually recognized that the objectivity of a psychological datum resides not in its having a muscular basis but in its status as a repeatable and verifiable event. It is also recognized that while a mentalistic psychology is indeed prone to the errors and failures which so concerned Watson, there nevertheless are many mental phenomena of legitimate scientific interest. Therefore, while much of contemporary psychology is concerned with an S-R, or strictly behavioral, analysis, there is increasing willingness to speculate about mediating processes within the organism and to seek subjectively apprehended data on central processes whenever this is a scientifically meaningful course.

The history of learning theory shows a continuing tension between two seemingly incompatible concerns—the desire to pursue a thoroughly objective analysis of learning on the one hand and the need to deal with processes arising within the organism on the other. This book will trace the erosion of a strict peripheral behaviorism in the face of repeated pressures to recognize the contribution of central processes. But we shall also attempt to make clear that the increased concern with central processes is in fact a major achievement of behaviorism in the sense that the determined pursuit of a peripheral stimulus-response analysis of learning has been invaluable both in delineating the nature of the central processes which must be postulated and in enabling psychologists to deal with those processes in an experimentally meaningful way.

ANIMAL PSYCHOLOGY

The nature of the kinship between human beings and other animals is an age-old question, but interest in this question did not take on a scientific character until the advent of Darwin's theory of evolution. Under the impact of evolutionary theory the issue of the degree of psychological continuity between humans and lower animals became more than a matter of philosophical inquiry. It became an empirical question of profound importance and one which urgently demanded the attention of natural scientists.

Early views on this question were based largely on uncontrolled observations of animal behavior. Darwin himself, hardly a disinterested critic in the matter, had cited numerous anecdotal reports of animal performance which appeared to require the attribution to animals of mental capabilities similar to those of humans (Darwin, 1871). Continuing in this tradition, Romanes published in 1881 a large and influential collection of natural observations and anecdotal reports on "animal intelligence." These reports were drawn from a wide variety of sources, and all purported to demonstrate in one way or another the operation of higher mental processes in animals. By present-day standards Romanes's collection is more valuable for documenting the dangers of the observational procedures involved and the strength of the tendency to anthropomorphism than for what is revealed about the comparability of human and animal intelligence. The salient characteristic of Romanes's observers, whether professional or lay observers, was their remarkable willingness to seize upon and to interpret *any* patterned or organized aspect of animal behavior as revealing the operation of complex psychological processes. For example, Romanes quotes the following observation of ant behavior from a certain Mr. Belt.

A nest was made near one of our tramways, and to get to the trees the ants had to cross the rails, over which the waggons were continually passing and repassing. Every time they came along a number of ants were crushed to death. They persevered in crossing for some time, but at last set to work and tunnelled underneath each rail. One day, when the waggons were not running, I stopped up the tunnels with stones; but although great numbers carrying leaves were thus cut off from the nest, they would not cross the rails, but set to work making fresh tunnels underneath them. [Romanes continues] . . . . the degree of receptual intelligence, or "practical inference," which was displayed is highly remarkable. Clearly, the insects must have appreciated the nature of these repeated catastrophes, and correctly reasoned out the only way by which they could be avoided. (Romanes, 1893, pp. 52–53)

Another anecdote reported by Romanes:

A cat sees a man knock at the knocker of a door, and observes that the door is afterwards opened: remembering this, when she herself wants to get in at the door, she jumps at the knocker, and waits for the door to be opened. Now, can it be denied that in this act of inference, or imitation, or whatever name we choose to call it, the cat perceives such an association between the knocking and the opening as to feel that the former as antecedent was in some way required to determine the latter as consequent? (Romanes, 1893, p. 59)

In these instances we see the readiness of Romanes and his correspondents to conclude the operation of complex mental processes in lower species. But the deficiencies of these observations, which are representative of the hundreds compiled by Romanes, are apparent. Neither observation can be reproduced by another observer, and each is open to a number of alternative, and simpler, interpretations. For example, the ants may have responded simply to chemical traces resulting from their use of the previously established trail; the cat's behavior may not have been problem-solving in nature but rather investigatory and controlled by the simple perceptual properties of the knocker, e.g., its brightness or noise-producing properties.

While the Romanes collection and other similar reports of the period invariably painted a very humanlike picture of animal psychology, they proved of little ultimate value to comparative psychology.

Animal psychology as a rigorous and programmatic venture begins with the writings of the English psychologist C. Lloyd Morgan. Morgan, like Romanes, was an evolutionist, but he held that reports of seemingly intelligent behavior by animals were of little value in themselves for the evolutionist position. Meaningful interpretation of such reports was possible, argued Morgan, only against a knowledge both of the way in which those behaviors came about and of the habits of the species involved. Thus, what appears to be an intelligent or insightful solution to a problem when viewed apart from preceding behavior may in reality be merely a chance outcome of a fumbling, hit-or-miss behavior sequence. Alternatively, a behavior which suggests a higher-order psychological process might prove to be an instinctive or invariant behavior pattern of the species involved. These arguments, which Morgan supported by his own observations and other data, were influential in fostering a more systematic and comprehensive approach to the study of animal behavior.

Morgan also contributed an important rule to guide interpretation of data in this area. Lloyd Morgan's canon, as this rule has come to be called, states "In no case may we interpret an action as the outcome of the exercise of a higher psychical faculty, if it can be interpreted as the outcome of the exercise of one which stands lower in the psychological scale" (1894, p. 53). In other words, always seek explanation in terms of the lowest possible psychological process. This rule, which continues to guide the study of animal learning, may well be questioned by the student. After all, in any given case involving decision between invoking a lower- versus a higher-order process, the latter might well be operative in nature. There is no reason to suppose that the simplest explanation should always be the true one. But Morgan's canon makes sense when considered against our strong tendency to interpret animal behavior in terms of our own psychological processes, a tendency well documented by the Romanes collection, if not, indeed, by our daily observation. Morgan's rule, then, serves as a safeguard against the error we are most likely to make in explanations of animal behavior.

Morgan also took the first steps toward an experimental analysis of animal learning. True to his own arguments, he sought information on the total course of development of acts frequently taken to indicate higher-order processes. His procedure was to require animals to solve problems and to learn tricks, generally within homey surrounds, while he made qualitative observations of their activity from start to finish. For example, one of his learning tasks required a dog to learn to open a garden gate by inserting its head under the latch and then lifting it. Morgan's observations of the ways in which such problems were solved led him to a relatively conservative view regarding the mental capacities of animals. Rather than indicating that animals reasoned, his observations indicated that solution was achieved on a trial-and-error or accidental basis with the correct actions appearing only gradually after considerable repetition. The underlying process, Morgan thought, was simple association by contiguity. Briefly, he assumed that as a result of the repeated, contiguous occurrences of the correct response, its antecedent stimuli, and its consequent stimuli, connections were established between the brain processes paralleling each of these events. As a result of these connections, the occurrence of any of these brain processes would tend to arouse the others. In this way, the antecedent stimuli (e.g., the latch in the above illustration) would gradually come to evoke an "idea" or anticipation of the correct response and of its stimulus consequences. Morgan further assumed that organisms were so built that the anticipated response would be facilitated if the anticipated consequences were pleasurable but would be inhibited if these consequences were unpleasurable. In short, Morgan's theory was that animals learned about the stimulus consequences of their actions through a simple association process and then acted in accordance with the hedonic nature of those consequences.

Morgan's studies were carried to a fully experimental level by Edward L. Thorndike in his famous doctoral dissertation published in 1898, *Animal Intelligence: An Experimental Study of the Associative Processes in Animals*. Thorndike's general method was to place hungry animals (cats, dogs, and chicks) in boxlike enclosures from which they could escape by some simple act, such as pulling at a loop of cord or pressing a lever. Food was left outside the enclosure but within the subject's view. Observations were made of the way in which the animal arrived at the escape response and of the total time required to escape. After the subject succeeded in escaping, it was allowed access to the food for a brief period. The procedure was then repeated until the subject became adept at the escape response. The training was carried out under controlled conditions, and, as a rule, sizable samples of subjects were tested within a given species. This work provided the prototype of the instrumental conditioning procedures discussed earlier in this chapter.

Thorndike also found trial-and-error performance in his subjects. The first escape response appeared by accident in the course of much fumbling and useless movement. A single escape or even several escapes seldom sufficed to produce consistently rapid escapes, and in fact later trials often required more time than earlier ones. Only after a considerable number of confinements and escapes did the subject develop a consistent, rapid, and efficient escape response. Like Morgan, Thorndike believed that such performance demanded an interpretation of animal learning in terms of simple association processes, rather than in terms of reasoning or other higher-order processes. However, Thorndike conceived of the associations underlying problem solution as direct connections between aspects of the learning situation and the particular movement that led to escape. Moreover, Thorndike felt that this association was not formed on the basis of contiguity alone but rather was "stamped in" by the outcome or aftereffects of the movement. Specifically, the pleasure experienced by the animal in getting out of the box and securing the food strengthened the connection between the learning situation and the specific response that led to this outcome. Movements that left the animal in the box (an unpleasurable outcome) were thereby weakened and became less likely to occur in the learning situation. With successive confinements and escapes, the association between the learning situation and the escape response would gradually become preeminent since it is the only association invariably followed by a pleasurable aftereffect. In essence, Thorndike believed that animals function in terms of associations between stimuli and specific movements and that these associations are strengthened or weakened by their consequences. Note that this differs from Morgan's interpretation, which assumed that associations are learned on the basis of contiguous occurrence alone and that the consequences of a behavior (as anticipated) serve merely to facilitate or inhibit response occurrence.

Much of Thorndike's later work was concerned with establishing and investigating his *principle of aftereffects*, later termed the *law of effect*. This law became so important as a reference for virtually all subsequent learning theories that it is worth quoting here in its entirety:

> Of several responses made to the same situation, those which are accompanied or closely followed by satisfaction to the animal will, other things being equal, be more firmly connected with the situation, so that, when it recurs, they will be more likely to recur; those which are accompanied or closely followed by discomfort to the animal will, other things being equal, have their connections with the situation weakened, so that, when it recurs, they will be less likely to occur. The greater the satisfaction or discomfort, the greater the strengthening or weakening of the bond. (From *Animal Intelligence: Experimental Studies*, 1911, p. 244)

At about the time of Thorndike's earliest experiments, Pavlov was beginning his studies on classical conditioning, although his methods did not become widely known outside of Russia until considerably later. Given the widespread applicability of the Thorndikian and Pavlovian paradigms, both to species and to processes, their introduction greatly facilitated a broad experimental and comparative approach to understanding animal behavior. Coincident with these developments, the exaggerated claims of the early Darwinians regarding the similarity of animal and human psychology were giving way to a general recognition that this question properly required systematic study of similarities and differences in behavior across the phylogenetic scale. The early 1900's saw a marked increase in research in this vein accompanied by a steady improvement in the methods employed. Examination of the leading text on animal psychology of the latter part of this period (Washburn, 1926) reveals that a formidable array of methods, data, and theoretical principles had rapidly accumulated. About half of this text was devoted to review of data bearing on the sensory capacities of species ranging from protozoa to the apes, much of this work stemming from the early interest of physiologists in the functions of the sense organs. But considerable space was also given to data and analyses bearing on perception, learning, memory, imitation, and attention. In short, by the 1920's animal psychology had become a thriving and well-respected scientific enterprise, boasting a broad body of reliable data, a relatively sophisticated methodology, and experimentally based principles of behavior.

There was also general appreciation of the fact that psychological research with animal subjects had a number of distinct advantages over research at the human level. Only with animals is it possible to fully control past history, to strictly regulate motivational states, to determine the effects of experimental manipulations over an entire life span, and to investigate the role of the nervous system in behavior through extensive physiological and surgical intervention. Consequently, research on psychological processes with animal subjects can achieve a degree of experimental control and incisiveness not attainable at the human level. This consideration should be related to the nearly unanimous view of the times that there are major continuities between the psychological processes of humans and animals. In particular, the pioneer investigators of animal learning, such as Morgan, Thorndike, and Pavlov, viewed animal learning as essentially a simpler version of human learning.

Thus, when Watson called for the extension of animal psychology to the study of humans, that call was heeded in large part because animal psychology had been accepted as a fully scientific discipline and because the methods, data, and principles of animal psychology were seen as directly relevant and useful to understanding human activity.

This completes our consideration of the historical circumstances which fostered the growth of modern learning theory. We have reviewed three interrelated developments in the psychology of the early 1900's—the discovery and elaboration of conditioning, the rise of behaviorism as an alternative to mental psychologies, and the emergence of animal psychology as a scientific discipline. By 1925 these developments had fused, along the lines envisaged by Watson, to provide the rationale and methodological framework for most of the work on learning over the next 30 years. Essentially, the idea was that broadly applicable principles of learning could be uncovered by intensive stimulus-response analysis of the performance of a few selected lower species under conditioning procedures.

In the immediately following chapters, we shall be concerned with the application of this approach—the S-R approach, as it came to be called—during the period 1925 to 1955. In this period, the approach was directed toward achieving a general theory of learning which would be applicable to all organisms, and to this end a considerable number of alternative global theories and principles were put forth. The bulk of the research in this period was concerned with assessing the validity of these conceptual treatments, and the merits of the major competing theories and supporting research were continually, often hotly, debated. Since the period 1925 to 1955 was marked by theoretical controversy, it is sometimes thought of as a period of disunity among learning psychologists. But at base the psychologists engaged in these theoretical controversies were more united than divided, and what united them was their belief in the efficacy of the S-R approach. They were all behaviorists, and they saw in behaviorism both a way to avoid the irresolvable issues which had plagued psychology in the past and a means to generate significant, testable questions about the learning process. They saw in conditioning a powerful research tool which promised (and had already provided evidence of) truly generalizable analyses of the learning process. They were strongly influenced by the theory of evolution and believed that the components of human psychological processes were present in animals but in simpler, more readily analyzable form. Therefore, they saw in animal psychology a means of studying the essential features of learning under rigorous experimental control. These views as to the fundamental rightness and value of the S-R method marked these psychologists as one and transcended their theoretical disagreements. The battles over the interpretation of data produced by that method were joined in earnest precisely because those data were widely accepted as valid and significant.

The period from 1925 to 1955 was in fact, then, one of singular agreement among learning psychologists as to the proper ends and means of their science. There was also general agreement about the significance of their work. Consistent with the tenets of behaviorism, they took a strong environmentalist position and viewed learning as the primary determinant of human activity. To these psychologists, the understanding of learning was prerequisite to advances in other areas of psychology, and thus they saw themselves at the forefront of the field.

The confidence of the S-R theorists in the efficacy and value of their approach was accompanied by optimism and a sense of high purpose, making this an exciting period in the history of learning. The aims of the S-R theorists were lofty. They took as their task nothing less than the explanation of all learned behavior, and they sought basic, universally applicable principles of learning. Their data came chiefly from studies of lower organisms, but they never doubted the usefulness of these data to the understanding of human activity. Application of their work to human psychology was central to their thinking, and no aspect of human activity was deemed too complex to profit from an S-R analysis. They envisaged solutions, in terms of conditioning principles, to such major problems as the reform of education and the treatment of abnormal behavior. Research on animal conditioning even provided the basis for the design, in fiction, of a new culture (Skinner, 1948b)! In short, the S-R theorists were firmly convinced that experimental and theoretical analysis of animal conditioning would significantly advance our understanding of learning, particularly of complex human learning. And they were right.

3
The Conceptual Treatments
of Tolman and Hull

An introduction to the S-R theorists properly begins with an introduction to the species which served as their primary data source—the species *Mus norvegicus albinus,* familiarly known as the white rat. While perhaps unprepossessing as a psychological model, the albino rat nevertheless presents many distinct advantages as an instrument for research on learning. It has relatively well-developed sensory, central nervous, muscular, and glandular systems which enable it to adapt to a wide range of environmental settings, thereby making it suitable for investigation of many different learning problems. Appetitive and aversive drives are readily induced and controlled, and even without special stimulation the rat tends to be an active, exploratory, and manipulative creature. It is small, clean, docile, easily handled, and disease-resistant; breeds in large numbers and rapidly (gestation is only 21 days); and can be inexpensively maintained in large colonies. It has a life span of about three years, and the psychologist is thus assured of a ready supply of sensitive, responsive, "standard" subjects whose life history can be fully specified and controlled. In essence, the white rat can be viewed as the learning scientist's equivalent of the test tube, a simple yet very useful means of observing and testing the nature of learning.

Research on learning has of course made considerable use of a number of other species (e.g., the dog, the monkey) selected for suitability to particular lines of investigation. Given the general view that there are fundamental principles of learning common to all species, the selection of a research subject becomes primarily a matter of the most favorable combination of degree of behavioral complexity, convenience of measurement and control, and ease of supply and main-

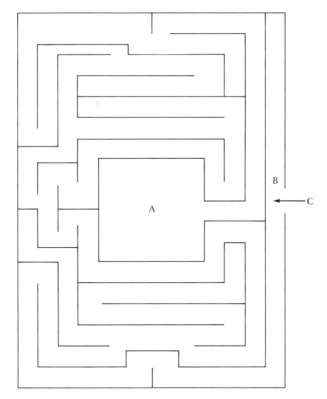

Figure 3-1. Floor plan of Small's adaptation of the Hampton Court maze. The food reward was located in the center compartment (A), and the subjects' home cage was placed at C flush against the entrance to the maze, B.

tenance. For the S-R learning psychologists of the 1930's, the white rat generally represented that most favorable combination.[1]

By 1930 considerable data had accumulated on the performance of animals within the simple classical and instrumental conditioning paradigms as developed by Pavlov and Thorndike, respectively. While S-R theorists relied heavily upon these data, they also made extensive use of the somewhat more complex problem of maze learning by white rats.

The maze was introduced as a research instrument in 1900 by Willard Small, who was also among the first to employ the white rat as an experimental animal. Impressed by reports of the natural burrowing and trail-making tendencies of the rat, Small reasoned that learning the true path through a maze should pose a problem appropriate to the instincts of this animal. He patterned his maze (see Figure 3-1) after the Hampton Court maze, a famous maze constructed of hedge-lined passageways in the gardens of Hampton Court Palace in England. A supply of food was placed in the center of the apparatus (point A in Figure 3-1), the animals were kept hungry, and their task was to learn the most direct route from the entrance (point B) to the food.[2]

Small sought to afford his subjects alternative modes of solution and to allow them to approach the problem in their own way with a minimum of restraint. By using wire mesh to make the partitions between alleys, he allowed the subjects to see across the pathways and thus possibly to learn by orientation to distant points in the maze. The subjects were allowed to traverse the maze in small groups, a condition which resembled their rearing experience and which afforded opportunity for learning by imitation. Finally, in order to minimize experimenter interference, the subjects were housed in a cage placed flush against the maze at point C and, by means of a sliding glass door, were allowed free access to the maze each evening until the following morning. Small did observe his subjects' initial passage from start to goal upon opening of the door each day, and he reported marked improvement in the speed and accuracy of this passage, amounting to almost perfect accuracy after about a ten-day period. But Small was not fundamentally interested in quantitative measures of correct path learning, and it is apparent that under the free-access procedure employed much of that learning may have been unobserved. Rather, much of Small's report is concerned with qualitative observations of such matters as the subjects' general adaptation to the maze, the emergence of exploratory activities, and the development of behavior patterns of a purposive, or goal-oriented, character.

A few years later Watson adopted Small's maze for a series of experiments on the relative importance of the different senses in learning by the rat (Watson, 1907). While agreeing that the maze problem is well-suited to the structure and habits of the rat, Watson made a number of changes in Small's apparatus and procedure. These changes are worth noting since they illustrate a fundamental difference in approach which was to pervade S-R research on learning. Noting that much of the maze behavior of Small's subjects went unobserved, Watson required his subjects to learn the maze through a series of daily trials, each trial consisting of one traversal from the start point to the goal. Between daily sessions the animals were kept in cages apart from the maze. While this procedure does enable a full record of the course of path learning, it greatly restricts observation of the animal's natural approach to the problem. It also serves to minimize whatever contribution general exploration of the maze might make to learning the correct path.

In another change, Watson increased the efficiency of training by giving his subjects a series of feeding experiences in the center portion of the maze prior to the start of maze training proper. Note, however, that such preadaptation precludes observation of how the animal responds when confronted with the maze problem as a novel whole, as in Small's procedure. In other changes, Watson replaced the wire mesh with solid, opaque walls and trained his animals individually. These changes enabled Watson to specify the stimulus situation more precisely, but they also precluded the possibility that an animal might learn by orientation to distant points in the maze or by imitation of another animal. Finally, Watson discarded Small's error measure as ambiguous (not all erroneous turns eventuated in a blind) and measured learning solely in terms of change in the time required to traverse the maze from start to goal. Little attention was given to qualitative observation of general activity.

It can be seen that Watson's changes emphasize objectivity, ease of measurement, and control and specification of the learning situation. Small, on the other hand, had sought to minimize behavioral constraints, to afford opportunities for alternative solution modes, and to make broad observations of activity in the learning situation. Each experimenter gained his priorities at the expense of the other's. There is no infallible best choice between these alternative approaches; research on learning has usually sought a balance between breadth of observation and precision of observation. The scientific temperament is biased toward the latter, but there is always the risk that precision and control may be secured at the price of a more veridical view of learning. As we shall see, more than one important theoretical issue has turned on this point.

In regard to the maze as a research instrument, however, Watson's view prevailed and the trend by later investigators was toward even greater simplicity of apparatus and experimenter control of the learning situation. Watson's prescribed trial procedure became standard and the Hampton Court maze plan was abandoned in favor of simpler mazes composed of segments which were identical or very similar to one another. One widely employed maze of this type is the T-maze (see Figure 3-2), which consists of one or more serially linked T-shaped units each of which confronts the subject with a choice between two alleys (the arms of the T), one leading to a dead end and one leading either to the next unit or to a rewarded terminus. As

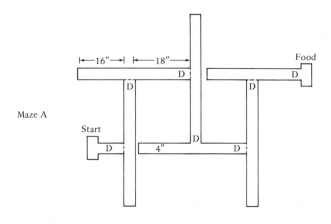

Maze A

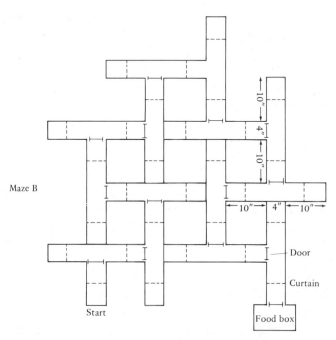

Maze B

Figure 3-2. Two T-mazes of varying complexity. Maze A is a six-unit maze (from Blodgett, 1929, p. 117); maze B is a 14-unit maze.
(From Elliott, 1928, p. 20. By permission.)

illustrated in Figure 3-2, by varying the number and pattern of T-units the experimenter can readily devise problems varying in difficulty and spatial organization properties while maintaining objective equality of the problem elements.

Such devices were used extensively during the next several decades, and these tasks, however simple in appearance, uncovered learning phenomena both basic and challenging to theory. Error patterns were observed which appear to characterize *all* serial learning, regardless of the nature of the response. Subjects exhibited intriguing response patterns which had the character of learning strategies. Choice point behavior was found to provide a rich source of evidence on the role of central mechanisms in learning. These and other behavior patterns were exhaustively studied in relation to a large number of conditions (readily varied in mazes like the T-maze) such as the number of choice points, relative lengths of correct and incorrect arms, direction of the blinds (goal-pointing vs. non-goal-pointing), availability of visual or other cues at the choice points, presence versus absence of opaque walls, and type and magnitude of goal incentive. In sum, the study of how rats learn to choose the correct path through a maze became a standard yet flexible and versatile research paradigm which provided much valuable information about the nature of selective and serial learning. It was within this paradigm that the early developments in S-R theory came about.

Those developments centered around the work of two major figures in the history of learning theory—Clark L. Hull and Edward C. Tolman. Both men were pioneers in advancing the cause of behaviorism, and each was fully committed to an experimental, stimulus-response analysis of learning. Both were convinced that the study of animal learning was the best route to understanding human behavior, and each pursued the strategy of seeking fundamental principles through the study of relatively simple situations such as maze learning. In Tolman's words,

What is it that we rat runners still have to contribute to the understanding of the deeds and misdeeds, the absurdities and the tragedies of our friend, and our enemy—*homo sapiens?* The answer is that, whereas man's successes, persistences, and socially unacceptable divagations—that is, his intelligences, his motivations, and his instabilities—are all ultimately shaped and materialized by specific cultures, it is still true that most of the formal underlying laws of intelligence, motivation, and instability can still be studied in rats as well as, and more easily than, in men (Tolman, 1945, p. 166).

And on another occasion he says, "I believe that everything important in psychology (except perhaps such matters as . . . involve society and words) can be investigated in essence through the continued experimental and theoretical analysis of the determiners of rat behavior at a choice point in a maze. Herein I believe I agree with Professor Hull . . ." (Tolman, 1938, p. 34).

But if these men were in whole-hearted agreement about methods and broad aims, they differed sharply in their explanations of learning. It is in their differences in theoretical orientation that many of the controversies, and major accomplishments, of the S-R approach are rooted.

TOLMAN'S VIEWS

General Orientation

Tolman began his formulation in a series of papers appearing in the early 1920's and continued to elaborate his system over the next 30 years. On the whole his theory of behavior and learning remained remarkably intact over this period despite numerous empirical and theoretical challenges.

We noted that Tolman was a vigorous advocate of behaviorism, but from the outset he implemented the behavioral formula in a way which was to make him unique among S-R theorists. Watson had argued that learning could be analyzed in terms of associations between small bits of the environment and behavior, each specified ideally in purely physiological terms. The stimulus-response relations of simple classical and instrumental conditioning provided the prototypical example. The former involves an association between a discrete aspect of stimulation, e.g., a tone of a given frequency, amplitude, and duration, and some particular glandular or muscular reaction, e.g., increased salivary flow. In instrumental conditioning, again a specific reaction, e.g., a lever press, is associated with a relatively discrete stimulus, e.g., sight of the lever. More complex forms of learning were seen as combinations of such S-R associations. Thus, running the correct path through a maze could be interpreted as a behavior chain composed of elemental simulus-response associations, e.g., left turn associated with the stimuli defining choice point 1, right turn associated with the stimuli defining choice point 2, and so forth. Under this view, the units of learning were essentially *molecular* stimulus and response events, and complex behavior was seen in terms of combinations of and interactions between these elemental units. But when Tolman looked at behavior he saw entirely different units of analysis.

For Tolman, behavior *occurs in acts* related to broad features of the environment. For example, consider the behavior of a rat placed in a maze for the first time. The rat will infallibly be observed (after some initial pause) to slowly traverse all or nearly all alleys, blind and open, with frequent pauses and retracings, and this locomotory activity will be accompanied by responses such as sniffing, rearing up, pushing against walls or doors, scratching at crevices, and biting or manipulating parts of the maze. Each of these reactions can be related to some momentary stimulus input—rearing up may be elicited by the end of a blind, biting by a protruding nail head. But note that all these responses can be seen to familiarize the organism with its new environment, and in this sense the rat's activity can be said to be not a concatenation of elemental responses but rather the single act of *exploration*. The new surround instigates the act of exploration, and the fact that exploration may be analyzed further into subordinate S-R units does not violate the conception that the association between "new surround" and "exploration" may be the primary and natural unit.

As another example, consider the behavior of a hungry rat placed for the first time in a maze which is similar in general respects to mazes in which the rat has previously learned to find food but different in the correct path pattern. Such an animal will reliably be observed to soon begin moving through the maze with little or no exploratory activity, quickly retracing from blinds and proceeding rapidly until it arrives at the goal box. Again, the behavior could be described either as a succession of specific S-R relations or as "looking for food." In this case, the act of looking for food is presumably related to those features of the environment which define the situation, for the rat, as one similar to those in which path-taking behavior eventuated in food.

Now it is apparent that, as in the foregoing illustrations, interactions between organism and environment can usually be analyzed either in terms of molecular stimulus-response units or in terms of the molar units of behavior acts. That is, the organism can be conceived of as emitting particular responses as a result of moment-to-moment changes in stimulation or as engaging in acts composed of

functionally equivalent elements and related to broad aspects of the
environment. These levels of analysis and the particular S-R associations selected within each are, at base, perceptions of the theorist,
and there is no a priori basis or absolute criterion to determine which
may be the more appropriate or meaningful. In this light, selection
of the unit of analysis is the first step which a behavior theorist takes,
and it is a most important one since it conditions all subsequent theorizing.

On this issue Tolman came down firmly on the side of a molar
behaviorism. Organisms are built, he assumed, to release behavior in
chunks, or sets, of responses, and evidence of the actlike structure of
behavior is, he felt, everywhere apparent in animal and human activity:

> A rat running a maze; a cat getting out of a puzzle box; a man
> driving home to dinner; a child hiding from a stranger . . . a
> pupil marking a mental-test sheet; a psychologist reciting a list of
> nonsense syllables; my friend and I telling one another our
> thoughts and feelings—*these are behaviors* (qua *molar*). And it
> must be noted that in mentioning no one of them have we re-
> ferred to, or . . . for the most part even known, what were the
> exact muscles and glands, sensory nerves, and motor nerves in-
> volved. For these responses somehow had other sufficiently iden-
> tifying properties of their own. (Tolman, 1932, p. 8).

The assumption that acts are the natural units of behavior gave
distinctive form and direction to Tolman's theorizing in that it led
him to explain the particulars of behavior in terms of the characteristics of acts, whereas other S-R theorists sought to comprehend the
characteristics of acts in terms of the properties of molecular S-R units.

Tolman asserted that when behavior is examined from a molar
viewpoint two major characteristics stand out—purpose and cognition. The reader may be surprised to see these terms within an
avowedly behavioral psychology, and an animal behavioral psychology at that. But Tolman was not reluctant to employ terms from
mental psychology or from everyday language, providing he could
give them behavioral referents. The use of such terminology, plus
Tolman's fondness for coining new word combinations, introduces
some ambiguity into his writings, and at some points the reader must
exercise considerable interpretative judgment. Fortunately, Tolman
typically supported and clarified his arguments by frequent reference

to maze behavior, and we shall follow suit here by concretizing his thoughts in this fashion.

In saying that behavior exhibits *purpose*, Tolman referred to the fact that acts are typically observed to *persist until* some specific end, or goal, is attained. For example, when a hungry rat is placed in a maze, it engages in trial-and-error path selection regardless of the number of errors and retracings involved until the goal and food are finally reached. This "persistence until" quality, as Tolman termed it, is reliably observed each time the animal is introduced to the maze and becomes a particularly salient aspect of path-taking behavior upon mastery of the problem. Should the true maze path be altered at that point or should obstacles be inserted on the true path, the path taking will be observed to persist while the particulars of behavior change to meet the new situational demands.

The purposive character of behavior is further attested to by the fact that acts conform to the most direct route to a goal. To put this in the context of maze learning, consider a maze which offers an animal three alternative paths to the same goal. Path 2 is identical in all respects to path 1 except that it is somewhat longer. Path 3 is the same length as path 1 but requires a brief delay en route. Various experiments have established that in such a situation rats would quickly exhibit a preference for path 1, within fairly narrow limits of path differences. Moreover, such preference would be likely to obtain even if there had been considerable prior training on path 2 or 3.

These objective features of maze behavior, a "persistence until" aspect and a tendency to conform to more efficacious means to the goal, are also observable, Tolman argued, in most behavior acts, including those of a nonspatial nature. While there is little in the way of experimental observation to directly support this assertion, general observation suggests that it is certainly plausible to assume that persistence and economy of goal-directed behavior characterize many human and animal activities. To Tolman, the apparent pervasiveness of these characteristics constituted *prima facie* evidence that behavior is by nature fundamentally goal-seeking, or purposive.

The *cognitive* aspect of behavior was also everywhere apparent to Tolman. He argued that behavior exhibits cognition (knowledge) whenever its occurrence can be said to presuppose facts about the environment. Cognition in this sense can be revealed in striking ways by sudden changes in a learning environment. For example, assume that a rat has thoroughly mastered a complex maze like that of Figure 3-2, so that if the animal is made hungry and placed at the entrance it reels off the correct sequence of turns in rapid succession, pausing only upon entering the goal box. Suppose, now, that one of the alleys on the true path is considerably reduced in length between trials. On the next trial, the animal runs full speed into the end wall of the shortened alley! It simply traverses the alley as if the old length were present. In Tolman's words, "His behavior postulates, expects, makes a claim for that old length" (Tolman, 1926, p. 356).

Another phenomenon of maze running which Tolman felt points to cognition in behavior is that as maze training progresses, animals are often found to make an *increasing* number of errors on choices involving goal-pointing blinds, at least over the initial stages of learning. This outcome reflects an increasing tendency to enter alleys pointing in the general direction of the goal box more frequently than those pointing in the opposite direction, or away from the goal. "Such a predominance of entrances obviously states, exhibits, an imputation that the food box lies in the general direction of such alleys . . . (the animal's) various behaviors . . . can be said, at any stage of learning, quite objectively to express his thereunto acquired object adjustments (cognitions) with respect to the 'getting-on-toward-food' possibilities of such maze parts" (Tolman, 1925, p. 292).

Note that the behaviors involved in each of these illustrations make sense *only* when they are related to certain facts or knowledge about the learning environment. Running into the wall makes sense when it is realized that an open alleyway at that point had been, up to that time, a fact of the learner's maze environment; an increasing number of errors on goal-pointing blinds as training progresses makes sense when it is considered that the goal is in fact located in the direction of those blinds. In this light, these observations constitute strong presumptive evidence that the organism does indeed act out of knowledge of the environment. More important, these observations help to establish the basic point that *any and all* behavior can be considered to be directly informative about knowledge of the environment. In Tolman's words, "Every behavior act in going off and being what it is expresses, implies, certain specific characters in the environment. And this is so because the continuance of its going off can be shown to be contingent upon there actually proving to be such characters in the environment. If these expected characters are not found, the act sooner or later ceases or modifies itself . . . practically all behaviors are thus cognitive or postulative" (Tolman, 1927, p. 434). Thus, as in the case of *purpose*, Tolman viewed cognition as an inherent, pervasive, and self-evident property of behavior.

It is to be emphasized that the words *purpose* and *cognition* have been used throughout here in a fully objective way. They are defined by experimental and natural observation. Despite the mentalistic tone of *purpose* and *cognition*, Tolman has no interest whatever in the mental processes, if any, which might accompany these behavioral phenomena. Both terms refer simply and only to features of behavior exhibited under particular stimulus conditions. The same point can be made about the terms *demand* and *expectation*, which Tolman frequently used interchangeably with *purpose* and *cognition*, respectively. *Demand* and *purpose* can be interchanged in the sense that the "persistences to and from" which define purposes ultimately reduce to persistences to and from states of bodily quiescence or bodily disturbance (demands). External environmental states (goals) are sought only as a means of modifying these bodily states. Therefore those environment-behavior relations which define purpose also define the demand character of organismic activity. Similarly, the equivalence of *expectation* and *cognition* is based upon equivalence in defining operations. But whether purpose or demand, expectation or cognition, or any of the other richly suggestive and frequently original terminology which Tolman favored, these concepts were always anchored in external stimulus-response correlations; such anchoring is the distinguishing mark of the S-R theorist.

To recapitulate, the main points of Tolman's approach are as follows: (1) While behaviorism is the route that psychology must take, the appropriate units of analysis are not particular responses but behavior acts, conceived of as clusters of functionally equivalent responses organized about goals or demands. (2) Behavior acts directly display purpose and cognition, the former in the form of persistence and economy of expression toward goals and the latter in the form of their demonstrable dependence upon facts of the environment. (3) The tendencies of organisms to behave in molar, purposive, and cognitive ways are not derived or acquired; rather, these are given, fundamental properties of all organisms, even of the lowly rat.

Essentially, Tolman is saying that we can see organization, purpose, and cognition manifest throughout behavior—this is simply how organisms are; therefore, we must seek to understand them in terms of these characteristics.

What general view of learning does this orientation imply? It implies that what an organism will initially do in any learning situation, what behavior acts appear, depends upon what purposes (demands) and cognitions the organism brings to that situation. Similarly, it implies that the general course of behavior throughout the learning will be a product of the working out of those purposes and cognitions in relation to the given environmental circumstances. And it implies that the critical changes which take place in learning are changes in purposes and cognitions, primarily the latter, as a result of the action of the given environmental circumstances. In short, it views purpose and cognition as the determinants of behavior and learning.

Major Features of Tolman's Theory of Learning

Tolman's theory essentially elaborates the general view that behavior is determined by purposes, or demands, (hereafter only the latter of these terms will be used) and cognition. The major assumptions specify types of cognitions and demands, the conditions governing their development, and the particular ways in which they influence behavior. Tolman reformulated his thoughts somewhat on these matters a number of times, but one can pull out a set of basic assumptions which constitute the influential, enduring core of his thinking about learning. These assumptions will be reviewed here in relation to instances of maze learning behavior which can be said to be consistent with them.

Demands and Behavior. As we have seen, demand considered as a general state can be defined in terms of "goal-oriented" properties of behavior, but specification of the presence and strength of particular demand states requires, in addition, the statement of some antecedent stimulus manipulation of the type generally assumed to create a biological need. For example, reducing an animal's food intake in a given way specifies a hunger demand, placing an animal in an unfamiliar but harmless surround specifies an exploratory demand, and depriving an animal of sleep creates a demand for sleep. Such demand states are assumed to influence behavior in several important ways. First, they may directly evoke innately organized behaviors which are instrumental to the satisfaction of the demand. Second, they are assumed to bring about selective attention to demand-related features of the environment. To illustrate these functions, consider the typical behavior of a rat maintained on a moderately reduced food intake and placed in a maze for the first trial of a training series, this also being its first exposure to the maze. These conditions specify two demand states, hunger and exploration, but behavior evoked by the latter is likely to predominate during the initial trials. On the first trial, the rat is likely to engage in a variety of investigatory behaviors throughout the maze with greater attention being given to the more novel aspects of the situation. During this phase the rat may pay little or no attention to relatively familiar features of the environment; it is not uncommon to observe a hungry subject totally ignore the food in the goal box while continuing investigation of the surround. As the rat becomes familiar with the maze over a number of trials, the hunger demand becomes dominant and there is a corresponding shift in attention and pattern of locomotory behavior. Task-extraneous features of the maze are now totally ignored and "food-searching" behavior emerges, i.e., path-taking behavior proper emerges.

Tolman's theory emphasized another important relation between demands and behavior. High levels of any demand state were assumed to impair learning by rendering organisms relatively insensitive to stimulus input. To illustrate, suppose that a group of rats first learns under moderate hunger demand to take maze path A to find food and then is given the option of either taking path A or a much shorter path, B. Suppose further that when path B is made available, one half of the subjects are placed on a feeding schedule defining intense hunger demand while the remaining subjects continue on the moderate-demand schedule. The likely outcome of this procedure is that the moderate-demand subjects would soon adopt path B while the high-demand subjects would persist much longer, perhaps indefinitely, on path A. In Tolman's view, the intense demand state has essentially blinded the latter subjects to the new opportunity in their learning environment. Throughout his writings, Tolman emphasized this hypothesized relation between demand intensity and perceptual processing as one of the primary sources of learning deficiencies and maladaptive behaviors.

The Nature and Development of Cognitions. The assumptions which lie at the heart of Tolman's theory of learning are those concerned with the nature and development of cognitions. As we have seen, Tolman argued the general propositon that cognition can be inferred whenever behavior can be said to presuppose facts of the environment. Within his theory of learning proper, cognitions were conceived of as units of knowledge stored in the organism and activated by particular stimuli. These units concerned either the outcomes of organism-environment interactions or the relationships between environmental events. Thus, in Tolman's final theoretical statement (1959) cognitions are said to take either the form "an instance of this *sort* of stimulus situation, if reacted to by an instance of this *sort* of response, will lead to an instance of that *sort* of further stimulus situation," or else the form "an instance of this *sort* of stimulus situation will simply by itself be accompanied, or followed, by an instance of that *sort* of stimulus situation" (Tolman, 1959, p. 113). Cognitions of the former type are referred to by the shorthand expression s_1r_1---s_2, and cognitions of the latter type by the expression s_1---s_2.

Now, although Tolman gave weight to s_1r_1---s_2 cognitions in his final theoretical statement and even in some of his early papers, it is nevertheless true that cognition of the s_1---s_2 form was the critical ingredient of his overall thinking about learning, particularly in regard to its influence on contemporary investigators. Therefore, Tolman's theory will be explicated here with reference only to s_1---s_2 cognitions.[3]

Tolman held that learning *is* cognition formation. Cognitions were assumed to develop primarily through repetition of environmental sequences. Specifically, repetition of an environmental sequence was assumed to develop an association whereby one element of the sequence, s_1, evokes an expectation of a subsequent stimulus, s_2. For example, in maze training the subject is typically restrained on each trial in the start box for a few seconds prior to opening of the start box door, and therefore as trials progress the cognition should develop that placement in the start box (s_1) will shortly be followed by opening of the start box door (s_2). Or to express this another way, placement in the start box will come to evoke the expectation "door opening." Similarly, after repeatedly traversing, say, a black alley which eventuates in the goal box, the cognition should develop that the black alley (s_1) leads to food (s_2), or, in expectation terms, sight of the black alley comes to evoke the expectation "food at the end." In general, exposure to a maze results in the formation of a number of such knowledge units expressing "what leads to what" in the maze. Consistent with Tolman's general orientation, the s_1, s_2 elements of these knowledge units are thought of as molar, obvious, and commonsense features of the learner's environment, features such as "start area," "goal box," "alley to the left," and "second choice point."

Cognition formation not only involves the differentiating out and strengthening of specific s_1---s_2 patterns but also involves the learning of these patterns across different situations. Suppose an animal is trained to find food in each of a number of mazes differing in path pattern, size, color, and other particulars of construction. While each maze would pose a number of situation-specific s_1---s_2 relations, certain s_1---s_2 relations would be invariant over *all* the mazes. Thus, as a consequence of the fact that traversal of each maze eventuates in food, cumulative maze training should develop the general cognition that mazes (s_1) afford food (s_2). Such general cognitions, applicable across a class of learning situations, were generally referred to as *beliefs* or *means-end readinesses* by Tolman, while the term *expectation* was usually reserved for situation-specific cognitions.

Whether specific or general in nature, cognitions in Tolman's system always derived ultimately from repeated sequences of organism-environment interactions. In this sense cognitions were tied to the observable stimulus, or input, side of the broad S-R formula. How were cognitions anchored to the response, or output, side? Tolman suggested that means-end readinesses could be tied to behavior by measuring the transfer between successive learning experiences which differ in regard to the *specific cognitions* they occasion. To illustrate, consider two mazes, A and B, differing markedly in the particular s_1---s_2 patterns they present, i.e., the mazes differ clearly in the motor patterns required for solution, in visual features, etc. Now if it is found that animals trained in succession on A and B learn B more rapidly than animals trained on B alone, such an outcome can be taken to index the operation of general cognitions acquired in the course of the A-B learning experience, since transfer of specific cognitions is precluded under the procedures employed. It is from such transfer experiments, with their positive or negative transfer effects, that Tolman would infer the kind and strength of the means-end readinesses established by a given learning experience or series of experiences.

Specific cognitions or expectations were to be indexed by means of "disruption experiments." In these, one simply observes behavior following a sudden change in the final term of a previously experienced s_1---s_2 relationship. In an early experiment frequently cited in this connection (Elliott, 1928), rats were trained for nine days in a maze under one type of goal reward (bran mash) and then, on the tenth day, were shifted to another known reward (sunflower seeds). On the trials immediately following this change, there was a decided increase in both errors and running time and, more important, a disrupted, searching sort of behavior was observed in the goal box. Elliott writes, "The change in performance appears to be more than a mere temporary disturbance, since it increases rather than decreases during the course of six days. On the day of the change of reward, the tenth day, the animals did not eat steadily while in the foodbox, but divided their time between eating and random searching" (p. 23). For Tolman, such disruption is direct evidence for an expectation of the previously obtained bran mash reward. If the change in rewards had not produced "surprised" and disrupted behavior patterns, then one must infer that the subject had not acquired an expectation of the bran mash reward. In general, experiments of this type, with their attendant disrupted behavior or lack of disruption, were held to provide a direct means of tying expectations to behavior in any learning situation.

Conditions Affecting Cognition Formation. Repetition of an s_1---s_2 relation was held to be the primary factor in strengthening an association between s_1 and an expectation of s_2. But from time to time Tolman discussed the action of other variables. Notably, cognition formation was said to be assisted by recency of the s_1---s_2 sequence, by short temporal intervals between sequence elements, and by proximity of the sequence to biological goals (rewards and punishments). The importance attributed to these conditions varied somewhat in different writings; in particular, Tolman gave little weight to the action of rewards and punishments in his early treatments, but he ultimately took the position that they always played some part in the acquisition of cognition (Tolman, 1959). In the main, however, the roles of recency, temporal interval, and biological goals were distinctly subsidiary to that of sheer frequency of occurrence of given s_1---s_2 relations.

Other variables stressed by Tolman in this regard appear to be more directly concerned with the question of *which* s_1---s_2 relations become associated. This question is a fundamental one, since in all but extremely simple situations the organism will usually be exposed to a myriad of concurrent and overlapping s_1---s_2 relationships. In a maze which poses a choice between a black alley leading to the goal and a white alley leading to a blind, an obvious s_1---s_2 relation which the subject might learn is that between sight of the black alley (s_1) and expectation of the goal (s_2). But the black alley would doubtless contain a number of other distinctive features (distinctive to the rat, at least) which also stand in an invariant sequential relation to the goal or, for that matter, to one another, e.g., features such as floor texture, odors, extramaze cues visible from the alley, and the stimulus patterns specifying the beginning or end of the alley. Given that such rich potential for cognition formation characterizes most learning situations, the question arises as to whether subjects learn all possible s_1, s_2 relationships and, if they do not, how the learner selects those particular s_1---s_2 relations which make up its cognitive map.

The general thrust of Tolman's thinking here is that s_1---s_2 learning *is* a highly selective process. While he did not provide a theoretical treatment of selectivity at the level of specific stimuli, he did suggest several broad conditions affecting the direction and scope of cognition formation. One such condition, the subject's demand state, we have already touched upon. Demands sensitize the learner to stimuli which are relevant to the satisfaction of the demand, and these stimuli are thus likely to enter into s_1---s_2 patterns. Intense demand states tend to focus the learner's attention almost exclusively on demand-related features and hence tend to produce what Tolman called "striplike maps." Moderate or low demand intensities, on the other hand, favor the development of broad cognitive maps, i.e., favor multiple s_1---s_2 learning.

Breadth of s_1---s_2 learning is also significantly affected by the physical arrangement of the learning situation. Two types of mazes which contrast sharply in this regard are elevated and alley mazes. The elevated maze presents the animal with a wall-less pattern elevated several feet above the floor so as to prevent the animal from straying off the track. The alley maze, in contrast, is a walled, enclosed structure. It is apparent that any point in the open elevated maze affords a subject more potential s_1---s_2 relationships both near and remote in nature than would a corresponding point in an enclosed alley maze with the same path pattern. The contrast of elevated and alley mazes represents an unusual degree of difference in this physical arrangement variable, but Tolman emphasized that even moderate variations in the layout of a learning situation could have important consequences for cognition formation. In general, Tolman's experiments favored "open" learning situations which afford the subject opportunity for multiple s_1---s_2 associations, and in this respect Tolman's thinking was similar to that of Small, the first maze psychologist.

Finally, selection in s_1---s_2 learning was also held to be determined by the salience, or attention-getting quality, of the various task stimuli, with stimuli of relatively high salience being more likely to enter into s_1---s_2 association. Tolman had little to say about just what properties were critical to stimulus salience, but such common sense considerations as intensity, magnitude, novelty, and subject preference were indicated.

Note that many of the conditions we have discussed as influencing cognition formation could be said to significantly affect the organism's perception of s_1---s_2 relations. The effects of demand and salience variables were explicitly assumed to be mediated by perceptual selection processes, and physical arrangement of the learning situation is patently a perceptual factor. In regard to the role of temporal interval between sequence elements, it is doubtless easier to detect that a particular stimulus is likely to follow another stimulus when the temporal interval between them is relatively short. And even rewards and punishments were said to act, at least in part, by emphasizing the difference between correct and incorrect routes to their loci (Tolman, 1932, pp. 344–345). In general, the conditions which facilitate or hinder learning in Tolman's theory are largely conditions which facilitate or hinder perception of relevant s_1---s_2 relationships. For this reason, Tolman's theory is sometimes referred to as a "perceptual learning" theory.

The Relations Between Demands, Cognitions, and Behavior. Since in Tolman's system what get learned are cognitions, not behaviors, the question arises as to how one gets from learning to behavior. What principles relate learning to action? There are two aspects to Tolman's treatment of this question. First, he assumed that behavior is a joint function of demands and cognitions, that is, that behavior is a product of the united action of the demands and cognitions operating at any time. Second, he assumed that behavior simply conforms appropriately to existing demands and cognitions.

It follows directly from the first of these assumptions that whether or not what is learned will be translated directly into behavior depends upon the nature of the organism's demand state. If an animal has both an accurate cognitive map of a maze and a demand for the food at the goal, there will ensue a rapid, errorless sequence of movements from start box to goal. But if the same animal is placed in the maze after being satiated for food, little or no goal-directed movement will occur. In the first instance, there is a direct correspondence between what the animal has learned about the correct route to the goal and its performance in the maze, whereas in the second instance this correspondence has been eliminated by the removal of the appropriate demand. Such considerations led Tolman to emphasize a distinction between learning and performance—between an organism's knowledge of s_1---s_2 relationships in the learning situation and what the organism actually does in that situation. In the light of this distinction, it can be seen that while correct, goal-directed performance can be used to infer knowledge of the relevant s_1---s_2 relations, absence of such performance cannot be used to infer absence of such knowledge. As we shall see, the learning-performance distinction has proved fundamental in the development of theory and research on learning.

Tolman's second assumption concerning the relation of learning to behavior, namely, that behavior conforms appropriately to demands and cognition, was his response to the problem of how moment-to-moment behavior is selected and shaped. Effectively, the assumption amounts to a denial of the significance of this problem since it simply asserts that, given any cognitive state, behavior will take care of itself. The final version of Tolman's theory (1959) sought to make the nature of the link between cognitive states and specific actions somewhat less mysterious by shifting to cognitions of the s_1r_1---s_2 type, that is, by emphasizing that organisms may acquire cognition of the s-r relations as well as of the s-s relations of the learning situation. But of course the "r" here has the status of a representation only, and so the link to actual behavior remains vague. Throughout the bulk of his writings, Tolman was content with the simple view that behavior automatically follows, or tracks, cognitive state.

Summary

To recapitulate briefly, Tolman's theory proceeded from the basic assumption that demands (motivational states) and cognitions (knowledge of the environment) are the major determinants of behavior. Demands and cognitions are conceived of as central, intervening events in a causal chain linking behavior to antecedent environmental conditions. But demands and cognitions are capable of objective definition in terms of observable stimulus-response relations. Learning itself is a matter of cognition formation, conceived of as the development of associations between present stimuli and expectations of subsequent stimuli. The primary factor in cognition formation is repeated experience of stimulus sequences, although other conditions play some part. The particular behavior emitted at any point in a learning situation is the outcome of the combined action of existing demands and cognitions, and behavior was assumed to adjust appropriately to these central determinants.

Before going on to consider Tolman's theory in application to some basic phenomena of learning, we shall first review the major features of Hull's approach to learning.

HULL'S VIEWS

General Orientation

Hull began his treatment of learning somewhat later than Tolman, publishing his first article directly on this topic in 1929. Over the next 20 years he wrote a series of papers and books which progressively elaborated a systematic treatment of learning. Like Tolman, Hull made some significant changes in his system in his later writings. But again as in Tolman's case, we can readily distinguish the most influential form of the theory.

From the outset Hull was attracted to conditioning, particularly Pavlovian conditioning, as a theoretical base, and he proceeded to construct his theory of learning on the phenomena of conditioning experiments. The feature of conditioning which so recommended it to Hull was its apparent simplicity. Interestingly enough, it was precisely this feature which led Tolman to reject conditioning in favor of maze learning as an observational base more appropriate to the purposive and cognitive nature of organisms. But Hull welcomed the simplicity of the conditioning experiment as better revealing the operation of basic principles which must, he argued, be involved in all behaviors, simple or complex. When Hull turned to maze experiments, as he frequently did in analyses of learning, it was not to find a somehow truer picture of learning, but rather because he was eager to show that principles derived from conditioning experiments could satisfactorily explain more complex behaviors.

Reliance upon conditioning as a source of explanatory principles was the hallmark of Hull's orientation, and he felt strongly the legitimacy and value of this approach. His views in this regard are worth quoting. In discussing the generality of principles discovered in conditioning experiments, he states:

> The compound adjective in the expression, "conditioned-reaction principles," . . . refers to the locus of *discovery* of the principles rather than to their locus of *operation*. Their original isolation in conditioned-reaction experiments was presumably due mainly to the relative simplicity of such situations; except for the practical difficulty of isolation they might quite as well have been isolated in any complex adaptive situation. Galileo is said to have made his original experiments concerned with the laws of motion by rolling a ball down a simple inclined plane, but one should hardly for that reason regard the laws of motion thus

isolated as restricted in their action to small spherical objects. Presumably the same principles of motion are also active in avalanches, but an avalanche would be a poor type of situation for their isolation and quantitative determination. (Hull, 1939, p. 233)

Consistent with this line of reasoning, one of Hull's earliest papers (1930a) sought to show that principles of simple conditioning were at work even in the higher mental processes. Specifically, the paper attempted to account for the origin and nature of knowledge and purpose in terms of conditioned habits. Since this paper nicely illustrates the major features of Hull's approach, it will be reviewed in some detail at this point.

Like any S-R theorist, Hull analyzed the flowing panorama of the organism's world into functionally discrete stimulus events. Under this view a naturally recurring chain of environmental circumstances can be described as a series of selected, recurring stimulus aspects (S's), each reducible ultimately to a pattern of energy at the organism's receptor surfaces. Figure 3-3 represents a sequence of environmental events schematized in this fashion, the sequence terminating in a biologically significant goal event, S_G. (Figure 3-3 and the following figures employing S-R schemas are patterned after those in Hull 1930a.)

$$S_1 \quad - \quad S_2 \quad - \quad S_3 \quad \cdot \; \cdot \; \cdot \quad S_n \quad - \quad S_G$$

Figure 3-3.

Suppose that into this situation an organism is introduced whose prior conditioning history has so shaped it to reliably emit characteristic responses to each stimulus component of the sequence, as schematized in Figure 3-4.

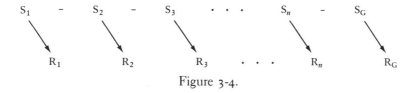

Figure 3-4.

Now any organism capable of movement has receptors embedded in its muscles, tendons, and joints, which respond to the displacement of these members. Response by these receptors initiates neural signals to the brain in the same manner that response by external receptors, for example, receptors of the eye or ear, triggers neural input to the brain. Each movement of the organism, then, internally generates a characteristic stimulus pattern which can function in the same ways as externally initiated stimulus patterns. In Figure 3-5 the internal stimuli produced by the organism's responses (R) to the stimulus events of the external environment are represented by s's. To simplify the exposition, assume that the time intervals between stimulus aspects of the world flux are such that each external stimulus event (S) occurs within a few seconds of the response-produced stimulus (s) arising from the organism's reaction to the preceding external stimulus. Thus, S_2 shortly follows s_1, S_3 shortly follows s_2, and so on.

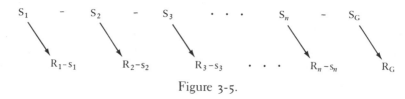

Figure 3-5.

To this point Hull's analysis has simply pictured in S-R terms the major links in an organism-environment interaction sequence. Successive aspects of the world flux produce particular reactions by the organism, and these reactions in turn result in both distinctive internal stimuli and new external stimuli. But this simple S-R description also reveals—and this is the point of the analysis—that such interaction sequences contain the conditions for classical conditioning. Recall that in Pavlovian conditioning an association is established between a stimulus and a response by the pairing of that stimulus with another which reliably evokes the behavior in question. As Hull invoked this principle it took the form "all the components of a stimulus complex impinging upon the sensorium (i.e., upon receptors) at or near the time that a response is evoked, tend themselves independently to acquire the capacity to evoke substantially the same response" (1930a, p. 513). Inspection of Figure 3-5 with this principle in mind shows that conditioning can be expected at each juncture of the sequence. The response R_2 is a characteristic reaction to S_2, and S_2 occurs close in time to s_1; consequently, s_1 should acquire power to evoke R_2. By the same logic, s_2 should come to evoke R_3, s_3 to evoke R_4, and so on. Figure 3-6 shows the associations expected to develop through conditioning in the form of dotted arrows between "s and next R" elements of the sequence.

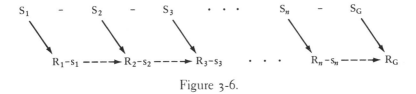

Figure 3-6.

After one or more repetitions of the world sequence, then, we can expect the elements of the behavior sequence to become knitted together by links forged within the organism itself. There would then exist two independent bases for the behavior sequence—first, the pattern of external stimulus events which give rise to the sequence and second, internal, response-produced stimuli associated with reactions in the sequence. To make this clear, suppose that the world sequence is now interrupted after the occurrence of, say, S_1. The resulting situation is shown schematically in Figure 3-7. The newly acquired associations, unless outweighed by some more potent stimulus influence, should continue the behavior sequence in much the same manner as when evoked by the world sequence itself.

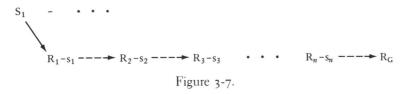

Figure 3-7.

To put this in general terms, conditioning theory indicates that repetition of any organism-environment sequence conditions an internally organized sequence which can operate independently of, yet faithfully corresponds to, the original interaction. In Hull's words, "Henceforth the organism will carry about continuously a kind of replica of this world segment. In this very intimate and biologically significant sense the organism may be said to know the world. No spiritual or supernatural forces need be assumed to understand the acquisition of this knowledge. The process is entirely a naturalistic one throughout" (1930a, p. 514).

Once the organism has acquired knowledge of the world in the form of such self-maintaining stimulus-response sequences, other conditioning processes can be expected to come into play with important behavioral consequences. One of these processes, which Hull was to emphasize throughout his writings, is the tendency for a conditioned reaction to anticipate the S-R association on which it is based.

For example, consider a salivary conditioning procedure in which the conditioned stimulus is a one-minute tone followed by one minute of silence and then by presentation of the unconditioned stimulus (e.g. food). Initially, of course, salivary flow will be elicited only by presentation of the unconditioned stimulus, but after this procedure has been repeated a number of times it is likely that salivary flow will appear during the one-minute interval preceding the unconditioned stimulus, and with increasing magnitude as trials progress. That is, the conditioned response will be observed to advance in time toward the conditioned stimulus in such a way as to consistently antedate presentation of the unconditioned stimulus. An anticipatory tendency of this nature is a characteristic feature of conditioning experiments. When this principle of anticipatory conditioned reaction is applied to parallel world and internally organized action sequences (Figure 3-6), it is evident that the terminal reactions of the internally organized sequence will tend to antedate the world stimuli which originally evoked them. In terms of Figure 3-6, s_n will tend to elicit the goal reaction, R_G, prior to occurrence of S_G.

A little reflection will make clear that such anticipatory responses have great adaptive potential. Suppose an animal engages in a behavior sequence which terminates in a painful stimulus; for example, suppose a rat traverses a maze, enters the goal box, and thereupon receives a painful shock. Suppose further that the goal reaction, R_G, is a successful flight or withdrawal response elicited by that shock at the goal. According to the present analysis, repetition of this sequence of events should result in the animal withdrawing from the goal region *prior to the appearance of the goal box itself,* as a result of the tendency of the conditioned sequence to run off faster than the world sequence on which it is based. If the painful stimulus in the goal is sufficiently intense, and therefore if conditioning is strong, the organism may seldom or never expose itself to the goal region again. The adaptive value of the anticipatory goal reaction in this case is obvious.

The point of greatest significance about the anticipatory mechanism illustrated here is that it enables one to resolve the problem of foreknowledge, or foresight, in strictly behavioral terms. The essence of the problem of foresight is to understand how an organism can react to an event which may be impending but which has not as yet taken place. Yet this is precisely the behavior expected under the anticipatory mechanism described above. In the particular illustration employed, the rat flees appropriately before the painful stimulus or its immediate surround is encountered. The particular value of Hull's analysis lies in showing that such foresightful behavior can be derived from the operation of simple conditioning principles and that we need not postulate foresight as a basic, "given" mental capacity of the organism.

To expand the applicability of Hull's analysis, it can be seen by a similar line of reasoning that successful attack reactions to aversive events could also become anticipatory. By means of an early attack reaction an organism might thwart a painful event, such as an attack by another organism. Thus, the two types of reactions elicited by aversive events, flight and fight reactions, have the potential of occurring anticipatorily with adaptive benefit to the organism. But what about sequences which terminate in rewarding, or appetitive, events? Would not the same mechamism tend to elicit anticipatory appetitive responses? Suppose, for example, that in our earlier maze illustration there were food rather than shock in the goal box. As training progresses, would not the responses involved in seizing and ingesting the food tend to occur prior to attainment of the food itself? Indeed they would, said Hull, and we can confidently expect to find evidence of such conditioned anticipation. However, he pointed out that there are limits to the form and magnitude of anticipatory appetitive reactions because these reactions if full-blown could interrupt the behavior sequence on which they are based. Therefore, Hull argued, anticipatory appetitive goal reactions are likely to involve only those *portions* of the goal reaction which do not conflict strongly with the antecedent behavior sequence. In our maze illustration, for instance, we can expect salivation and incipient chewing movements to occur anticipatorily since these responses can take place concurrently with movement to the goal. Anticipatory responses of this nature are frequently observed during the course of conditioning instrumental behavior sequences with food reinforcement (e.g., Cowles, 1937; Wolfe, 1936).

Hull's analysis, then, can account for the appearance of anticipatory, or foresightful, behaviors in the presence of stimuli associated with either aversive or appetitive events. But if this approach is to have relevance for the long-range foresight which appears to characterize the behavior of higher organisms, it is also necessary to show that conditioned anticipatory goal reactions need not be restricted to the locus of the reinforcing event but can occur early in the behavior chain leading to reinforcement. Hull met this problem by a simple extension of his analysis, and perhaps the reader can now anticipate the general form of the explanation. If foresightful behavior is held to consist of conditioned goal reactions, what is required if these conditioned behaviors are to be observed at points removed from the locus of their conditioning? Since conditioned behaviors appear only upon occurrence of their controlling conditioned stimuli (or closely similar stimuli), conditioned goal reactions can appear at points distant from the goal only if their conditioned stimuli appear at those points. This approach, then, appears to require the existence of stimuli which are present both at the goal region and at far removed points if long-range foresightful behavior is to be generally expected. Is there any basis to assume that such stimuli exist? Hull argued that, in fact, most organism-environment interaction sequences are likely to involve stimuli which persist throughout the action sequence from start to finish. He had in mind primarily those internal stimuli associated with the organism's need state. For example, Hull assumed that food deprivation creates distinctive "hunger stimuli" (e.g., the recurring crampings of the digestive tract) that persist through all the organism's actions until hunger is eliminated by sufficient food intake. When a hungry animal is trained in a maze, then, the distinctive hunger stimuli will be present at the goal and, like any stimulus at the goal, will become conditioned to elicit portions of the goal reaction. But since these hunger stimuli are carried about by the animal, there is now a basis to expect appearance of the goal reactions at virtually any point in the maze-behavior sequence, even in the start box itself. As a result of this conditioning process, the subject will appear to "know what is in the goal box" and to "know where it is going" at the very outset of the behavior sequence.

Internal drive stimuli had yet another important role to play in Hull's analysis. Since these stimuli persist throughout a behavior sequence they would be expected to become conditioned to each reaction in the sequence under the same principle whereby response-produced stimuli become conditioned to the reactions which they precede. Figure 3-8 extends our previous schematic analyses by incorporating the conditioning of the drive stimulus component, here symbolized by s_d, throughout the behavior sequence. Hull further assumed (with some empirical justification) that s_d becomes conditioned more strongly as the final reactions in the behavior cycle are approached; consequently the conditioned anticipatory tendency between s_d and R_G would be stronger than that between s_d and any other reaction in the sequence, even though the s_d gains some association with every element of the behavior chain.

Of course, in natural life an organism obtains food not through any single behavior sequence but rather through a number of different behaviors in a variety of environmental settings, and as the organism learns various ways to obtain food the hunger stimuli should

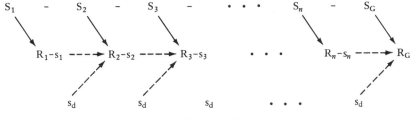

Figure 3-8.

become conditioned in the manner of Figure 3-8 throughout each of the effective food-procuring behavior chains. Since the s_d component thus becomes associated with the set of food-procuring behaviors, the organism tends, whenever hungry, to run through this repertoire of acts, at least to the degree permitted by the existing external stimulus situation. In this way, need states can be expected to give rise to behaviors which exhibit an organized, goal-seeking, purposive character.

The foregoing analyses represent some of Hull's earliest efforts to explain phenomena relating to knowledge and purpose in behavior. These analyses by no means cover all of his thinking on these problems, but our purpose at this point is not to provide such coverage nor even to assess the adequacy of these accounts. We shall return to these matters at length in a later chapter. Our purpose now is simply to illustrate Hull's general approach to learning, particularly as it compares to Tolman's approach. Let us recapitulate the main points. Repeated sequences in the outer world produce parallel reaction sequences in the organism. Through conditioning the organismic sequences acquire a tendency to run off by themselves independently of, yet in correspondence with, the world sequences. In this way the organism acquires what Hull termed "an intimate functional copy of the world sequence, which is a kind of knowledge" (1930a, p. 523). The organismic sequence tends to run off at a faster rate than the world sequence with the result that the knowledge becomes foreknowledge, or foresight. By this means, organisms come to respond in adaptive anticipatory fashion to impending aversive and appetitive events. Since the organism's drive state constitutes a distinctive, persistent stimulus aspect in most learning situations, goal behaviors become conditioned to drive stimuli along with other stimuli at the goal. When the link between drive stimuli and goal reactions is well developed, goal reactions (foresightful behavior) may then appear early in the reinforced sequence. Finally, since drive stimuli are present throughout reinforced behavior sequences, they will tend to become associated with the constituent behaviors, giving rise to a tendency for need states to evoke sets of need-serving (purposive) acts.

This account of knowledge and purpose illustrates the most significant aspects of Hull's approach. First, it can be seen that Hull's units of analysis are molecular relative to those of Tolman. There is a strong implicit assumption that behavior at base is a matter of short-term glandular and muscular actions under the control of the immediate stimulus input. The conception underlying the schema of Figures 3-3 through 3-8 is that of an organism interacting with its environment by means of a series of overlapping yet discrete associations between aspects of the peripheral stimulus input and specific movements. In this conception the particular movements composing an activity assume far greater importance than in Tolman's view, since it is particular movements, not acts in Tolman's sense, that become the units of learning. In Hull's system the molecular S-R units may combine to give rise to molar behaviors, but they retain in combination their identity as basic functional and explanatory units. For example, Hull held that as maze training progresses each locomotory movement becomes conditioned to the internal stimuli arising from the previous movement, the result being that maze traversal may ultimately take on the appearance of a cohesive, goal-oriented action with an "inner-directed" quality. But despite this actlike structure, the organism's actions at any point in the maze will nevertheless remain a matter of the particular stimulus-response associations operating at that point. In this connection, recall Tolman's observation that a rat is likely to run head on into the end wall of a previously much-traversed but now suddenly shortened alleyway. For Tolman, the rat bumps into the wall because it has acquired an expectation, or knowledge, of a longer alley and hence reliance upon peripheral stimulus input is reduced; rapid forward movement becomes the appropriate goal-oriented behavior to the cognitions evoked at that point. But in Hull's view the rat is literally driven into the wall by the specific "move forward" response which has been conditioned to the stimuli at that point in the maze-behavior sequence. The rat is not acting out of knowledge of the coming stimulus; it is responding as conditioned to the stimulus input immediately present.

Another feature of Hull's approach illustrated by the analysis of knowledge and purpose is description of complex phenomena in terms of *peripheral events*. In the context of psychological theory, peripheral events are defined by action of the receptors and effectors and they are to be distinguished from *central events*, or those conceived as intervening between receptor-effector activities and ultimately definable by brain processes. Hull's and Tolman's conceptions of knowledge provide a striking contrast of peripheral and central explanations. As we have seen, Tolman explicitly conceptualized cognition as a central process, although to be sure he anchored cognitions in observable S-R relations. Given his definition of cognition as the evocation by a stimulus of an expectation of a subsequent stimulus, it is an easy step to conceive of Tolman's cognitive events as some form of central representation of a stimulus (e.g., a memory) which moves forward in time to antedate occurrence of that stimulus. In contrast, Hull defined knowledge strictly in terms of peripheral response processes. Specifically, knowledge consists of responses to external stimuli which become conditioned to contiguous antecedent internal stimuli and in this way come to antedate, in conditioned form, the external stimuli. In essence, Tolman views knowledge as a central copy of perception of the world, whereas Hull views knowledge as a peripheral copy of behavior to the world.

Peripheral events appear more amenable to objective study than central processes, particularly from the viewpoint of the neurophysiology of 1930, and this claim to greater objectivity underlies Hull's preference for peripheral activities as explanatory mechanisms. Watsonian behaviorism of course provided a major impetus to peripheralism in theory construction, and Watson himself had earlier contributed perhaps the most striking instance of peripheral explanation by his conception of thought as implicit speech, detectable as movements of the vocal cords. Hull's translation of knowledge and purpose into internally linked conditioned movements and fractional anticipatory goal responses falls directly in the line of Watson's theorizing. And like Watson, Hull was also to make frequent use of peripheral reactions which were not observed but merely postulated to occur. But from the viewpoint of orthodox behaviorism, postulated covert peripheral reactions could more readily be defended as real and potentially measurable physical events than could central reactions. In this light, Hull's peripheralism can be considered a further expression of the Watson-inspired zeal for a behavior-based objectivity.

We noted earlier that Hull differed from Tolman in his reliance upon the conditioning experiment as a source of explanatory principles, and this feature of Hull's approach is evident in the analysis of knowledge and purpose. From a broader perspective the use of conditioning principles and of molecular and peripheral units of analysis can be seen as aspects of a basic *modus operandi*, namely, derivation of the complex from the simple, of the higher-order from the basic.

Note that Hull's account of knowledge and foresight is essentially a working out of the implications of two assumptions, one concerning the appropriate units of behavioral analysis, the other concerning generality of conditioning principles. By logical analysis Hull shows that if molecular S-R associations obtain and if the principles of conditioning experiments operate throughout organismic activity, then we can expect, indeed, then we must expect, that behavior will anticipate environmental events and that anticipatory behavior will be appropriate and adaptive in relation to aversive and appetitive events. Tolman felt that observations of foresightful behavior warranted the postulation of cognition as a primary capacity of the organism, but Hull's analysis argues that foresight can be seen as a learned phenomenon, and a relatively simple form of learning at that. This feature of Hull's approach, the deduction of complex phenomena from the operation of simple S-R processes, accounted for a good deal of its attraction to contemporary behaviorists.

Much of our discussion of the broad quality of Hull's approach as compared to that of Tolman can be succinctly summarized by simply considering the titles of two of their earliest papers on learning. In 1925 Tolman published a paper which outlined his major ideas for the explanation of behavior and learning. Its title was "Purpose and Cognition: The Determiners of Animal Learning." In 1930 Hull published the paper we have considered here at length. Its title was "Knowledge and Purpose as Habit Mechanisms."

Basic Features of Hull's Theory of Learning

One of Hull's major goals was to improve the quality of psychological theory, which he felt had generally failed to meet the criteria of scientific explanation. Using the theoretical systems of the natural sciences as exemplars, he early and repeatedly emphasized what he discerned as the essentials of sound scientific theory (Hull, 1930b, 1935a, 1937), and much of his own theorizing can be seen as an effort to demonstrate that psychology was capable of generating theoretical systems as rigorous as those of the natural sciences.

Hull maintained that scientific theory in the best sense was characterized by a three-phase procedure. A theory should begin with formulation of a relatively small number of carefully defined postulates, or basic assumptions. Essentially, the postulates are the theorist's "best guesses" as to the most fundamental factors operating in the phenomena in question. Then, considering the implications of these postulates in combination, the theorist should deduce by a rigorous and clearly detailed logic a series of theorems, or testable statements of what should be observed under specific conditions. In psychological theory, a theorem should take the form of an hypothesis about the outcome of a concrete experiment. Theorems may refer to already-known facts and experimental outcomes, but of particular importance in the development of theory is the deduction of theorems which predict the outcomes of experiments never before conducted. Finally, there should be a continual program of theorem testing. Observational confirmation of theorems constitutes evidence for the truth of the system; disconfirmation is evidence of its falsity. If the theorems cannot be unambiguously confirmed or disconfirmed, the system is scientifically meaningless.

Construction of a valid theory by these procedures is not an all-at-once affair. Postulates formulated in the early stages of any discipline would be expected to yield a number of theorems which are disconfirmed or incapable of clear test. But by revision of postulate sets in the light of experimental findings and by recurring cycles of the postulation-deduction-test sequence, there should evolve postulates increasingly capable of deducing the known and unknown facts of behavior. In Hull's words, "Thus theoretical truth is not absolute, but relative" (1935a, p. 514).

Hull's own work shows such a programmatic progression. As our review of the 1930 "Knowledge and Purpose" paper revealed, his earliest theoretical efforts showed a strongly deductive character. Over the next decade he published a series of papers, each attempting to deduce particular phenomena of learning (chiefly of maze learning) from a few postulates. His presidential address to the American Psychological Association (Hull, 1937) was a first attempt to organize his previous work into a single postulate set, or general theory of learning. These efforts culminated in the publication of his book *Principles of Behavior* (1943), which presented a set of 16 postulates intended to provide a comprehensive account of learning.

Hull made further revisions of his postulates after 1943, but it is the *Principles of Behavior* and his earlier papers that present his most influential views on learning, and our characterization of Hull's learning theory will be culled from these writings.

We shall first present and briefly discuss a number of the postulates from *Principles of Behavior*, selecting those that best communicate the substance of the theory. Then we shall consider some of the ways those postulates were applied to deduce major phenomena of learning and behavior.

At the heart of Hull's theory lies the so-called *reinforcement postulate* (postulate 4 in *Principles*). Paraphrased to remove the symbols and technical terms used in *Principles*, this postulate reads:

> Whenever an effector activity and a receptor activity occur in close temporal contiguity and this conjunction is closely associated with the diminution of a need or with a stimulus which has been closely and consistently associated with the diminution of a need, there will result an increment to a tendency for that afferent impulse on later occasions to evoke that reaction. The increments from such successive reinforcements summate to yield a combined habit strength (or S-R association strength) which is a simple positive growth function of the number of reinforcements. The upper limit of this curve of learning is the product of (1) a positive growth function of the magnitude of need reduction, (2) a negative function of the delay in reinforcement, and (3) a negative growth function of the degree of asynchronism (nonsimultaneity) of stimulus and response.

This postulate gives Hull's position on the two most important questions in learning theory—what is learned? and how is it learned? To put these issues more formally, what is the nature of the organismic changes that define learning and what conditions are necessary to bring those changes about?

The postulate holds that what is learned are receptor-effector connections. But in what sense can receptors and effectors be said to be "connected" in nervous functioning? While there are instances of direct linkages between receptor and effector structures (in monosynaptic reflexes), the representative mode of relating receptor and effector activities is by way of central (brain) structures. Thus, what is learned in Hull's system must be conceived of as a change within the central nervous system, a change which has the effect of making particular effector activities more probable upon recurrence of particular receptor activities. The precise nature of this change was left open by Hull: "It exists as an organization as yet largely unknown, hidden within the complex structure of the nervous system" (1943, p. 102). But that he conceived of this central process as a relatively simple hookup mechanism is indicated by his speculation that it is "a change in the state of the conduction structure of the nervous system whereby the propagation of the [afferent] nerve impulse, s, is routed into the efferent fibers leading to the organs which execute the reaction, r" (1942, p. 66).

Hull did not actually work at the neurophysiological level as these conceptions of learning imply. Rather, his choice of terms here devolves from his general conviction that "it would seem wisest to keep the causal segments [of behavior analysis] small, to approach the molecular, the fine and exact substructural details, just as closely as the knowledge of the substructure renders possible" (1943, p. 21). One may well question whether knowledge of neural substructures at that time permitted any very meaningful hypothesis about the neural correlates of learning, but the issue of the validity of Hull's neurophysiologizing is really of little importance. The significant point to note about the definition of learning as "receptor-effector connections" is that for Hull, as for Tolman, the critical change in learning becomes an unobserved central event the nature of which must be inferred from behavior. It is also worth noting that defining learning as a receptor-effector connection biases one to conceive of the stimulus and response elements of learning in molecular units, i.e., in units small enough to permit their ready conceptualization as particular afferent and efferent processes.

As regards how learning takes place, the postulate specifies two conditions. There must be temporal contiguity between the stimulus and response processes involved, and this conjunction must be followed closely in time by reduction in a need or by a stimulus that has consistently accompanied need reduction. By needs (or drives— we shall use these terms interchangeably) Hull referred to bodily states that deviate materially from conditions optimal for survival of the organism or of the species, such states as pain, hunger, thirst, and sexual deprivation. It is the specification of need reduction as the primary condition for learning that most clearly distinguishes Hull from other learning theorists. To Hull this proposition was an eminently reasonable extension of the doctrine of evolution, which he felt constrains analysis of organismic activity from the viewpoint of survival. If survival is taken as the paramount principle of organic functioning, it follows that behavior must function to satisfy needs. Casual observation alone suggests that needs are strongly associated with behaviors that appear to serve the needs. Thus, pain reflexively evokes vigorous movement away from the source of pain, hunger is associated with increased locomotor activity (searching) in many animals, cold produces shivering, and so forth. A review of the experimental evidence on such need-behavior relationships led Hull to formally postulate that organisms possess innate receptor-effector connections, established by the processes of organic evolution, which facilitate the satisfaction of particular need states (postulate 3 of

Principles). But it is clear that a large proportion of the needs of higher organisms are likely to be poorly served by innate, ready-made behavior patterns alone. Rather, higher organisms require, in addition, a system of widely adaptable behavior if they are to meet the highly variable and unpredictable situations of need which characterize their world. Given these considerations, it is a small step to the hypothesis that organisms may have evolved a system of behavioral adaptation that is governed primarily by the need-serving consequences of behaviors. Such was the general reasoning that led Hull to postulate need reduction as the primary reinforcing (strengthening) event in learning. The survival value of this principle is obvious; it provides that, as a rule, behavior will change in the direction of maximizing need fulfillment.

Postulate 4 also states that any stimulus that regularly accompanies need reduction will acquire the power to strengthen receptor-effector connections. Stimuli which thus acquire reinforcing capacity were termed *secondary reinforcers* by Hull, and need-reducing events themselves were termed *primary reinforcers*, "primary" in the sense that their capacity to strengthen receptor-effector connections is presumably innate.

In support of this aspect of postulate 4, Hull was able to cite a number of experimental demonstrations of acquired reinforcement, although evidence on the generality of the phenomenon was slight at the time. In an early study of secondary reinforcement, Bugelski (1938) first trained rats to press a bar for a food pellet reward, each food reinforcement being accompanied by a distinctive clicking sound from the pellet-release mechanism. The bar press response was then extinguished by adjustment of the apparatus so that presses were no longer followed by food delivery. During this extinction treatment one half of the rats continued to receive the customary click after each bar press while the other half did not. Bugelski found that the click-extinction group gave about 30 percent more bar presses before extinguishing than did the no-click group, thereby demonstrating that the clicking sound acted to maintain the response in the absence of primary reinforcement, i.e., that it acted as a primary reinforcer itself would act.

Cowles (1937) used a learning paradigm rather than an extinction procedure to demonstrate secondary reinforcer formation. Two chimpanzees were first trained to insert small colored disks into a slot machine which delivered a food reward (raisins) for each disk inserted. Subsequently the subjects were trained on a series of choice tasks in which the only differential consequence of choice behavior was whether or not the subject received a disk. On each trial of a choice task the subject was allowed to open one of an array of five discriminably different lidded boxes, only one of which (the same on all trials) contained a disk. On the initial trials of such a problem, the subject would of course be expected to choose on a chance basis, but if the disks have acquired reinforcing power by virtue of their association with food in the earlier training, the subject should come to choose consistently the "correct" (disk-containing) box. Upon completion of 20 choice trials with a given box designated as correct, the disk was shifted to a new box and another 20 choice trials were given, and so on through the five boxes. Cowles found that the average score of his subjects on the second half of each 20 trials on a given problem (where by chance 20 percent success would be expected) was 74 percent correct. This figure compares very favorably with the 93 percent correct behavior observed during a comparable training series with food reward for correct choice. Clearly, receipt of the disks had an effect on choice behavior much like that of receipt of the food itself.

Hull cited several additional studies of secondary reinforcement, but the Bugelski and Cowles experiments suffice to convey the general nature and importance of the phenomenon. These studies demonstrate, in Hull's terms, that stimuli associated with need reduction can be expected to contribute to the maintenance of receptor-effector connections (Bugelski experiment) and to strengthen new receptor-effector connections (Cowles experiment). Moreover, Hull argued further (although without data) that stimuli associated with a *secondary* reinforcer also take on reinforcing powers. That is, he speculated that the transfer of the reinforcing power of need-reducing events could be progressively extended to additional stimuli in a chainlike process. These principles greatly increase the range and flexibility of the basic need-reduction postulate since they mean that virtually any stimulus in the organism's environment can come to play the critical role ascribed to need reduction. Furthermore, they mean that receptor-effector conjunctions may be strengthened at points far removed temporally and spatially from actual need reduction. These points are of particular importance to the application of Hull's theory to human behavior, where the great bulk of learning appears to occur in the absence of any apparent biological need satisfaction. Note, however, that biological need reduction remains the paramount principle of learning in Hull's system in the sense that secondary reinforcers ultimately owe their existence to association, direct or indirect, with need reduction.

The remainder of postulate 4 specifies a number of conditions influencing the strength of receptor-effector connections, or, to use Hull's alternative term, habit strength. Since habits, or receptor-effector connections, are not observable, habit strength must be inferred from relationships between particular antecedent stimulus conditions and consequent behaviors. On the antecedent, or input, side, habit formation was held to depend chiefly upon the number of reinforcements, i.e., upon the number of times the stimulus-response conjunction was followed by need reduction or stimuli associated with need reduction. More specifically, Hull postulated that habit formation proceeds incrementally with successive reinforcements up to some physiologically determined limit of habit strength. The postulate also specifies amount of reinforcement, delay of reinforcement, and stimulus-response asynchrony as important determinants of habit strength and defines the nature of their influence. On the behavioral side, Hull anchored habit strength by several measures, including response magnitude, response latency, and resistance of the reaction to extinction.

The remaining postulates we shall consider here do not have the central, defining character of postulate 4 in Hull's system. Accordingly, these postulates will simply be listed at this point in paraphrased form, along with brief interpretative comments.

The Stimulus Trace Postulate (from postulate 1 of *Principles*). After termination of the action of a stimulus on a receptor, the afferent neural impulse generated by that stimulus continues its activity in the central nervous system for some seconds, gradually diminishing to zero.

There is neurophysiological evidence to support the notion that neural activity initiated by a stimulus may persist after cessation of that stimulus, but it is not clear that such neural traces are as directly related to the stimulus input or as long-lasting as Hull supposed.

Generalization Postulates. The reaction involved in the original conditioning becomes connected with a considerable zone of stimuli other than, but adjacent to, the stimulus conventionally involved in the original conditioning; this is called *stimulus generalization*. The stimulus involved in the original conditioning becomes connected with a considerable zone of reactions other than, but related to, the reaction conventionally involved in the original reinforcement; this may be called *response generalization* (from *Principles*).

These assumptions simply acknowledge the well-known tendencies for conditioned responses to occur to stimuli similar to the stimulus involved in the conditioning, and for conditioning to strengthen not only the particular response involved but a range of similar responses as well.

Postulates Specifying Relations Between Motivation and Learning. Associated with every drive (need state) is a characteristic drive stimulus (S_D) whose intensity is an increasing function of the drive in question (postulate 6 of *Principles*).

The most important aspect of this postulate is the idea that drives can function as stimuli, an idea we encountered earlier in Hull's theorizing about knowledge and purpose. In support of this postulate, Hull showed (1933) that rats could learn to take one maze path to obtain water when thirsty and to take another path to obtain food in the same goal box when made hungry on alternate days. Leeper (1935) later showed that this discrimination between drive stimuli could be accomplished with considerable facility if the food and water reinforcers were placed in spatially distinct goal boxes.

Habits ($_sH_R$ units) are "sensitized into reaction potentiality" by all drives active within an organism at a given time, the magnitude of the reaction potential being a product of habit strength multiplied by the strength of drive (postulate 7 of *Principles*).

In saying that habits are sensitized into reaction potentiality by drives, Hull was saying essentially that drive brings about a readiness to respond according to existing habit structures, the strength of this readiness, or predisposition to respond, increasing with the strength of drive. The sensitization of habits into reaction potential by drive was viewed by Hull as a necessary step in response evocation. However, prediction of precisely which learned responses will be evoked on a given occasion requires consideration of additional variables, notably the nature of the stimulus situation in relation to the organism's existing $_sH_R$ units.

Two other aspects of this postulate should be noted. First, it is assumed that different drives exert a combined, common sensitizing influence on habits. One implication of this assumption is that learned responses would be expected to persist even after the drive under which they were established is reduced, if other drives remain to sensitize the habit. Second, the postulate embodies the logical counterpart of

Tolman's learning-performance distinction in that it holds that behavior is a multiplicative function of habit strength and drive strength.
Thus, an organism may have acquired very strong receptor-effector
connections which are fully appropriate to a given situation, yet these
habits would not be reflected in behavior if drive were at zero strength.

> *Postulates Dealing with Extinction.* Whenever a reaction is
> evoked there is created a primary negative drive which has the
> capacity to inhibit the reaction potentiality of that response, the
> amount of inhibition increasing with the number of response
> evocations and with the amount of work involved in response
> execution. This reactive inhibition spontaneously dissipates dur-
> ing periods of nonresponding (postulate 8 of *Principles*).

The central idea of this postulate, that the sheer act of responding
creates a tendency against repeating a response, is the basis of Hull's
treatment of extinction. Since he assumed that the negative drive and
associated inhibition produced by responding are usually slight relative to the strengthening effect of reinforcement, it follows that reactive inhibition will be difficult to detect when behavior is undergoing
reinforcement. But when reinforcement is consistently withheld, inhibition from repeated responding will accumulate to progressively
weaken (extinguish) the behavior. Note that the postulate sees extinction not as a matter of unlearning the habit itself, but rather as a
matter of offsetting the effects of the drive to make the response, i.e.,
as reducing the reaction potential of the habit in question. Finally,
since reactive inhibition is assumed to dissipate with time, it follows
that extinguished responses can be expected to reappear when the organism is later reintroduced to the learning situation, as in the commonly observed phenomenon of "spontaneous recovery" (see p. 13).

The concept of reactive inhibition provides only for temporary
weakening of behavior through nonreinforcement, but it is known
that learned behaviors are permanently weakened by repeated extinction treatments. Long-term extinction is accounted for by postulate 9
of *Principles*, as follows:

> Stimuli closely associated with the cessation of a response be-
> come conditioned to the inhibition associated with the evocation
> of that response, thereby generating conditioned (learned) inhi-
> bition. Conditioned inhibition and reactive inhibition summate
> against the reaction potentiality to a given response.

The first part of this postulate follows from the drive reduction principle of postulate 4. Recalling that responding induces a negative drive, it is obvious that cessation of nonreinforced responding will be followed by reduction of this drive state. Consequently, stimuli contiguous with the response of "stopping responding" should become conditioned to evoke that response through the action of drive reduction. As a result of repeated extinction treatments, then, the organism eventually learns not to respond in its former manner, and this learning, like all learning in Hull's system, is governed by the drive reduction principle.

Application of Postulates in Explanation of Some Basic Phenomena of Learning

Classical and Instrumental Conditioning as Instances of a Single Principle of Learning. We noted earlier that classical and instrumental conditioning are *the* two fundamental operations for producing behavior change in the laboratory. Hull argued that despite the apparent procedural differences between classical and instrumental conditioning, both forms of learning follow postulate 4.

The application of postulate 4 to instrumental conditioning is straightforward since in this procedure learning is typically brought about by the following of a response with a stimulus that is either patently need-reducing or likely to have been associated with need reduction. To illustrate classical conditioning as an instance of the same principle, Hull considered a conditioning experiment involving the following representative circumstances. A dog is restrained in a standing position on a laboratory table. An electric grid is attached to the dog's foot, and current to the grid passes through a switch which is opened by upward motion of the foot. A buzzer is sounded two seconds before a shock is delivered to the foot grid. The shock elicits as its dominant reaction a reflex lifting of the foot, which breaks the circuit, thus terminating the shock. After a number of such conditioning trials, the dog is observed to lift its foot at the onset of the buzzer, i.e., prior to the delivery of shock. In Pavlov's terms, the buzzer comes to elicit foot withdrawal as a consequence of its pairing with a stimulus (shock) which unconditionally elicits that response. In this view, presentation of the shock is the event which reinforces the bond between buzzer and foot withdrawal. Hull's analysis, however, points out that each trial ends in reduction of the pain-fear drive induced by shock. Thus, each trial arranges temporal contiguity between a new stimulus-response conjunction (buzzer–foot withdrawal) and drive reduction. From this viewpoint, the critical reinforcing event is not shock presentation but shock termination, or more precisely, the drive reduction brought about by shock termination. Hull believed that a similar analysis could be applied to any Pavlovian procedure employing an aversive stimulus as the unconditioned stimulus. Similarly, cases of classical conditioning with appetitional UCS's were also seen as instances of the drive reduction principle. Thus, in Pavlovian salivary conditioning, the unconditional stimulus (meat powder) at once elicits the response to be learned (salivation) and serves as a reinforcer by reducing hunger drive. In short, Hull argued that classical and instrumental conditioning and their various derivative procedures contain at the core the conditions of postulate 4—a stimulus-response conjunction followed closely in time by need reduction or by a stimulus which has accompanied need reduction.

Deduction and Elaboration of the r_g Mechanism. The concept of the fractional anticipatory goal reaction, or r_g, as introduced in the 1930 "Knowledge and Purpose" paper, was based upon empirical principles of conditioning. In Hull's later writings, the r_g appears as a clearcut deduction from the postulate system. Specifically, by postulate 4, need reduction or a stimulus that has accompanied need reduction will strengthen any contiguous S-R conjunction, and one such conjunction in learning situations is that between the persisting drive stimulus (postulate 6) and the goal reaction. Moreover, by postulate 2, traces of stimuli in a sequence leading to a reinforcer can also be expected to persist and overlap in time with the goal reaction. It follows directly from these postulates that drive stimuli as well as other stimuli occurring prior to the point of reinforcement will tend to elicit the goal reaction, or, more typically, those fractions of the goal reaction which can occur in the absence of the goal object itself. Finally, by postulate 4, the strength of these anticipatory tendencies should increase with nearness to the goal.

In his 1930 "Knowledge and Purpose" paper, Hull employed the r_g primarily in explanation of foresightful behavior. Later papers progressively extended the r_g mechanism in analysis of learning phenomena, with the result that the r_g came to bear a major explanatory burden in Hull's system. Several of the more important applications of the r_g will be considered at this point.

Figure 3-9 illustrates Hull's conception of the conditioning of the r_g throughout a well-practiced behavior sequence eventuating in re-

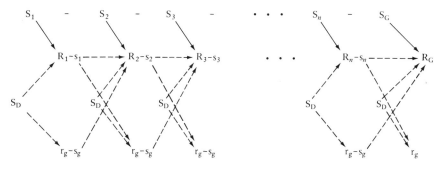

Figure 3-9.

inforcement. Here R_G designates the total goal reaction, r_g those portions of the goal reaction which can occur simultaneously with antecedent reactions, S_D the drive stimulus, and the dotted arrows associations expected to develop through conditioning under the assumptions of postulate 4. Since the S_D will be conditioned to elements of the reaction sequence with an intensity proportional to the proximity of each to the goal, it follows that S_D will become most strongly conditioned to the goal reaction itself. With sufficient reinforcement of an S-R series, S_D should eventually come to evoke r_g through the entire sequence, as depicted in Figure 3-9.

Like any other reaction r_g causes characteristic proprioceptive stimulation to arise from the movements involved. Stimulation arising from the r_g, designated by the symbol s_g, was termed by Hull the *goal stimulus*. Now, since r_g persists throughout the behavior sequence, s_g must also do so and, like the drive stimulus, would be expected to become conditioned to every reaction of the sequence (see Figure 3-9). Moreover, s_g, by virtue of its proximity to the primary terminal reinforcement, should acquire strong secondary reinforcing power; consequently, s_g provides a means of strengthening S-R conjunctions at points far removed from the terminal reinforcer.

As elaborated above, the r_g-s_g mechanism can be expected to play a critical role in the organization of behavior. Its presence throughout an S-R series exerts a dual integrative influence. As a conditioned stimulus component it contributes to evocation of the successive reactions, and as a secondary reinforcer it strengthens and maintains antecedent reactions. Therefore, it can be expected that as training progresses, i.e., as s_g gains in conditioned excitatory and reinforcing power, the series of S-R relations will become organized into a tightly cohesive, goal-oriented action. And, in fact, maze-running behavior is commonly reported to undergo such a qualitative organizational change under extended training. As one description has it,

> The new behavior is often characterized as appearing to be more "purposeful" than at the beginning; [the animal] acts as if he were "going somewhere." Beneath this rather vague characterization there may be discerned certain fairly concrete and definite behavior tendencies, the most notable of which involve anticipatory movements. When an animal is approaching his goal (the food box) he is apt progressively to speed up his pace. Another significant observation is that when an animal is approaching a familiar 90° turn in a maze, he quite generally begins his turning movements some time before he reaches the corner. . . . As a food goal is neared the organism tends to make mouth movements of a masticatory nature (Hull, 1931, p. 488).

The integrating and goal-directing functions of r_g-s_g can perhaps best be realized when it is considered that this mechanism is literally a case of anticipation serving to bring about the end anticipated. The r_g-s_g both represents the goal and, by its associative connections, provides a mechanism for attainment of that goal. In this light, the r_g functions as what Hull termed a *pure stimulus act*, that is, an act whose function is to serve as a stimulus for other acts. It is apparent that the occurrence of r_g during a behavior sequence has no instrumental value by itself. But when it is considered that r_g produces s_g, which is linked to only those reactions which invariably lead to the goal, then r_g is seen to play a very useful role indeed. Namely, it produces the stimulation that organizes and directs behavior toward the realization of reinforcement.

Considered as an achievement of deductive reasoning, the r_g mechanism is a thing of beauty. Beginning with empirically based assumptions about the nature of learning, Hull deduced as a necessary consequence of these assumptions the development through learning of a purely physical mechanism that automatically functions to guide behavior adaptively to the procurement of reinforcement. Not surprisingly, the r_g concept, consistent with its first usage in the 1930 paper, became Hull's primary means of accounting for those purposive and cognitive aspects of behavior stressed by Tolman and others.

In addition to its explanatory possibilities in relation to behavior under reinforcement, the r_g-s_g also enables an account of some puzzling aspects of behavior during extinction. One of these concerns the fact that withholding the terminal reinforcement from a learned behavior sequence tends to produce disintegration of the habit sequence as a whole, rather than merely a weakening of the final response members. The question for the learning theorist is, why should an event taking place at the goal so strongly influence habit organization at points remote from that event? In terms of Figure 3-9, why does extinction at S_G cause the animal to behave at S_1 as if it knew that reinforcement at S_G is unlikely? From the viewpoint of the r_g-s_g mechanism, the answer is clear. Omission of the terminal reinforcer weakens the full goal reaction and, simultaneously through generalization, the r_g's throughout the sequence. But when r_g is eliminated, s_g is eliminated also along with its conditioned excitatory and reinforcing powers. Since s_g by virtue of these powers functions as the "glue" of the behavior chain, rapid and widespread deterioration of the sequence necessarily follows.

The r_g-s_g mechanism also explains why extinction of the goal reaction will weaken its antecedent behavior series alone, with the organism then likely to continue to strive for reinforcement by alternative behaviors. This follows because an r_g is peculiar to the behavior sequence(s) of which it forms part of the terminal reaction. And since extinction of an r_g has no effect on the S_D which evokes it, the S_D will persist to evoke, in conjunction with the given environmental stimuli, any other behavior sequence conditioned to it in the past.

Another phenomenon to which the r_g is applicable concerns the effect of a sudden change in reward. We previously cited (p. 59) Elliott's observation that the substitution of one type of food reward for another produced a temporary decrement in the maze performance of rats and a disruption of their behavior at the goal. Similarly, Tinklepaugh (1928) found that monkeys would sometime leave untouched a surreptitiously changed but otherwise acceptable food reward. Such disruption of goal behavior was taken by Tolman as evidence for the existence of central expectations. But the r_g-s_g concept suggests that the disturbance results from interference with a pattern of ongoing glandular and muscular reactions, namely, the r_g. If anticipatory goal reactions are occurring with particular vigor in the goal region, as expected in Hull's theory, then the sudden withdrawal of their normal stimulus supports could be expected to produce considerable disruption and hence interference with the usual goal reaction to the newly substituted reward.

Still other important roles of r_g will be discussed in the immediately following sections and, at a more particular level, in the following chapter.

The Goal Gradient Hypothesis. Postulate 4 holds that habit strength varies inversely with the time interval separating the reaction being conditioned from the reinforcing state of affairs. Specifically, Hull (1943) interpreted the existing data on learning under delay of reinforcement as showing that the increment to habit strength from a primary reinforcement decreases sharply with delays of the order of only a few seconds. On this basis one would expect that in behavior sequences of any appreciable length the effect of terminal reinforcement would be limited to the latter response members. For example, in a maze like that of Figure 3-10, if traversal of any arm requires, say, five seconds, then food reinforcement at the goal would exert progressively smaller strengthening effects on correct responses at choice points 8, 7, 6, etc., with no strengthening effect at all accruing to points 2 and 1. However, as Hull pointed out, the effects of delay of primary reinforcement are inevitably coupled with the operation of secondary reinforcement. Again by postulate 4, stimuli at choice point 8 should by virtue of their association with need reduction take on power to strengthen antecedent reactions. Similarly, stimuli at choice point 7 should then become secondarily reinforcing through association with the acquired reinforcers at choice point 8; stimuli at point 6 will then acquire reinforcing power from the point 7 stimuli, etc. Moreover, with the development of anticipatory goal responses, s_g becomes available as an additional potent reinforcer for points earlier in the sequence. By these means, acquired reinforcing power is transferred backward through the sequence, eventually making possible the learning of correct responses early in the sequence. Since each point of reinforcement is necessarily temporally separated to some degree

Figure 3-10. Multiple T-maze. The stems and arms of each T-segment are of equal length.

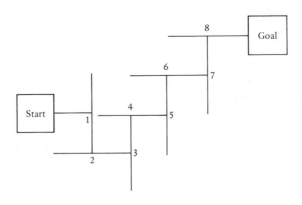

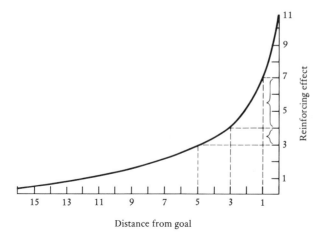

Figure 3-11. The gradient of reinforcing power extending backward from a goal. (After Hull, 1932, p. 33. Copyright 1932 by the American Psychological Association. Reprinted by permission.)

from its "parent" reinforcement, the amount of reinforcing power transferred should progressively decrease with distance from the primary reinforcer.

These considerations, together with some direct experimental evidence, led Hull to hypothesize the existence of a declining gradient of reinforcing power (compounded of primary and secondary reinforcement) extending backward from the goal throughout the sequence. Hull further theorized that this gradient is a negatively accelerated function of the general form depicted in Figure 3-11. In this illustration, magnitude of reinforcement and distance are specified in arbitrary units and distance from the goal refers to either spatial or temporal distance. The important point about this conception is that as distance from the goal increases, the decline in reinforcing power is seen to proceed at a lower rate. This means that the difference in reinforcement consequent upon two spatially (or temporally) separate responses will be greater if that degree of separation obtains near the goal than if it obtains farther from the goal. This relation can be illustrated by determining the reinforcing consequences of correct versus incorrect choice at choice points 8 and 6 of Figure 3-10. To simplify the illustration, assume that following choice at these points the animal proceeds directly to the goal and that each arm of the maze equals 1 unit of distance on the scale of Figure 3-11. Note that an incorrect response *at either choice point* will add 2 units of distance (traversal and then retracing of the blind arm) to the distance consequent upon correct choice. Hence correct and incorrect choices are separated, in terms of distance to the goal, to the same degree at points 8 and 6. From Figure 3-10 we see that correct choice at point 8 occurs 1 unit of distance from the goal (i.e., one arm to be traversed), whereas incorrect choice at point 8 occurs at 3 units of distance from the goal. Correct choice at point 6 is 3 units distant from the goal, and incorrect choice is 5 units from the goal. Reference to Figure 3-11 shows that the differential in reinforcement consequent to correct and incorrect choice will be about 3 units at point 8 but only about 1 unit at point 6. In general, the magnitude of the difference in reinforcement for correct and incorrect responses will decrease with increasing distance of the choice point from the goal.

The goal gradient hypothesis, the general features of which follow from Hull's basic postulates, leads in turn to a number of quite specific deductions about maze learning. Some of these deductions as formulated by Hull in an early treatment of the subject (1932) are:

1. That animals will tend to choose the shorter of two alternative paths to the goal.

Consider the alternative pathways abcd and efgh in Figure 3-12. Both pathways eventuate in the goal, but it is clear that the response of entering alley e will be followed by primary reinforcement sooner than will the response of entering alley a; and according to the goal gradient hypothesis the cues of alley e will acquire greater reinforcing power than those of alley a.

2. That the greater the difference between the length of the paths, the more readily will the shorter path be taken.

The greater the difference in path length, the greater the difference in reinforcement favoring choice of the shorter path.

3. That animals will come to choose the direct path to a goal rather than enter any blind alley.

The fact that blinds are eliminated in maze learning is not a trivial theoretical problem, as might first appear. Since animals retrace out of blinds and eventually reach the goal, food reward should strengthen the erroneous as well as the correct choice behaviors. But from the viewpoint of the goal gradient hypothesis, the problem is readily resolved when it is considered that entering any blind alley creates a longer path to the goal and hence is less reinforced than taking the direct path.

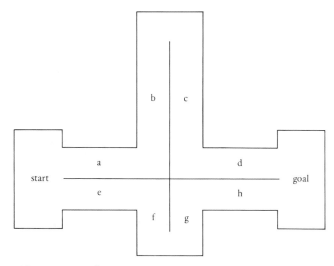

Figure 3-12. Alternative pathway maze.

4. That long blinds will be eliminated more rapidly than short ones.

Since animals tend to traverse to the end of a blind in the initial stages of learning, long blinds essentially have the status of a longer path to the goal.

5. That the order of elimination of blind alleys during maze learning will tend to be in a backward direction from the goal.

This follows because (a) the absolute magnitude of reinforcement following correct choice increases as the goal is approached and (b) as seen in Figure 3-11 the difference in reinforcing effects following correct and incorrect choice increasingly favors correct choice as the goal is approached.

6. That long mazes will be learned with greater difficulty than short mazes with the same path patterns.

This follows from the considerations that (a) the formation of conditioned reinforcement diminishes with increasing distance from the goal and (b) the difference in reinforcing effects following correct and incorrect choice decreases with increasing distance from the goal (Figure 3-11).

7. That a maze may be too long to be learned at all.

The differential in conditioned reinforcement consequent upon correct and incorrect choice behavior early in the sequence may be insufficient to make the correct choice response dominant.

8. That animals traversing a maze in the final stages of training will move at a progressively more rapid pace as the goal is approached.

Since the goal gradient acts to increasingly favor strengthening of correct habits over incorrect habits as the goal is approached, and since one measure of habit strength is speed of response upon onset of the stimulus member of the S-R association, it follows that the gradient of reinforcement should be paralleled by a speed-of-loco-motion gradient. As Hull noted, the twists and turns of a maze pattern like that of Figure 3-10 might well obscure observation of a locomotion gradient. But a gradient of response speed should appear clearly in a direct-alley maze, where the animal simply learns to move from one end of a straight alley to the other in order to find a food reward.

9. That of two alternative paths to a common goal the animal will traverse the early section of the shorter path faster than it will traverse the parallel section of the longer one.

Assume that in the maze shown in Figure 3-12 an animal is first trained to take path abcd to the goal and food reward with the door to path efgh locked throughout training. The animal then receives equal training on path efgh with the door to path abcd locked. At the end of training the experimenter determines the length of time required by the animal to traverse alleys a and e. Given the less favorable reinforcement history for responses in alley a as compared with alley e, a slower pace would be expected in a.

10. That the final parallel sections of two alternate paths of different lengths leading to a common goal will be traversed at approximately equal speed.

Since alleys d and h (Figure 3-12) are the same distance from the goal, the goal gradient should be the same along d and h and hence the speed on these pathways should be about the same.

Additional deductions could be cited, but the foregoing should be sufficient to convey the impressive scope of the goal gradient concept. The phenomena deduced range from the general tendency of organisms to elect the most economical means to a goal, which Tolman took as evidence of the fundamentally purposive nature of behavior, to predictions of detailed aspects of locomotion in different parts of a maze. Moreover, several deductions are predictions of the behavior to be expected under previously untested conditions (e.g., the speed-of-locomotion gradient in a straight alley). It is clear that Hull's theorizing here goes well beyond mere description of behavior in S-R terms or the simple drawing of analogies to conditioning experiments.

The goal gradient hypothesis stimulated many experimental tests, which in itself is testimony to its fruitfulness. By and large, considerable empirical support was found for the foregoing deductions as well as other deductions from the hypothesis, although the research frequently pointed to other factors in maze learning which required refinement or qualification of the theorems. For example, Spence and Shipley (1934) did observe a tendency for blinds to be eliminated in the backward order suggested by the goal gradient hypothesis, but the strength of this tendency depended upon both the stage of learning and whether the blinds were goal-pointing or non-goal-pointing. They argued that interpretation of their data required consideration not only of the goal gradient but also of inhibitory effects from frustration of approach tendencies in goal-pointing blinds and of the development of interfering anticipatory turning responses near the goal in the later stages of training. In other analyses of maze learning, Hull (1935b) did incorporate a frustration, or inhibitory, principle, and the role of anticipatory responses is indicated by his theory on other grounds. Of course, such corrective interaction of theory and experimental test was precisely the route that Hull advocated. The goal gradient hypothesis, providing as it did for guidance of research through deduction of a broad range of phenomena from a few basic assumptions, nicely illustrates Hull's theoretical approach and its possibilities.

The goal gradient analysis also enables some insight into another facet of Hull's work. The function in Figure 3-11 can be said to constitute a crude quantitative model of behavior in relation to a goal. The logical next step toward rigorous quantification would be to determine the mathematical characteristics of the goal gradient, i.e., to seek a mathematical equation specifying the goal gradient curve. In fact, Hull made several efforts to quantify the gradient in this way. A logarithmic function was first proposed but was later rejected in favor of an exponential equation (Hull, 1943) which appeared to fit a considerable range of empirical findings.

Hull sought to interpose such mathematical description between the basic postulate and test stages of his overall program wherever possible. As a rule we shall not consider Hull's mathematical efforts since in general they did not prove to substantially advance the predictive power of his verbal postulates. But it would be remiss to fail to note that Hull's pioneering work in this vein did much to establish the possibility of a systematic mathematical treatment of learning and behavior.

The Habit Family Hierarchy. Behavioral adaptation can be generally thought of as the resolution of some problem state by the attainment of a particular goal, or end state. An organism becomes hungry and must find food, or danger signs appear and the organism must remove itself from those signs, or the organism becomes confined and must find a way out, and so on. In the natural life situation, it is seldom the case that a frequently encountered problem state can be consistently resolved by a single behavior. Given the variability of the environment, organisms ordinarily are reinforced for a wide variety of behaviors in relation to given problem states and the attainment of the appropriate goal. Therefore, Hull argued, we can expect that for any characteristic problem state the mature organism is likely to be equipped with a number of alternative behaviors, each effective for achieving the desired end but differing in the number of reinforcements received. Such clusters of behaviors having in common the initial stimulus situation and the final state of affairs were termed by Hull *habit family hierarchies.*

Perhaps the simplest examples of habit families are found in alternative locomotor sequences between points in space. If while following a footpath we encounter a pool of water, several ways to solve the problem would immediately suggest themselves. We might detour to the right or to the left or perhaps try to jump over the water. Which alternative would occur would depend (other factors constant) upon how often each has been reinforced in previous similar situations. In like manner, Hull thought, any commonly encountered problem state comes to evoke a number of response alternatives, all productive of the same end yet differing in their probabilities of occurrence in that situation.

An important feature of habit families is that, by deduction from the postulates governing development of fractional anticipatory goal reactions, components of behaviors evoked by the end state will eventually become conditioned throughout each of the action sequences leading to it and consequently will be present everywhere throughout all members of a given habit family. This means that whenever this

r_g is evoked, it will tend to evoke in turn, through its s_g, all members of the habit family.

The idea that organisms acquire families of habits integrated by a common anticipatory goal reaction was used by Hull to account for a variety of phenomena indicative of cognition in learning. As applied to maze learning, the habit family hierarchy was conceived of as a set of alternative locomotor habits, acquired through general loco-motory experience, which the animal brings to the maze situation, "any one of which, in free space, would mediate transition of his body from the starting point to the goal" (Hull, 1934, p. 41). To put this in more familiar terms, Hull is saying that at the outset of the experiment the animal knows many different ways of moving through the space between the start and goal points of the maze. Over the initial trials, of course, the animal must discover the goal region and reinforcement through exploration and trial-and-error path selection. As training progresses fractions of the goal reaction become condi-tioned back to the start region of the maze by the S_D mechanism we previously described. Now, it must be borne in mind that the goal reaction includes not only the consummatory response but also per-ceptual and other reactions elicited by the goal region. Consequently, some portion of the reactions defining "being at that place" will be evoked at the start point. But since the stimulation arising from this fractional goal reaction is already conditioned to all members of the habit family mediating movement to that place, all of these locomo-tor sequences will tend to be evoked at the start point.

This analysis has several interesting implications. Once the r_g com-mon to the habit family becomes conditioned to the start point, it tends to evoke the strongest members of the habit family. As a rule, these will be the shorter (and hence more reinforced) paths to the goal point. Since blinds pointing in the direction of the goal are the beginnings of relatively shorter paths, entrance into these blinds will therefore tend to be evoked after a number of successful movements to the goal by more circuitous routes. In this way, Hull accounted for the paradoxical finding of increasing errors on goal-pointing blinds as training progresses (previously discussed in relation to Tolman's theory, p. 54).

The habit family concept was also useful in accounting for goal striving behavior. If a well-learned pathway to the goal is blocked, the r_g-s_g mechanism will immediately evoke the next-strongest member of the habit family directed to that goal. Should several learned path-ways be blocked in succession, the organism would be predicted to shift its behavior systematically through the remaining possible alternatives in accordance with their position in the habit hierarchy. In this way, the organism would exhibit a persistent, directed striving toward the goal suggestive of the operation of purpose and foresight, but in reality mediated by a hierarchy of previously learned simple habits attached to a common stimulus.

In general, the habit family hierarchy was Hull's means of freeing his theoretical interpretation of learning and problem solving from a rigid dependence upon the particular behaviors observed to have been reinforced. In essence, Hull reasoned that the particular behaviors reinforced in any learning situation are likely to be linked through previous learning to other similar behaviors. Because of that linkage, those previously learned behaviors may transfer appropriately to the new learning situation.

SUMMARY

The foregoing set of postulates, concepts, and deductions conveys the essential features of Hull's theory. Like Tolman, Hull sought to build a general theory of behavior, and both men devoted a quarter-century of productive effort to this task. We have not attempted to provide an exhaustive description of these efforts here; the theoretical writings of both men have received comprehensive coverage in a number of other texts. Our main aims have been, first, to make clear the different ways in which Tolman and Hull implemented the behavioral formula and, second, to define each theorist's assumptions about the nature and conditions of learning.

The major differences between Tolman and Hull can be said to stem from their choice of different behavioral units of analysis. Tolman assumed that the natural units of behavior are *acts*, or sets of functionally equivalent movements. Given these molar units of analysis, he perceived knowledge and purpose as fundamental properties of organismic activity. Under this orientation he sought to understand the particulars of learning and behavior as the working out of the organism's knowledge and purposes in relation to environmental context. Consistent with this approach, he stressed the contribution of central processes to learning and viewed peripheral events primarily as a means of inferring the nature of central processes. Hull, on the other hand, analyzed behavior in terms of the elemental stimulus-response units which characterize the conditioning experiment. Under this orientation he sought to show that cognitive and purposive aspects of activity are acquired through conditioning of molecular S-R units. In correlation with these views, he stressed the contribution of peripheral rather than central processes in learning.

In regard to assumptions about the nature and conditions of learning, Tolman saw learning as the acquisition of knowledge about the environment through the development of associations between recurring stimuli. Organisms learn "what leads to what," and the primary condition for this learning is repeated experience of environmental sequences. Hull saw learning as the modification of behavior through the development of associations between stimuli and responses. Organisms learn to behave in particular ways in particular situations, and the primary condition for this learning is the occurrence of need reduction.

Consistent with the theme of continuity in the development of learning theory, Tolman's and Hull's theories can be seen as elaborating earlier viewpoints in some basic respects. Both theorists believed that organisms change by a process of associating events in their experience, and in this regard these men expressed the view of the earlier mental psychologists and of their philosopher predecessors. It is apparent that Tolman's views resemble those of Morgan, particularly in the ideas that organisms learn to associate stimulus events and that the formation of these associations is independent of the consequences of behavior. Hull is clearly in the tradition of Thorndike's pioneering work, both in conceiving of learning as the establishment of associations between stimuli and movements and in regarding the after-effects of behavior as the critical event in the formation of these associations.

Tolman's and Hull's conceptual treatments of learning, which we shall hereafter term *cognitive theory* and *response-reinforcement theory*, respectively, comprise the framework for the most important work on learning theory during the 1930's to 1950's. The conceptual oppositions of these approaches generated a number of intensely disputed issues which went to the heart of a the problem of learning. These controversies, which in many respects constitute the cream of S-R theory, are the subject matter of the next chapter.

4

Controversies Between Cognitive
and Response-Reinforcement Theories

The cognitive and response-reinforcement orientations were not the only theoretical avenues open during the 1930's to 1950's, but they were surely the most attractive and galvanizing to psychologists dedicated to establishing a general theory of learning. As we have noted, the two viewpoints expressed, in S-R terms, long-standing perspectives on the nature of learning. Moreover, to espouse one of these theories was almost necessarily to be concerned with efforts to disprove the other, differing diametrically as they did on such basic questions as the role of the response and the role of reward in learning.

In the best scientific tradition, the dispute between the cognitive and response-reinforcement theories was waged in the laboratory. Specifically, the dispute took the form of experimental challenges and counterchallenges within several research paradigms that were designed as critical tests of the nature and conditions of learning. In our review of these issues, Tolman and Hull will appear as the dominant figures, largely because their theoretical writings provided the frame of reference wherein the labors of many investigators were effectively channeled. Our review will be selective, and we shall have occasion to cite the research and names of a relatively small number of investigators. But it should be noted at the outset that these issues attracted the efforts of a great many creative theoreticians and ingenious experimenters whose collective labors, over a quarter of a century, contributed significantly to our understanding of learning.

We shall begin with the issue of latent learning, the most protracted and in many ways the most significant of the research exchanges.

LATENT LEARNING

The Blodgett Effect

The latent learning experiments sought to deal with what many have regarded as the major difference between cognitive and response-reinforcement theories, namely, the role of reward and punishment (reinforcement) in learning. As we have detailed, cognitive theory held that learning proceeds through the sheer experiencing of environmental sequences. Rewards and punishments, such as food, termination of painful stimuli, and other biologically significant events, might serve to emphasize environmental sequences leading to them, but such consequences were not held necessary for the formation of associations between stimulus events. Response-reinforcement theory, on the other hand, maintained that stimulus-response associations are not strengthened unless followed by stimuli that serve organismic need.

Evidence on this issue was sparse at the outset of the S-R program. Earlier investigators had displayed great ingenuity in devising tests of animal learning capacities, but reward and punishment were almost invariably a part of these procedures; and reinforcing stimuli clearly play a major procedural role in the instrumental and classical conditioning paradigms. A response-reinforcement theorist could argue that the prevalence of rewards and punishments in the procedures which had evolved for studying animal learning testifies to the critical role of reinforcement. But this circumstance could also be accounted for on the purely practical grounds that rewards and punishments are necessary for the efficient *production* of the changes one wishes to study. The issue remained, then: are the rewards and punishments employed in animal studies necessary to the *learning* that takes place or are they merely necessary for the *performance* of that learning?

To attack this issue requires a procedure that affords an animal clear opportunity to learn new ways of behaving *without* being reinforced for that behavior. The multiple-unit maze appeared well suited to this end. As previously noted, a hungry animal is likely to roam throughout its local environment, and a hungry rat can thus be counted on to move throughout a complex maze, entering all alleyways from start to goal. If an animal is given a number of such maze exposures and if all traces of food are kept out of the maze during this period, then the conditions for a test of learning without reward might be arranged. Following such a series of nonrewarded movements through the maze, food is placed in the goal and the hungry animal allowed to traverse the maze once again. On this traversal, as on the preceding ones, the animal proceeds in a desultory manner, roaming into blinds and true pathways alike. But unlike on preceding traversals, the animal for the first time finds and consumes food in the goal box. The question is, what would the two theoretical positions predict about the animal's behavior on its *next* placement in the start box?

To the cognitive psychologists who devised this test the predictions were straightforward. Under Tolman's theory the repeated experience of the maze sequence during the nonrewarded runs should produce learning of the s_1---s_2 relationships of the maze and hence learning of the true pathway. This learning would not be expressed in behavior during the preexposure period since the animal has no demand for the goal box, but once this demand is created by feeding at the goal we can expect the animal to utilize whatever knowledge of the pathway it has acquired. Therefore, on the trial following the first reward, cognitive theory predicts a substantial reduction in the number of errors (blind alley entries) as the animal traverses from start to goal. In contrast, under Hull's theory, correct-path learning cannot begin until the first rewarded trial, and hence only a slight reduction in errors would be expected on the following traversal, a reduction presumably comparable to that following any single rewarded trial.

The first clear test of these conditions was made by Blodgett (1929). Utilizing the maze shown in the upper portion of Figure 3-2, Blodgett compared the performance of hungry rats trained under three conditions. Group I subjects traversed the maze individually once a day for seven days and were allowed to eat for three minutes in the food box at the end of each run. During each traversal, doors located at the choice points (see dotted lines marked D in Figure 3-2) were closed behind the animal to prevent retracing to preceding maze sections. Rats in Group II were given one daily run through the maze in the same manner except that for the first six days these subjects found no food in the goal box and were kept in it without reward for two minutes at the end of each run. On day 7 and on two subsequent days they were treated exactly like Group I, that is, they found food in the box for three minutes at the end of each run. Group III consisted of animals trained in the manner of Group II except that food reward was introduced at the end of day 3 and each day thereafter. Throughout training Blodgett recorded the number of errors by each animal, defining an error as one or more entries into any given blind alley.

Figure 4-1 presents the error scores of each group throughout training. It can be seen that as long as Groups II and III were without food reward their performances were inferior to that of Group I. But

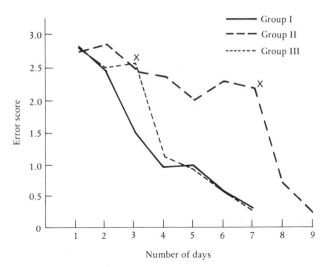

Figure 4-1. Daily error scores of Groups I, II, and III in the Blodgett experiment. (After Blodgett, 1929, p. 120. By permission.)

on the run *following* the first rewarded run (marked by x), the error scores of Groups II and III plummeted, and by the second day after food reward their error rates were fully comparable to that of Group I.

Blodgett interpreted the large drop in errors by Groups II and III following the introduction of reward as showing that these subjects had learned a good deal more about the maze during nonrewarded trials than was indicated by their performance during that period, and he introduced the term *latent learning* to denote such sudden performance shifts following a period of nonrewarded exposure to the elements of a learning task.

Blodgett's findings were confirmed by Tolman and Honzik (1930) and by Kanner (1954) in studies that replicated the essential conditions of Blodgett's experiment. In addition, the concept of latent learning as indicated by experiments of the Blodgett type was further buttressed by some related studies showing significant facilitation of maze learning when rewarded trials are preceded by a period of unrewarded free exploration of the maze, rather than by successive runs from start to finish (Thistlethwaite, 1951).

At face value these observations seem convincingly in favor of the cognitive viewpoint. But an issue as basic as the role of reward in learning is not to be so easily resolved. In fact, the Blodgett-type experiment, as dramatic as its result appeared, was to prove indeterminate to this issue. The reasons are instructive.

The critical issue in interpreting data like those of Figure 4-1 is whether the drop in errors upon introduction of reward for subjects given unrewarded runs (hereafter called the Experimental subjects) is greater than can be expected under response-reinforcement theory. In treating this question, cognitive-oriented investigators were usually able to establish by appropriate statistical tests that the Experimentals' error drop was greater than (1) that following the first rewarded run by the regularly rewarded subjects (Controls), (2) the largest drop following any single reward in the Controls, and (3) the drop following a reward administered at the same level of performance by Experimentals and Controls. To illustrate the last criterion, note that in Figure 4-1 the performance of Group II Experimentals at the time of their first reward (day 7) is roughly equivalent to the performance of the Controls (Group I) on day 2, yet the error drop in the Experimentals on the next run is almost twice that of the Controls on their next run. By such criteria, it was argued, the error drop of the Experimentals must reflect hitherto latent learning of the maze pathway, that is, learning without reward that was reflected in performance only when reward was introduced. But this line of argument imputes to response-reinforcement theory the view that the effectiveness of the food reward in reducing error scores cannot be expected to vary with prior maze exposure. As research progressed on this problem, it became apparent that there were a number of grounds on which to question this supposition.

One consideration is that the rewarded trials of the Experimentals and Controls are coming at different points in their adaptation to the test situation. Two aspects of this adaptation process may be distinguished. One concerns reactions to the routine handling necessitated by the experiment. From the rat's eye view, removal from the home cage, manual handling by the experimenter, and placement in a new environment constitute a massive change in stimulation, and this routine handling characteristically elicits such "emotional" reactions as defecation, increased muscle tension, immobility, and pronounced startle response. These reactions subside considerably with repeated introductions to the experimental situation, and experimenters typically put their subjects through the handling sequence for several days prior to the start of experimentation in order to reduce influence of the "handling reaction" on test performance. But the possible persistence of the handling reaction during the experiment proper is indicated by such observations as that hungry rats may continue to ignore food in a novel, harmless surround even after many daily placements in that surround (Evans and Hunt, 1942; Hall, 1934).

A second aspect of adaptation to the test situation concerns an animal's tendency to orient to and explore novel stimuli. The strength of this tendency in the maze setting was early suggested by Dennis's observations of path-taking behavior in the absence of food rewards (Dennis, 1935; Dennis and Sollenberger, 1934). In Dennis's maze three identical elevated pathways radiated out from a central choice point at equal angles in a Y-configuration. In such a unit, whenever an animal returns to the choice area it may enter any of three alternative paths differing only in history of entry. In several studies with both single and multiple Y-mazes, Dennis found that whenever rats faced such a choice, they were very likely to take a pathway not previously entered or, in the case of a choice between pathways already taken, they tended to take the one less recently occupied. These findings should be related to a report by MacCorquodale and Meehl (1954) that when hungry rats were given extensive rewarded training in the Blodgett maze with the blinds blocked off to prevent entry throughout the training runs, subsequent unblocking of the blinds resulted in entry into nearly every blind by nearly every subject, even though the animals presumably knew the correct path to food. Taken together, such observations suggest that systematic exploration of blinds (and of true pathways) is an essential part of maze learning and occurs regardless of the presence or absence of reward.

The significance of these emotional and exploratory reactions for interpretation of the Blodgett design is that they are likely to be considerably stronger when reward is begun for Controls than when reward is begun for Experimentals. From the viewpoint of Hull's theory, this circumstance could account for a larger reward-induced error drop by Experimentals than Controls. That is, the increased tendency to take the correct path that results from the initial rewards faces more interference from competing reactions in Controls than in Experimentals. It is also possible that the strengthening effect of the food reward is reduced in proportion to the magnitude of the emotional and exploratory reactions.

Another complication in the Blodgett design is that the Experimental group often shows some reduction in errors during the nonrewarded period. Such an error drop is clearly discernible in Group II Experimentals of Figure 4-1. The suggestion is that even under the condition of a single run per day without food reward, rats nevertheless acquire some disposition to stay out of the blinds. Intrigued by some similar observations, MacCorquodale and Meehl (1951) tested animals under conditions that might allow fuller development of this disposition. In one experiment MacCorquodale and Meehl placed hungry rats one at a time in a replica of the Blodgett maze and allowed them to explore freely for 30 minutes on two successive days. Throughout the experiment all maze doors were locked in the open position and there was no food in the maze. At the end of each of these extended exploratory periods the rat was placed in a small box for 30 minutes and then in an empty neutral cage for two hours prior to return to its home cage. This procedure was followed to ensure that removal from the maze was not followed by anything that might be construed as a rewarding state of affairs. The rats then received a test trial by being placed singly in the start of the maze and allowed to move freely to the empty goal. On this test run the animals exhibited a striking tendency to choose "correctly," making only 20 percent of the possible number of errors (blind entries). Such results make clear that there are factors operating *within the maze structure itself* that dispose toward selection of the correct pathway. The nature of these factors is not clear; MacCorquodale and Meehl speculated that rats may dislike having to turn about in the confined space of a blind alleyway.

But even if there is some error reduction in the Experimental group during the nonreward period, would not comparison of the impact of reward on Experimentals and Controls at a point of equal errors still provide a valid measure of latent learning? An attempted replication of Blodgett's experiment by Reynolds (1945) is pertinent to this question. In addition to keeping the usual record of errors during the nonrewarded runs, i.e., the number of entries into blinds, Reynolds also recorded how *far* his subjects penetrated into the blinds. He found that both errors *and* depth of blind entry were significantly reduced as nonrewarded runs progressed. This means that the Experimentals' errors at time of first reward are likely to involve minimal or incomplete penetration of the blinds. Since the point at which Controls make an equivalent number of errors will necessarily be earlier in training, the Controls' errors are likely to involve more complete penetration of the blinds. This difference in the *form* of the errors at time of the initial rewards would clearly favor more rapid elimination of errors in the Experimentals. In essence, reward may impact on different behavioral baselines in Experimentals and Controls with respect to strength of blind-taking tendencies.

Finally, observations of decreasing errors during nonrewarded runs also provided an opening for the argument that need-reducing reinforcing events may not, in fact, have been eliminated from this training period. Granted, there are no *obvious* forms of need reduction. But it was not unreasonable for the response-reinforcement theorist to suggest that the drop in errors may have been brought about by less conspicuous forms of need or drive reduction, such as reduction in an "activity drive" as a consequence of unimpeded movement through the maze. Perhaps, too, if placement in the maze arouses a mild fear drive, then movement to the goal may have been rewarded by the reduction of this drive consequent upon removal from the maze and return to the home cage. Such reinforcements would presumably be of smaller magnitude than that involved in the food reward, and under Hull's 1943 postulates, which held that the buildup of habit strength depends upon magnitude of reinforcement as well as number of reinforcements, this difference in reinforcement magnitude would account for the findings of a much smaller drop in errors by Experimental subjects than Controls over the initial portion of training, as seen in Figure 4-1. But since under this view the smaller magnitude of reinforcement operating on the Experimentals would develop relatively weak tendencies to avoid the blinds, the large drop in errors by Experimentals following the first food reward would be totally unexpected. However, this error drop could be accounted for if the reinforcement postulate were revised so that the buildup of habit strength is made to depend solely upon number of reinforcements (regardless of magnitude) and so that magnitude of reinforcement is held to determine the degree to which habit strength is reflected in performance. Under this revised view, Experimentals could be expected to learn the maze pathway in the absence of food reward provided that some other form of drive reduction, however weak, followed movement to the goal or withdrawal from the blinds. The large increase in reinforcer magnitude represented by introduction of the food reward would then stimulate full performance of the previously learned truepath choices. Hull did make just such a revision of the reinforcement postulate in a later version of his theory (1952).

Additional complications in the Blodgett design could be noted, but the foregoing are the major difficulties and are sufficient to show that the response-reinforcement theorists were not without recourse in the face of the Blodgett effect. Can it be concluded that the factors noted above, considered either singly or in combination, enable a full account of the large error drop by Experimental subjects as observed by Blodgett and others? Or can it be concluded that the Blodgett effect reflects, at least in part, true latent learning of the maze pathways in the absence of reward, as cognitive theorists maintained? To answer these questions convincingly would require quantitative estimate of the contribution of these factors. That is, it is not sufficient to simply establish that there is likely to be a difference between Experimentals and Controls at time of first reward in, say, adaptation to handling and the maze situation; one must also be able to estimate *how much* this difference contributes to observed differences between these groups in error reduction under reward. This estimate requires, in turn, at least a general knowledge of how the adaptation process varies with presence versus absence of the food reward, amount and type of pretraining, complexity of the maze, exposure time on rewarded runs, depth of blind alleys, etc. But knowledge of this basic type was lacking, not only in regard to the adaptation process but in regard to the other factors we have discussed as well. Indeed, it was only as work on the Blodgett effect proceeded that there developed proper appreciation for the complexity of this experimental situation, and it was the theoretical controversy itself that revealed the need to know more about the role in learning of such factors as adaptation, exploratory behavior, and the subtler forms of motivation and reward. In short, there simply was not an adequate empirical base to permit quantitative evaluation of alternatives to Blodgett's latent learning interpretation.

Indicative of the uncertain state of knowledge were several failures to produce the Blodgett effect under conditions that appeared to preserve the essential features of Blodgett's design (Kanner, Experiments I and II, 1954; Meehl and MacCorquodale, 1951; Reynolds, 1945). These failures were doubtless the result of procedural variations which seemed innocuous at the time. For example, in the studies just mentioned preexperimental handling generally extended over considerably longer periods than in the studies showing the Blodgett effect, and the more extensive handling would tend to equalize adaptation to the maze situation in Experimentals and Controls at the time of first reward.[1] But the noted inconsistencies in experimental outcomes made many investigators pessimistic about the possibility of achieving a clear resolution of the opposing theories in this situation. The controversy over the Blodgett effect eventually assumed the status of an impasse in which alternatives to the cognitive interpretation could be neither convincingly established nor convincingly dismissed. In short, the Blodgett design, which had initially appeared to provide a clean and dramatic demonstration of learning without reward, was found to provide neither a clear measure of learning nor firm control over the absence of reward.

Some Lessons from the Blodgett Design

The reader may wonder why we have recounted at some length a set of experiments that ultimately proved indeterminate on the issue to which they were directed. We did so in part because the Blodgett-type experiments taught us a good deal about the complexities of research on learning theory. In particular, these studies underscored the difficulty of clearly distinguishing learning from performance. But we also reviewed these experiments because the history of research on the Blodgett effect embodies an important point about the nature of progress in learning theory, one that it is well to note at the outset of our review of the controversies between cognitive and response-reinforcement theorists. The point is that major issues in learning theory are not likely to be resolved by a single piece of research, however compelling that research may appear initially.

Learning theorists have always been attracted by the idea of the *experimentum crucis*, that is, the perfectly conceived, conclusive test of a basic issue. Buoyed by the promise of the new S-R methodology, the early S-R theorists were particularly optimistic about the possibility of resolving theoretical controversies by *experimenta crucis*. At first impact the Blodgett design appeared to have the character of such an experiment; it was accepted by cognitive and response-reinforcement theorists alike as a serious challenge to the law of effect. It was only as a result of the further studies spurred by the Blodgett effect that the methodological and theoretical complexities of this situation became apparent.

The fact that initially compelling evidence on a theoretical issue often later proves less convincing should not be cause for concern. An *experimentum crucis* is extremely rare in any science. This circumstance simply tells us that in the final analysis the truly persuasive evidence on fundamental issues will almost invariably be *cumulative* in nature, perhaps spanning many years of research. This is particularly true in considering research on learning theory. The student must learn to read evidence in larger units than the single experiment or even series of related experiments. The Blodgett effect is a case in point. While itself inconclusive as a demonstration of latent learning, it led directly to other studies of latent learning that, in the totality of evidence produced, did have an important bearing on cognitive and response-reinforcement interpretations. It appears that in science, as in mazes, a good deal of true-path learning is the result of defining blind alleys.

Spence's Designs

Kenneth Spence, a major figure in the history of learning theory and in the development of the response-reinforcement position in particular, approached the issue of latent learning from a somewhat different perspective than Blodgett. Blodgett had sought to determine whether an animal learns the layout of an environment in the absence of a reward, whereas Spence asked whether an animal learns the whereabouts of a reward in the absence of a need for that reward. In Spence's general procedure, an incentive (e.g., food) for which the animal is fully satiated is placed in one arm of a single-choice maze and the animal is given a number of runs through *both* arms. Since the subject has no demand for the incentive, it simply sees but does not partake of it, and hence choices of the incentive-bearing arm are not rewarded by consumption of the incentive. Subsequently a need state is created for the incentive, and the question is whether the subject will now choose the appropriate incentive-bearing arm, thereby evidencing a latent learning of how to get to that reward. Within this procedure, the exposure to the incentive can be given either in the presence or in the absence of a strong need for a different incentive. This variation proved to have important consequences, and hence the two experimental designs (irrelevant need present vs. irrelevant need absent) will be considered separately.

Design A: Incentive Present with a Strong Irrelevant Need. In a design involving presence of a strong irrelevant need, Spence and Lippitt (Experiment I, 1940; 1946) trained rats in a Y-maze under a two-stage procedure. First, the subjects received five trials on each of 12 days under strong thirst motivation (18-hour water deprivation) but satiated for food, which was constantly available in the living cages. The right arm of the maze led to a goal box containing water, which the subjects were allowed to drink, and the left arm led to food for half of the animals and to an empty goal for the other half. Under these conditions, the thirsty animals would be expected to quickly learn to choose the right (water-bearing) arm. To ensure that all animals had ample experience in entering the left alley and finding the goal box containing food (Group F) or the empty goal box (Group O), a number of forced runs (only the left arm open) were distributed among the five daily trials, so that at least two trials and not more than three were to the nonwater side. Upon completion of this phase of training, all animals were placed on a 24-hour food deprivation schedule while thirst was now satiated by making water continuously available in the home cages. Under these altered motivational conditions the subjects then received five free-choice trials daily with food reward in the left arm.

Spence and Lippitt made several predictions concerning performance in the second phase. According to cognitive theory, they maintained, Group F subjects should have learned the "left alley---food" sequence during the first phase of training as a result of their repeated experience of that sequence. Therefore, when now placed in the maze motivated for food and satiated for water they can be expected to take the left alley. In contrast, Spence and Lippitt predicted from the response-reinforcement position that Group F subjects would continue to respond on the first hunger trials to the right (water) alley because this response has been strengthened by the reinforcements received in the first phase. Finally, it was argued that even if the subjects of Group F do not shift response appropriately on the first hunger trial, cognitive theory predicts they should learn to choose the left alley for food more readily than Group O subjects, who had hitherto experienced only an empty goal on the left.

On both counts the data favored response-reinforcement theory. On the first hunger trial all subjects continued to take the right (water) arm, and there was no difference in the rate at which Groups F and O learned to choose the food-bearing arm.

A considerable number of studies have been carried out employing procedures similar to those of Spence and Lippitt, and the outcomes have generally (though not invariably) been in favor of the response-reinforcement position (see MacCorquodale and Meehl, 1954, for review of these experiments). But again, the issue was not to be so readily resolved. Cognitive theorists could fairly argue that the presence of a strong irrelevant need during the latent period renders this design questionable as an adequate test of cognitive theory. Tolman, it will be recalled, had explicitly postulated that as demands increase in strength there is a channeling of cognition formation to stimulus relationships relevant to the need. Hence in the foregoing Spence and Lippitt experiment, for example, the thirst drive might have acted against formation of S_1---S_2 associations informative of the location of food, i.e., against latent learning.

Such considerations led Johnson (1952) to undertake a direct test of the role of drive strength in latent learning. He trained food-satiated rats in a T-maze under either 0, 6, 12, or 22 hours of water deprivation. Throughout training one goal was empty and the floor of the other goal box was covered with food pellets so that the animal could not avoid coming into contact with them. A mixture of free and forced trials ensured equal experience with both goals. After several days of such training (during which no animal ever ate food in the maze), the animals were satiated for water and tested under 22-hour food deprivation. It was found that on the initial hunger trial only animals that had been trained under the lower thirst drives chose the food side at an above-chance level; specifically, the percentages of rats choosing the food side were 93, 79, 50, and 57 for the groups arranged from lowest to highest thirst drive during training.

These data clearly indicate the importance of the strength of irrelevant drives in tests for latent learning of incentives, and they reinforce the suspicion that Spence's design A is of limited value as a test of cognitive versus response-reinforcement interpretations. Moreover, Johnson's data, in conjunction with Tolman's postulates, suggest that latent learning of an incentive might be commonly observed whenever exposure to the incentive occurs in the absence of a strong, competing need state. This condition, of course, is the essential feature of the second variant of Spence's general approach, and we now turn to consideration of this class of experiments.

Design B: Incentive Present with No Competing Irrelevant Need. This design can be illustrated by a second Spence and Lippitt study (Experiment II, 1940; Spence, Bergmann, and Lippitt, 1950) that imposed essentially the same conditions as their foregoing experiment with the exception of a strong irrelevant need. Rats were trained in a Y-maze while fully satiated for food and water. Food was always present in one goal of the Y and water in the other. Training comprised four choice trials a day for seven days, with equal exposure to the two incentives accomplished by a mixture of free and forced choices. The experimenters initially had some difficulty getting the satiated subjects to repeatedly traverse the maze, but they solved this problem by instituting a weak "social" motivation-reward condition which consisted of returning the rat at the end of each run to a cage containing other rats. At the end of the first phase of training, then, the subjects had 14 experiences of "arm_A---food" and "arm_B---water" sequences in the absence of any strong, competing need state. At this point, the animals were made either hungry or thirsty through appropriate deprivation and tested for choice of the maze arm appropriate to the newly imposed need state.

The results were clear. On the first trial with motivation for one of the maze incentives there was a significant increase in choice of the appropriate maze arm, showing that the subjects had learned something about the location of the incentives during the exposure period.

Similar outcomes with this design were forthcoming from other laboratories. In fact, of the eight studies that can be said to have met the criteria of design B, all were positive.[2] Moreover, cognitive theorists and response-reinforcement theorists alike found the evidence from these experiments adequate to the conclusion that animals in some sense learn the location of incentives for which they have no demand. A set of conditions yielding a form of latent learning had clearly been isolated.

Of course, the uniformly positive outcomes observed under design B fit cognitive theory nicely and reinforce the cognitive interpretation of the failure of design A to yield generally positive outcomes on latent learning. But how did response-reinforcement theorists respond to the set of design B experiments? First, they were quick to point out that these data do not demonstrate learning in the absence of motivation and reward. Spence, Bergmann, and Lippitt (1950) made much of the point that they had found it necessary to employ a social motivation-reward condition (i.e., placing their satiated subjects with other animals at the end of each run) in order to maintain maze traversals during the training phase. This meant, they argued, that the "arm_A---water" and "arm_B---food" sequences were not experienced without the occurrence of a reward or goal situation following the sequence. The reward might be a relatively weak one, to be sure, but then latent learning was commonly thought to be less robust than normally rewarded learning. While not all studies employing design B indicated the need for additional weak incentives to keep the subjects going in the latent period (e.g., Thistlethwaite, 1951), the response-reinforcement theorist could always hold that the mere fact of continued performance argues the presence of *some* form of motivation, such as a mild activity drive, that could provide a basis of reinforcement. From the viewpoint of response-reinforcement theory, then, Spence's design B, like Blodgett's design, does not bear critically on the issue of whether learning occurs without reward. However, Spence's design B *was* accepted as establishing that exposure to a "nondemanded" reward enables the later location of that reward in the presence of appropriate motivation. And this finding does require explanation by the response-reinforcement theorist.

The explanatory mechanism was the fractional anticipatory goal reaction (r_g). As we have seen, it can be deduced from Hull's postulates that stimuli associated with a goal reaction will come to elicit fractions of the goal reaction in anticipatory fashion. In the life history of an organism, such basic goal reactions as ingestion of food are most closely associated with a limited number of stimulus complexes composed of perceptual properties of the appropriate goal objects, e.g., visual pattern and odor of particular foods, together with the drive stimulus arising from the need for those goal objects. Therefore, these stimulus complexes should become most evocative of anticipatory ingestion reactions, or r_g's. Now when an organism encounters a familiar goal object for which it is satiated, an important aspect of this r_g-evoking complex is missing (the drive stimulus) but the remaining perceptual aspects should still tend to evoke the r_g as a consequence of their long history of association with the goal reaction. The response-reinforcement analysis of Spence's design B proceeds from this proposition, namely, that when a satiated organism encounters food or water, these objects weakly but reliably evoke portions of the consummatory responses normally made to them.

Consider a design B experiment in which food has been placed in the left arm of the maze and water in the right. Suppose further that the left and right arms are distinctively and differently marked (e.g., one black and one white), as was commonly the case in these studies. The above proposition indicates that from the outset of phase 1 training, fractional eating responses, $r_{g(e)}$, will be evoked in close relation to the stimuli of the left arm (or to their traces) and fractional drinking responses, $r_{g(d)}$, will be evoked in relation to stimuli of the right arm. Associations should develop between these stimuli and responses if they are followed by a reinforcing event, and, as we have noted, Spence felt that reinforcement was provided by reduction of such subsidiary drives as social motivation. Alternatively, Hull (1952) suggested that the reinforcement in this situation might be secondary in nature, instigated by the mere sight of the goal objects theselves.

As these conditionings progress and the r_g's become evoked by the pathway stimuli; the proprioceptive stimulation arising from each of these fractional goal reactions, $s_{g(e)}$ and $s_{g(d)}$, will occur in conjunction with the responses of traversing the left and right arms, respectively. Since these S-R conjunctions also occur in temporal proximity to the reinforcing event, these, too, will become associated. That is, stimuli arising from the anticipatory goal reactions become conditioned to the left and right choice responses.

This analysis indicates that stage 1 of training actually involves the conditioning of different stimulus-response chains to the left and right maze arms, with the critical link in each chain, the r_g-s_g, being covert. Designating the stimuli of the left and right pathways as S_{left} and S_{right} and the choice responses as R_{left} and R_{right}, the S-R chains established in phase 1 can be schematized as shown in Figure 4-2. A verbal description of these chains would be "stimulation by left-alley cues evokes fractional eating responses which in turn tend to evoke left-going responses" and "stimulation by right-alley cues evokes fractional drinking responses which in turn tend to evoke right-going responses."

$$S_{left} \text{---} r_{g(e)} \text{--} s_{g(e)} \text{---} R_{left}$$

$$S_{right} \text{---} r_{g(d)} \text{--} s_{g(d)} \text{---} R_{right}$$

Figure 4-2.

Since the animal receives equal exposure to the maze arms during stage 1, these competing behavior chains should be organized at equal strength at the end of that training, and so the subjects' free choices should be about equally divided between the two pathways. But if *one* of the need states is now suddenly made strong, there will be a sharp increase in the drive stimulus arising from that need and, consequently, in the r_g's that are linked both to that drive stimulus and to one of the choice responses in the maze. Thus, if the animal is deprived of food, the hunger drive stimulus will augment fractional anticipatory eating responses in the maze and, concomitantly, the stimulation arising from these covert responses. The result is that stimuli favoring the left-going response are now more strongly represented at the choice point and so the animal goes left to food. Similarly if the animal is made thirsty but not hungry, the balance of stimulation in the maze would be tipped in favor of elicitation of the right-going response. In short, the frequency of left- versus right-going response will be a function of the relative strengths of the two fractional anticipatory goal reactions, and these in turn depend upon which drive is present and which is absent. The result is that the animal chooses appropriately to its drive state.

This explanation of the latent learning observed under Spence's design B follows clearly from Hull's postulates, clearly enough to have been independently deduced on several occasions (Hull, 1952; Meehl and MacCorquodale, 1948; Spence, Bergmann, and Lippitt, 1950). Moreover, it enabled an account of some related findings. For example, Maltzman (1950) failed to find latent learning in an experiment that was similar to that of Spence, Bergmann, and Lippitt except that the social-reward cages were attached to the maze itself, just beyond the compartments containing the food and water. Maltzman argued that the presence of the irrelevant social incentive in the maze itself produced relatively strong r_g's for this reward which interfered with, if not precluded, the fractional anticipatory eating and drinking responses necessary to mediate successful test trial performance. This line of analysis also suggests how the r_g concept might account for the generally negative outcomes under Spence's design A. That is, the strong irrelevant drive in these studies could be expected to interfere with conditioning of r_g's associated with the incentive for which the subject was satiated.

But about the time the r_g analyses of Spence's designs were being promulgated in defense of response-reinforcement theory, there appeared yet another type of latent learning experiment that seriously challenged the adequacy of the r_g construct.

Separation of Sequence and Incentive Learning

This development is the logical next step in the latent learning research. As we have seen, response-reinforcement theory handled latent learning of the whereabouts of a goal object by arguing that fractional responses to that goal object become conditioned to the stimuli and responses leading to it. The response-reinforcement theorist's reliance upon the goal stimulus in this account naturally led to the question of whether it might be possible to demonstrate latent learning of a stimulus sequence that does not involve a goal stimulus (incentive). Of course, cognitive theory maintained that organisms might learn such sequences, but it also maintained that the learning would be reflected in performance only if the sequence is motivationally significant for the organism. The problem, then, is how to demonstrate that the learning of a stimulus sequence has proceeded in full independence of the sequence-incentive linkage inevitably required to measure that learning.

In the context of maze research, this problem was handled by physically separating pathway training and goal training and then testing for integration of these experiences. More specifically, the subject is first run through both arms of a T containing highly differentiated endboxes; in this phase of training the distinctive endboxes are not related to reinforcement in any way. Subsequently, the subject is *placed directly* in one of the endboxes where for the first time it experiences a motivationally relevant incentive unique to that endbox. Finally, the subject is then placed at the choice point and tested for choice behavior appropriate to the experienced incentive.

This design was first applied in an effective way to the latent learning problem by Seward (1949) and by Tolman and Gleitman (1949). In Seward's procedure rats were trained in a maze fitted with perceptually distinctive goal boxes (e.g., black vs. white interiors) positioned so they could not be seen from the choice point. In the first stage the subjects were made hungry and allowed to explore the T for 30-minute periods on each of three days. During this time the goal boxes were empty and the rats could move freely throughout the maze. Under cognitive theory this training should allow learning of the two arm---endbox sequences. On the fourth day the rat was placed directly into one endbox through the top of the endbox where for the first time it found the door to the maze closed and the food cup filled. The subject remained in the endbox for one minute or until it started to eat, and this training provided the only link between a goal stimulus and this particular endbox. The subject was then lifted out and placed in the start box to make a free choice of the maze arms. If the rat had latently learned the arm---endbox sequences, then it presumably would choose that arm leading to the reinforced endbox. Of the 32 rats in the experiment, 28 chose the arm leading to the box where they had just been fed.

Before analyzing Seward's findings, let us consider the essentials of Tolman and Gleitman's experiment. Their study is one of the few latent learning experiments involving a negative reinforcer. As in Seward's study, training was carried out in a simple T with markedly different goal boxes invisible from the choice point. In the first phase, hungry rats made one food-rewarded run to each goal box on each of nine days. Note that food was given equally often in each endbox, so neither this goal stimulus nor reactions to it could provide a basis for differentiation of the two arm---endbox sequences. On the tenth day, the two goal boxes were detached from the maze and placed in another room, where each subject was shocked in one of the endboxes but given a food reward in the other. Finally, about two hours later the animals were tested in the maze with the two goal boxes returned to their usual places. Latent learning of the maze pathways would be indicated if the rats now chose the arm leading away from the box in which they had been shocked. Of the 25 animals in this experiment, 22 chose the arm leading to the goal box in which they had not been shocked.

On the critical test trials of both studies the subjects clearly behaved as if they knew which arm led to which of the two distinctive endboxes. Since the pathway-endbox sequences were never differentially associated with the relevant goal stimuli, it is difficult to see how the r_g mechanism could possibly account for test performance. But it is seemingly crucial experiments like these that most sharply reveal the difficulty of anticipating and controlling all theoretically significant details. For despite the separation of sequence and goal training in these studies, postexperimental scrutiny defined in each case a basis whereby r_g's could have influenced test performance.

Let us first consider Tolman and Gleitman's experiment in this respect. As part of the experimenters' effort to make the goal boxes dissimilar, the subjects were required to jump over a small hurdle upon entering one goal box and to twist around several baffles upon entering the other. Consequently, during the initial pathway training under food reward, these responses, as distinctive motor reactions occurring in conjunction with the goal, could be expected to become conditioned in anticipatory fashion to the preceding pathways. Then in the shock-training phase, the animals would again make these motor reactions as they entered the area where shock was administered. Therefore, traces of the proprioceptive stimuli arising from the motor reaction would become conditioned to evoke fractions of the fear response elicited by the shock. The end result of these conditionings is that when the rat next enters the choice area of the maze, fear reactions and their associated s_g's will be evoked by the anticipatory motor response conditioned to the arm leading to the endbox in which shock was given. As it stands, this analysis is not sufficient to account for the observed test performance since at no point in the experiment were the fear s_g's conditioned to choice response. But successful test performance would follow if the further assumption is made that s_g's arising from reactions to punishment elicit avoidance behavior, i.e., the animal avoids those stimuli or discontinues the behaviors that produce those s_g's. The subject would then simply move in such a fashion as to terminate and avoid those s_g's, i.e., would enter the safe alleyway.

Lest the theoretical usage of the differential approach responses appear an after-the-fact matter, it should be noted that response-reinforcement theorists had long maintained that motor reactions in the goal region are likely to become anticipatory. In fact, Miller (1935) early stressed the role of such anticipatory reactions in mediating the effect of punishment throughout a behavior chain. Essentially, Miller found that shock given in a previously food-rewarded runway goal box depressed subsequent approach to that goal to the degree that the reactions accompanying receipt of shock, which included a distinctive motor component, were similar to those accompanying receipt of food in the earlier approach training. Miller interpreted his experiment in precisely the terms outlined in the preceding paragraph, i.e., anticipatory motor reactions instated under food reward mediate fear s_g's in the runway to the degree that these anticipatory reactions resemble the reactions accompanying shock.

In Seward's experiment, r_g's may have entered in a different but equally subtle way. Since Seward gave goal training with the goal box in its usual place on the maze, it is possible that test performance was mediated by r_g's conditioned to *extramaze* stimuli common to the goal and its pathway. To illustrate how this might come about, suppose that through the open top of the maze a particular overhead visual pattern is perceptible from the arm and endbox on one side of the maze and a different visual pattern is perceived from the other side. During rewarded goal training, $r_{g(e)}$ could then become conditioned to the visual stimulus on the reinforced side of the maze. Consequently, on the later test trial when the organism is stimulated at the choice area by the visual stimuli from both sides of the maze, $r_{g(e)}$-$s_{g(e)}$ will be evoked only in the arm leading to the reinforced endbox. Successful test performance would then follow if it is assumed that s_g's arising from reactions to reward elicit approach behavior, i.e., the animal approaches those stimuli or further engages in the behaviors that produce those s_g's. The animal would then move in the direction of cues evoking $r_{g(e)}$-$s_{g(e)}$ and consequently would be led down the arm leading to the reinforced endbox.

Seward sought to eliminate extramaze cues in his experiment, but since it is difficult to ensure that some stimulus differences of this nature do not remain (e.g., subtle differences in odor, temperature), he conducted two control experiments to check the possible operation of such cues. In the first, the procedures of the main experiment were repeated except that identical endboxes were used. The idea was that if test performance is based solely on extramaze cues, then it should be unaffected by removal of the cues distinguishing the endbox interiors. On the test trial of this experiment, choice of the arm leading to the reinforced goal box fell to the chance level, thereby indicating that correct test trial choice required differential intramaze cues. In the second control experiment, the original differentiated endboxes were again used and the procedure was the same as that of the main experiment except that during the goal-training phase the endbox was detached from the maze and placed just behind the start box. If test performance is independent of cues outside the maze, then displacement of the goal box during feeding should have no effect on the animal's ability to choose the arm leading to the reinforced goal box. But again test choices fell to chance level, suggesting that correct test trial choice was *not* independent of extramaze cues. The seemingly paradoxical results of these control experiments compel the conclusion that test trial choice in the main experiment was influenced by *both* within-maze and extramaze cues. Unfortunately, these control experiments do not enable conclusions about the relative influence of within- versus extramaze cues or about the way in which extramaze cues operated. Thus, while the control experiments do suggest that test trial choice in the main experiment was based in part upon a knowledge of "what leads to what" within the maze, they leave open the possibility that test performance was mediated in a primary way by r_g's conditioned to extramaze cues common to the goal box and its pathway.

It is possible, then, that in both the Seward experiment and the Tolman and Gleitman experiment, s_g's directly associated with the goal incentive were evoked in the pathway leading to the incentive. But relatively fine-grain analyses were required to establish this possibility. The very fact that interpretation now centered on consideration of such seeming minutiae of experimental detail indicates that with these studies the latent learning research had moved close to a revealing test of the cognitive versus response-reinforcement explanations. The Seward study and the Tolman and Gleitman study both pointed the way. Application of this design with greater precautions against the possible presence of stimuli common to the goal and pathway would appear to narrow the explanatory options in telling fashion.

The experiment which best appears to have met this requirement is that of Strain (1953). Rather than the usual T-maze, this study employed a straight runway divided into six equal-sized compartments separated from one another by opaque swinging doors. Each compartment had distinctive, salient perceptual properties achieved by variations in (1) patterns painted on the walls, (2) floor construction, and (3) designs on the swinging doors that allowed exit from each end. The top of each compartment was covered with a translucent glass which prevented the subject from perceiving extracompartment details. The apparatus was so constructed that the four inside compartments could be shifted in position within the runway. For any given animal, the sequence of compartments (i.e., position of the compartments) was the same throughout the experiment, and hence "what compartment led to what compartment" was invariant. But different compartment sequences were employed with different subjects to ensure that the behavioral effects observed were not dependent upon a particular perceptual sequence.

On each of five days each subject was allowed to explore in the runway for five minutes. During these periods the maze was empty and the animals moved freely from compartment to compartment through the swinging doors. Following another 2½ minutes of free exploration on the sixth day, each subject was placed directly in an end compartment with the door locked and administered two 3-second electric shocks. To eliminate olfactory cues that might result from

Table 4-1. Number of rats going initially toward or away from shock compartment on test, depending upon nearness to shock compartment (Group 1 closest, Group 4 farthest) (Strain, 1953)

Group	Total number of rats	Number going toward	Number going away from
1	20	3	17
2	20	6	14
3	20	6	14
4	20	12	8
Total	80	27	53

shock administration, this end compartment was replaced by a fully duplicate compartment for the shock trial only. Fifteen minutes after the shock experience the rat was replaced in the runway and again allowed free exploration. On this test trial subjects were placed in compartments at different distances from the shock box location. The question of major interest is, in which direction did the rats move when they left the compartment in which they were placed? Did they go toward or away from the shock box? It must be remembered that the rats could not see from one compartment into the other. Therefore, movement away from the shock end would indicate they had learned during the earlier exploration the sequences leading toward the shock compartment and toward the safe (opposite end) compartments.

Table 4-1 shows the number of rats placed in each of the four inside compartments who left in the direction of the shock box and the number who left in the direction away from the shock box. In the table, Group 1 refers to subjects placed in the compartment immediately adjacent to the shock compartment, Group 2 those placed in the compartment next removed, etc., with increasing numbers indicating increasing distances from the shock compartment. The data are clear. The rats exhibited a significant overall tendency to exit away from the shock compartment, and this tendency was greater the nearer the animals were placed to the shock compartment.

Strain's procedures leave little room to argue that this avoidance behavior was mediated by r_g's operating in the manner claimed for the Seward and the Tolman and Gleitman experiments. There was no opportunity for establishing different motor reactions to the stimulus sequences during the latent period, as in Tolman and Gleitman's study, since movement in the two runway directions of course involves the same motor behaviors. And Strain took elaborate precautions against the operation of extramaze cues. For example, the runway was located in a sound-insulated chamber, a uniform white sheeting was draped above and on the sides of the runway to eliminate extramaze visual cues, the animals were carried in and out of the experimental chamber in a covered cage, the runway direction was reversed for each training trial, and the subject's orientation upon introduction to the runway was systematically varied. In addition, for one half of the subjects the runway direction and the introduction orientation on the test trial were reversed from that obtaining on the shock trial. If reactions to shock were associated with some extramaze cue that later directed test performance, then these reversed subjects, in moving away from that extramaze cue on the test trial, would in fact approach the shock compartment, i.e., these subjects would show a higher number of test trial exits in the direction of the shock box. But their test trial performance did not differ from that of the remaining subjects.

In sum, Strain's procedures appear to justify the conclusion that the animals had no cues as to their direction of movement in the runway other than the visual stimuli distinguishing the inside of each compartment. Given this conclusion, it follows that test trial behavior was based on some internal representation of the sequence of those visual stimuli, a representation acquired during the initial exploration.

Strain's observations are supported by similar later experiments (Gonzalez and Diamond, 1960; Seward, Jones, and Summers, 1960) and, by implication, indicate a cognitive interpretation of the Seward study and the Tolman and Gleitman study as well.[3]

Is this set of experiments then fatal to the r_g construct? No, at least not in its essential aspects. The capacity of a theory is best revealed by stringent experimental challenge, and while these studies do force significant modifications in the response-reinforcement interpretation, the defining features of the theory could still be preserved by recourse to one of its early but seldom invoked provisions.

Interpretation of these experiments in terms of fractional anticipatory response formation proceeds from the idea that *any* reaction can become anticipatory. In practice, response-reinforcement theorists had tended to limit application of the fractional response mechanism to the goal reaction (i.e., to a primary appetitional or aversive reaction) and to motor responses in the vicinity of the goal. But in principle there is no difference between, say, a distinctive turning response as the animal enters the goal box and the more subtle responses involved in simply perceiving stimuli in the goal box, or for that matter, elsewhere in the maze. Hence perceptual responses ought to become anticipatory also. The belief that it is impossible to have a perceptual awareness of any sort without concomitant peripheral responses and the assumption that there are differentiating response accompaniments for all perceptual processes were part of Watsonian behaviorism. This viewpoint, along with the idea that fractions of perceptual responses may become anticipatory, was certainly strongly implicit in Hull's original conceptions (pp. 65–67, 95).[4] But the response-reinforcement theorists' penchant for dealing with behavior at its most palpable level had led them to eschew such lightweight, molecular responses in interpretations of learning. Observations like those of Strain, however, indicated the necessity of including perceptual responses in the category of fractional anticipatory behaviors.

Given the proposition that perceptual responses may become anticipatory, the response-reinforcement position can account for positive outcome of *any* experiment employing separation of sequence and incentive training. In the initial pathway exposure, fractions of the distinctive perceptual response evoked by the goal, or end, compartment would become conditioned to the pathway leading to it. This conditioning would occur through the action of food reward in Tolman and Gleitman's study, while some inconspicuous form of drive reduction, such as reduction of an activity or exploratory drive, would have to be assumed for Seward's and Strain's experiments. During the incentive-training phase, fractions of the appetitional or aversive reactions evoked by the incentive would become linked to internal stimuli arising from perceptual responses concurrently evoked by the end, or goal, compartment. Consequently, on the subsequent test trial the path or sequence eventuating in the reinforced endbox would evoke fractions of the perceptual responses to that endbox, and the stimuli arising from these anticipatory perceptual responses would in turn evoke r_g-s_g's of the goal incentive. Again, an approach-avoidance steering effect of s_g's would be assumed to account for the final selection of the appropriate choice behavior.

Status of the Latent Learning Issue Following the "Separation of Sequence and Incentive" Design

With the theoretical exchange resulting from this design, research on latent learning had essentially run its course, owing primarily to the difficulty of devising tests of the positions that had now evolved. Both sides now agreed that animals could acquire covert representations of organism-environment interactions, representations that might lie latent with respect to influence on overt behavior but that could serve to direct behavior given appropriate incentive conditions. But the two sides retained their essential differences in views of the nature of this internal representation and of the conditions for its acquisition. Consistent with our earlier description of Tolman's and of Hull's orientations, the cognitive theorist remained firm in the belief that these representations (or knowledge) consisted of "copies" of successive perceptions of the world, and the response-reinforcement theorist retained the view that they consisted of "copies" of behavior to the world. Since the postulated fractional responses of the response-reinforcement theorist generally had the unobserved status of central cognitions, it may seem that the issue of "what is learned" in latent learning had become a semantic matter. Indeed, many investigators now saw little pragmatic difference between explanation in terms of a fractional anticipatory response, particularly an anticipatory perceptual response, and explanation in terms of a cognition. But the difference in conceptualization does have important consequences for learning theory and research. As long as these covert processes are seen as directly derivative of peripheral responses, as persistently maintained by the response-reinforcement theorist, their link to observables is firm and they can be assumed to follow the laws governing conditioning of overt behaviors. But as derivatives of perception or as central processes, the conception favored by the cognitive theorist, their status in these fundamental regards is uncertain. Only under the response-reinforcement conception can we confidently seek to know unobserved psychological processes by studying observed behaviors. The practical significance of this differential line of reasoning for explanation of learning will become more apparent in the treatment of the remaining issues in this chapter.

While the latent learning experiments did not decide between the opposing viewpoints in a definitive "stand-or-fall" fashion, this body of research nevertheless had a number of important influences on both the theoretical positions at issue and the course of learning theory in general. Some specific effects will be apparent to the reader, such as altered conceptions of what might serve as a reinforcing event, but there were other, less obvious and broader influences on theory. Since similar effects were exerted by the other (and concurrent) controversies between cognitive and response-reinforcement theories, the influence of the controversies will be treated jointly following a separate review of the various research issues. Accordingly, we turn at this point to another issue separating cognitive and response-reinforcement theories, an issue that can be regarded as an offshoot of the work on latent learning.

LATENT EXTINCTION

Extinction refers to the abolition of learned responses by the withdrawal of reinforcement. Consider an animal who has been given a series of food-reinforced trials in a runway and, in consequence, has learned to rapidly traverse the runway whenever placed in the starting section. If this animal is now given continued runs through the alleyway but to an empty box, its movement through the runway will become progressively slower, both in starting time (latency) and in time taken to traverse the runway. Ultimately, the animal will show little running in the alleyway and will probably revert to a seemingly aimless meandering, much like an animal who had never received food in the goal section. How long it would take the animal to reach such a preconditioning level would depend upon a number of factors, most notably, the number and scheduling of previously rewarded runs. But in general, sustained withdrawal of reinforcement from a conditioning sequence results in progressive deterioration of the learned behavior toward the preconditioning level.

As detailed earlier (pp. 82–83), Hull explained extinction in terms of the buildup of response-produced inhibition. The essential idea is that repeated responding creates a fatigue-like state which eventually inhibits the response. In Tolman's theory, extinction is simply another case of cognition formation. Just as the animal learned to expect food in the goal box by experiencing the sequence "runway---goal box with food," it now learns to expect an empty goal box by experiencing the sequence "runway---goal box without food." Under Tolman's assumptions, the development of this new cognition, like the learning of any cognition, requires only that the animal perceive the prevailing stimulus sequence; it is not necessary that the animal actually make the learned response. In contrast, under Hull's theory the inhibition that produces extinction can develop only through performance of the learned response.

This difference in the response requirements for extinction suggests a simple test of the two interpretations. Suppose that following a series of food-rewarded trials in the runway but just prior to beginning normal extinction treatment, the animal is placed *directly* in the empty goal box for several minutes. From Tolman's viewpoint, this experience will conflict with the previously acquired cognition that the goal contains food and, since behavior is assumed to mirror cognition, we can expect some immediate weakening of the approach response on subsequent runs. Under Hull's theory, however, the direct goal placement should not have decremental effects on subsequent approach to the goal since the subject has not engaged in the nonreinforced response performance necessary to extinction.

Seward and Levy (1949) were the first to apply this test. Their apparatus consisted of an elevated straight pathway separated by doors at each end from starting and goal platforms. The latter platform contained a food cup so positioned that its contents were not visible from the runway. A group of hungry rats first received ten food-rewarded runs from the start to the goal platform. This training was sufficient to develop rapid runway traversal in all subjects. The following day, one half of the subjects (the Experimentals) were given five 2-minute placements on the goal platform with the door closed and the food cup empty. The other half of the rats (Controls) spent equivalent periods on a neutral platform located a short distance from the training apparatus. Regular extinction trials were then given the next day; on these trials each animal was repeatedly placed on the start platform and allowed to approach the empty food cup at the goal until the animal attained a criterion of two successive refusals to leave the start platform within three minutes. Prior to each extinction trial, each Experimental animal also spent two minutes on the goal platform and each Control subject spent comparable periods on the neutral platform.

Table 4-2 shows the effects of the goal placement on the number of trials required to extinguish the runway response and on latency and running speed on the first extinction trial. It can be seen that the Experimentals required less than half as many trials as the Controls to reach the criterion of extinction, a statistically significant difference. Moreover, the Experimentals took significantly longer to traverse the runway than the Controls did on the first extinction trial. Since this trial is uncontaminated by any effects of previous unrewarded approach responses, the decrement in running speed of the Experimentals must be fully independent of occurrence of the response itself. Interestingly, on the first extinction trial the Experimentals and Controls had about the same starting times, i.e., time to leave the start platform. This outcome, in relation to the Experimentals' first trial decrement in runway traversal time, suggests that the

Table 4-2. Effects of goal placement on extinction in the Seward and Levy (1949) experiment.

Group	Mean number of trials to extinction criterion	Running times (*sec.*)	Latency (*sec.*)
Experimentals	3.12	7.70	7.30
Controls	8.25	2.55	6.22

preextinction placement had a greater decremental effect on components of the habit sequence near the goal than on those early in the sequence.

Seward and Levy's experiment, then, demonstrates reduction in response strength that cannot be attributed to occurrence of the response. As Seward and Levy noted, their findings can be seen as the parallel, in extinction, of latent learning, and hence they introduced the term "latent extinction" in relation to their data. Seward and Levy's results were replicated in a sizable number of experiments involving a variety of learning situations (Kimble, 1961), and latent extinction was quickly established as a robust phenomenon.

Latent extinction is, of course, consistent with Tolman's theory but cannot be explained by Hull's inhibition interpretation of extinction. However, latent extinction can be considered as basically an extension of the controversy underlying latent learning research in the sense that the problem for theory in each case is to account for observations of change in behavior induced by the organism's perception of an altered relationship between a stimulus and a reinforcer. Given this perspective, it is not surprising that response-reinforcement theorists sought to extend to latent extinction the r_g concepts that had proved useful in analysis of latent learning.

An account of latent extinction in terms of r_g follows directly from Hull's conception of the role of r_g-s_g in the organization of behavior. As detailed in an earlier section (pp. 84–87), Hull deduced from his postulate system that with repetition of a behavior sequence terminating in reinforcement, fractions of the goal reaction will come to be evoked throughout the sequence, primarily as a result of the conditioning of r_g to the drive stimulus. In consequence, s_g, the proprioceptive stimulation arising from r_g, will become part of each of the stimulus complexes evoking the responses leading to the goal. Since r_g-s_g also occurs close in time to the primary reinforcer, s_g can also be expected to acquire the power to secondarily reinforce behavior earlier in the sequence. The presence of r_g-s_g throughout the S-R sequence thus contributes to both evocation and strengthening of the successive responses. In this way, Hull argued, r_g-s_g serves to direct and maintain the habit sequence as a whole. Given this conception of r_g-s_g as the "glue" of a behavior series, it follows that anything that weakens r_g will also weaken the entire behavior chain.

This line of reasoning leads naturally to the question of whether latent extinction might be interpreted in terms of the action of r_g's. Building on Hull's analysis and the idea that r_g's follow the laws of overt behaviors, Moltz (1957) showed how the procedures of the latent extinction experiments can indeed be expected to strengthen and weaken r_g's in a manner that would yield a latent extinction effect. First, during the initial rewarded training, r_g should become conditioned throughout the S-R sequence in the manner suggested by Hull. Then during the latent extinction trials, that is, during the nonrewarded direct goal placements, r_g should be evoked by those stimuli in the goal box, including the drive stimulus, to which the goal reaction was conditioned during the rewarded training. Since these stimuli are constantly present during the goal placement, the latent extinction trials should occasion many nonreinforced repetitions of the r_g. Just as nonreinforced emissions of any overt response will reduce the strength of that response's association with the eliciting stimulus, so, too, these nonreinforced emissions of r_g will weaken r_g both at the goal region and, primarily by means of the drive stimulus, throughout the antecedent S-R sequence. Reduction in the strength of r_g in turn reduces the eliciting and reinforcing conditions (s_g) favoring behaviors leading to the goal, and the result is that the organism manifests a decreased tendency to approach the goal on subsequent extinction trials.

The foregoing r_g analysis accounts for the basic fact of latent extinction and in this respect is on a par with the cognitive interpretation. But the r_g construct also enables prediction of new and more specific effects of latent extinction procedures. The power of the r_g construct in this respect derives from the identification of r_g's with readily observed behaviors, since this allows the response-reinforcement theorist to analyze r_g processes in terms of well-established empirical principles of overt behavior. In this vein, Moltz and Maddi (1956) sought to test and extend the r_g analysis of latent extinction by drawing upon principles that had emerged from research on extinction of overt responses. They cited two such principles of extinction: (1) the amount of reduction in response strength is primarily a function of the number of nonreinforced evocations of the response (e.g., the greater the number of nonreinforced repetitions of a previously food-rewarded bar press response, the less likely is the animal to engage in bar pressing in a subsequent exposure to the situation), and (2) the greater the strength of the drive during extinction trials, the greater the rate of response emission (e.g., if the animal is made hungrier at the outset of extinction of a previously food-rewarded bar press, then bar presses will be emitted at a higher rate during extinction). Coupling these empirical principles to the assumptions of the r_g analysis of latent extinction leads to the prediction that a higher drive level during the nonrewarded goal placements will result in a greater decrement in performance on the subsequent extinction trials. Specifically, Moltz and Maddi reasoned that if latent extinction is due to the weakening of an r_g and if r_g's function like overt behaviors, then increasing the drive level should increase the number of nonreinforced r_g's emitted during the goal placements, thereby resulting in a greater weakening of r_g and, concomitantly, of the power of s_g to support goal-oriented behaviors.

This hypothesis was tested by Moltz and Maddi in a three-step experimental procedure. In the first phase, rats were given ten trials a day for four days in a straight-alley runway with a single goal box while under a 22-hour hunger drive. On seven of the ten daily runs the subjects received food at the goal. Three nonrewarded runs were randomly interspersed with the daily rewarded runs, and distinctly different goal boxes were used on the rewarded and nonrewarded trials. In the next phase of the experiment, one half of the animals (the Experimentals) received the latent extinction treatment, which consisted of their being placed directly in the previously positive goal box without food for four 1-minute periods. Different subgroups of the Experimental subjects received the latent extinction placements under 0, 22, or 44 hours of food deprivation. The other half of the subjects (the Controls) was divided into three subgroups that were subjected to the same deprivation conditions but without any latent extinction treatment placements. In the final phase, all animals were returned to a 22-hour deprivation state and given 15 free choices in a T-maze, one arm of which led to the previously rewarded runway goal box and the other arm of which led to the previously nonrewarded runway goal box. Cloth curtains were placed in front of the doors to the goal boxes so that the animals could not tell from the choice point which arm led to the formerly positive box and which to the formerly negative. During these free-choice trials, food was never present in either goal box.

From the viewpoint of response-reinforcement theory, what outcomes would be expected on the T-maze choice trials? First, all Control groups would be expected to choose the arm leading to the formerly positive goal box more than 50 percent of the time. This prediction follows from the consideration that r_g's conditioned to the positive goal box during the runway training will persist in full strength to the start of T-maze testing in all Controls, since none were exposed to the goal box at any time during the interval between runway training and T-maze testing. Therefore, whenever Controls enter the formerly positive goal box, the evoked r_g-s_g reactions will secondarily reinforce and thereby strengthen the choice response leading to that goal. The expected outcome for Experimental subjects, on the other hand, depends upon their drive state during the latent extinction placement. Experimentals given latent extinction under food deprivation, particularly the 44-hour deprivation group, would be expected to exhibit little, if any, preference for the formerly positive goal. This prediction follows from the assumption that during latent extinction placements, r_g's conditioned to the positive goal box are evoked at a rate proportional to the strength of the hunger drive, thereby extinguishing the r_g-s_g reaction necessary for strengthening the antecedent choice response. But Experimental subjects who received the latent extinction placement under zero hours of food deprivation *would* be expected to choose the formerly positive rather than the negative goal. It will be recalled that under Hull's postulates, response evocation is a joint, multiplicative function of habit and drive. Under this view habit strength is not translated into behavior, whether covert or overt, when drive strength is zero. Therefore, latent extinction placements under zero hours of food deprivation would produce little if any extinction of the goal box r_g's, and hence the zero-hour Experimental group would be expected to perform much like the Controls.

Table 4-3 shows the mean number of choices of the positive goal by each group. These data are in accord with all of the response-reinforcement expectations. The three Control groups did not differ significantly from one another, but each chose the arm leading to the positive goal significantly more often than the chance expectation of 7.5 choices, thereby attesting to the secondary reward value of that goal box and its independence of the drive variation per se. The mean number of choices of the positive goal by Experimental subjects as a whole (7.57) was significantly lower than that of the Controls (9.67), showing that the latent extinction treatment did reduce the secondary reward power of the positive goal box. Among Experimental animals the zero-hour deprivation group made significantly more positive goal

Table 4-3. Mean number of choices of the formerly positive goal box by each group

Deprivation condition	Controls	Experimentals
0-hour hunger	9.30	9.30
22-hour hunger	9.90	7.10
44-hour hunger	9.80	6.30
Overall mean	9.67	7.57

box choices than either the 22- or the 44-hour group, and choice behavior of the latter two groups did not differ. Finally, note that the zero-hour Experimentals had the same mean number of positive goal choices as their Control counterparts.

It is particularly important to note that this experiment confirms prediction of a previously undemonstrated empirical relationship, namely, that latent extinction depends upon the strength of the drive state. This and other predictions of the experiment, as well as the procedures for their test, flowed from consideration of basic assumptions that make up the response-reinforcement position, such assumptions as that extinction proceeds by nonreinforced emission of behavior (in this case, r_g's), that r_g's are secondarily reinforcing, that r_g's follow the laws of observable behavior, and that habit strength and drive interact multiplicatively. The Moltz and Maddi experiment, then, illustrates how the postulate set and deductive logic of response-reinforcement theory could generate not only a counter explanation of a seemingly cognitive phenomenon but also a more specific description of the phenomenon and prediction of new behavioral aspects. This theoretical capacity is again evidenced in the response-reinforcement treatment of the next research controversy to which we now turn—that of sensory preconditioning.

SENSORY PRECONDITIONING

If an organism experiences repeated pairings of two neutral stimuli, such as a light and a tone, and if one of them is subsequently conditioned to evoke a specific behavior, will the other then be found to also evoke that behavior? This is the question to which the sensory preconditioning experiments were addressed. Implicit in this research question, of course, is the theoretical question of whether associations are formed between stimuli simply by virtue of their contiguous occurrence, for only if the organism does form some bond between the paired stimuli could one expect the later conditioning training of one to transfer to the other.

Brogden (1939) is generally credited with the first effective study of sensory preconditioning. In a three-step experimental procedure, he first gave eight dogs (the Experimentals) 200 two-second presentations of a bell and a light simultaneously, the 200 pairings being distributed in blocks of 20 over a ten-day period. In the second phase, one half of the animals were conditioned to make a leg flexion response to the two-second bell, and the other half of the subjects were similarly trained with the two-second light as the conditioned stimulus. Electric shock was used as the unconditioned stimulus, and CS-UCS pairings were presented at the rate of 20 per day until the subject met a criterion of 20 conditioned responses in a single session. The final phase was a test for occurrence of leg flexion to the stimulus (bell or light) not used in the conditioned phase. This was accomplished by 20 daily presentations of that stimulus without shock until the first session without any leg flexion.

Brogden also trained another group of eight dogs (the Controls) in only phases 2 and 3 of the above procedure, i.e., they were simply conditioned to one of the stimuli and tested for transfer to the other, with half of the animals being conditioned to the bell and half to the light. The phase 3 performance of the Controls provides a measure of generalization of conditioning from one of the experimental stimuli to the other in animals who did not experience the stimuli in paired presentations. Thus, the difference between the phase 3 performances of Controls and Experimentals measures the contribution of the preconditioning pairings to the transfer of conditioning.

The essential aspects of Brogden's experimental procedure are illustrated in Table 4-4.

Brogden found that during the test phase the Control animals made a total of only four leg flexion responses to the stimulus that was never associated with shock, whereas the animals who had received the preconditioning pairings made a total of 78 flexion responses to the stimulus never associated with shock. Clearly, the prior stimulus pairing led to significant transfer of the conditioned flexion response from one of the paired stimuli to the other. On this basis, Brogden concluded that the preconditioning pairing had resulted in "a sensory conditioning between the bell and light . . . [that is, a] sensory precondition-

Table 4-4. Illustration of the basic aspects of Brogden's experimental procedure

Group	Phase 1	Phase 2	Phase 3
Experimentals	Repeated pairings of bell and light	Conditioning of leg flexion by Pavlovian procedures: Light --→ Shock → leg flexion	Presentation of bell alone to test for appearance of leg flexion
Controls	No treatment	Same as above	Same as above

ing" (1939, p. 330). The implication is that the pairing operation had established *an association* between the two sensory events.

The reader may have noted that Brogden's control procedure is not complete in that Controls and Experimentals differ in total exposure to the test stimulus at the outset of phase 3. However, clear evidence of sensory preconditioning has also been obtained in experiments which give Controls unpaired presentations of the stimuli during phase 1, equal in number to the presentations given Experimentals (e.g., Silver and Meyer, 1954).

In the long line of experiments which have now stemmed from Brogden's study, sensory preconditioning has been successfully demonstrated in species ranging from rat to human, with both classical and instrumental conditioning procedures, and with a variety of voluntary and involuntary behaviors (Thompson, 1972). The one qualification in these demonstrations is that the phenomenon is frequently reported to be weak and highly variable across subjects. On the other hand, there has been a report of sensory preconditioning as strong as that observed under conventional classical conditioning (Bitterman, Reed, and Kubala, 1953), suggesting that we do not yet fully understand the optimal conditions for the effect (cf. Seidel, 1959).

Overall, it is clear that sensory preconditioning is a reliable and widespread phenomenon, and however strongly it may appear in any given instance, it is surely a potent datum for theory. As in the case of latent learning and latent extinction, it is a phenomenon that appears to pose problems for the response-reinforcement position but not for the cognitive position.

Consider, first, the application of cognitive theory. As a result of the initial pairing of the stimuli, presentation of one stimulus (S_1) should evoke a representation (expectation) of the second (S_2). Then in the conditioning phase, S_2 acquires a similar signalling function with respect to the shock stimulus. Accordingly, when the subject is then tested with S_1, there is evoked a representation of S_2 which in turn leads to an expectation of the shock. Since cognitive theory assumes that behavior adjusts appropriately to cognition, it follows that leg flexion will appear upon presentation of S_1. The outcome of the sensory preconditioning experiment is exactly what would be expected under a theory which assumes that organisms acquire knowledge of the relations between stimuli and act in accordance with that knowledge.

In contrast, sensory preconditioning constituted a double-barreled threat to response-reinforcement theory, double-barreled because it posed an apparent instance of learning without either response or reinforcement. During the stimulus-pairing phase, no response is required of the subject, who simply passively experiences the paired stimuli. The stimuli themselves typically are relatively mild events without apparent biological significance. Yet under these conditions some type of link is developed between the paired stimuli, as indicated by the behavior during the later transfer phase.

The response-reinforcement theorist's response to this challenge was, of course, to attempt to show that response and reinforcement are in fact present in the preconditioning phase and are at the basis of the observed transfer effect. The analysis proceeds as follows.

Granted that the tone and light (to take Brogden's conditions for illustration) do not produce any readily discernible responses, there must nevertheless be evoked *some* behavior unique to these stimuli, if only at the perceptual level, else these events would not constitute distinctive stimuli for the organism. Therefore, it is reasonable to assume that the tone and light each produce some distinctive recurrent attending response, such as a particular pattern of receptor-orienting reactions. Once the existence of such stimulus-attending reactions is acknowledged, it can be seen that the phase 1 stimulus-pairing operation provides a possible occasion for conditioning in the classical, Pavlovian mode, as shown in the S-R diagram in Figure 4-3. In the figure, what we have termed the attending reaction to the light is designated by r(L), and s(L) denotes the proprioceptive stimulation generated by and unique to the attending reaction to light.

Since the tone and attending response to light are contiguous S-R events, they will become associated, under the assumptions of Hullian theory, if followed by a reinforcing state of affairs. As we saw in

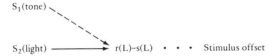

Figure 4-3. In this and subsequent figures, solid arrows indicate unconditioned S-R associations and broken arrows indicate associations formed through conditioning.

discussing the latent learning research, the response-reinforcement theorist was adept at suggesting possible sources of reinforcement within seemingly neutral stimulus sequences. In the case of sensory preconditioning procedure, the prime candidate for reinforcement was the offset of the preconditioning stimuli. The assumption here is that the preconditioning stimuli, presented as they are in these experiments against a sparse and otherwise constant environment, may induce a drive, presumably a mild aversive drive, that is reduced by their removal. While it is true that the conditions of the sensory preconditioning experiment literally force the response-reinforcement theorist to postulate some such reinforcing mechanism in this situation, it should be recalled that Hull's later theorizing held that number of reinforcements, not magnitude of reinforcement, is the critical factor in the strengthening of associations. Hence, it is entirely consistent with the theory to argue that a seemingly weak reinforcing event could produce the conditioning posited here.

According to response-reinforcement theory, then, the stimulus-pairing operation actually functions to condition some fraction (r_g) of the attending response elicited by each of the paired stimuli to the other member of the pair. With reference to Figure 4-3, the tone would also be expected to produce a unique r and s pattern that would become conditioned to the light by the processes described above. However, these aspects are omitted from the figure to simplify the illustration.

Figure 4-4 shows the next step in the response-reinforcement analysis. It diagrams the second, or conditioning, phase of the sensory

Figure 4-4.

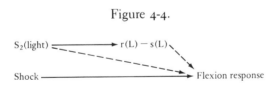

$S_1(\text{tone}) \dashrightarrow r(L) - s(L) \dashrightarrow \text{Flexion response}$

Figure 4-5.

preconditioning procedure, and the important point illustrated is that the leg flexion response occurs in contiguity not only with the light CS but also with the stimulation, $s(L)$, arising from the attending response to the light. The reinforcing action of the shock will therefore instate an association between the leg flexion and the stimulus compound of "light + $s(L)$."

Finally, Figure 4-5 indicates the behavior expected in the final phase of the experiment. As a consequence of phase 1 pairings, presentation of the test stimulus, the tone, will elicit portions of the attending reaction to light, $r(L)$-$s(L)$, which will in turn evoke leg flexion as a consequence of the phase 2 pairings. Since $s(L)$ is only part of the stimulus compound conditioned to leg flexion in phase 2, the flexion response will not be evoked as strongly as in the conditioning phase. Hence the tendency toward relatively weak response to the test stimulus generally noted in these experiments.

This explanation of sensory preconditioning can be seen as a variant of the basic r_g account of cognitive phenomena. In effect, it denies that "sensory conditioning" takes place, that is, sensory conditioning in the sense of a direct association between representations of the paired stimuli. Instead, pairing is seen as conditioning to each stimulus implicit responses evoked by the other, with the result that the stimuli come to evoke a common r-s component. Consequently, any behavior established in the presence of the r-s element will tend to appear in the presence of either stimulus.[5]

The criticism usually leveled against this response-reinforcement reply to the challenge of sensory preconditioning (and to its accounts of other cognitive phenomena) is that it in essence postulates an unobserved response for an unobserved cognition. While there is some justice to this criticism, due weight should be given to the consideration that the response-reinforcement explanation comes equipped with the means of its assessment. That is, since its hypothetical processes are conceived of as actual *behaviors*, they can be investigated and tested in terms of overt behavioral analogs. Assessment of the response-reinforcement interpretation of sensory preconditioning drew primarily upon two behavioral models of the postulated covert processes—mediated stimulus generalization and classical conditioning.

The Mediated Generalization Model

Mediated stimulus generalization was early demonstrated by Shipley (1933). Using human subjects, he first repeatedly paired a faint flash of light with a sudden tap on the cheek which reliably evoked a readily observable eye-blink response. Next, the tap on the cheek, and hence the blink response, were repeatedly paired with a shock evoking withdrawal of the finger from a flat surface. Finally, when the light was then presented alone, the finger withdrawal occurred in a majority of the subjects, even though the light had never been paired with shock or with finger withdrawal. This procedure is said to demonstrate mediated generalization because the generalization of the withdrawal response cannot be accounted for on the basis of similarity of the light and shock but can be accounted for in terms of the blink reaction acting as a connecting link, or mediator, between the light and the withdrawal response. That is, light evoked finger withdrawal because it had earlier come to evoke eye blink (by its pairing with tap on the cheek), and eye blink, in turn, had come to evoke finger withdrawal by its pairing with shock. Note that this mediational sequence is composed entirely of *observable* behaviors established by explicit conditioning.

Shipley's mediated generalization can be seen as a parallel, at the level of overt behaviors, of the mediating process postulated by the response-reinforcement theorist to occur covertly in the sensory preconditioning paradigm. In fact, Wickens and Briggs (1951) suggested that sensory preconditioning can be seen as a special case of mediated stimulus generalization. This perspective helps make plain that the transfer in sensory preconditioning experiments can be considered as arising not because the two stimuli have been presented together, as cognitive theory would maintain, but because a common, mediating response has been learned to them, as in mediated generalization. Under this view, the pairing operation is not really required for transfer of the conditioned response in a preconditioning paradigm; all that is necessary is that the subject acquire the same response to the two stimuli before the conditioning phase.

The idea that a common response rather than temporal contiguity is the truly critical factor in sensory preconditioning was tested by Wickens and Briggs (1951) at the level of overt behavioral mediation. In the preconditioning phase of their study, four groups of college students each received 15 presentations of a tone and light. For Group I the tone and light were paired together and subjects were required to say "now" when the stimuli appeared. As a result of this training these subjects should be likely to say (or think) "now" whenever light or tone is presented in the later experimental phases. For Group II the two stimuli were presented independently of one another (separated in time) and the same response ("now") was given to each stimulus. Thus Group II subjects should also be likely to think "now" upon presentation of either stimulus. For Groups III and IV the verbal response was given to one but not the other stimulus, light or tone, respectively. In the second phase of the procedure, all groups received 30 conditioning trials with the tone as CS, shock to the finger as UCS, and finger retraction as the conditioned response. During conditioning, the subject's hand rested on a flat surface and a slight upward movement of the finger closed a switch which terminated the shock or, if the subject responded in the brief interval which separated CS and UCS onset, prevented delivery of the shock. In the third phase of the experiment, transfer of finger retraction from the tone CS to the light was tested by ten presentations of the light alone.

Under the common response hypothesis, Groups I and II should show good transfer of conditioning from phase 2 to 3, since the CS and test stimulus evoke a common r-s component ("now") in these subjects. More important, if phase 3 transfer depends upon the subject's having made the same response to both stimuli in the preconditioning phase, rather than upon association of the stimuli through contiguity, then Groups I and II should display *equal* transfer since they had equal pretraining in making the common response to each stimulus. The fact that the pretraining stimuli were contiguous for Group I and discontiguous for Group II should make no difference. Finally, Groups III and IV should display little or no transfer since they were not pretrained to make the same response to the CS and test stimulus.

The results are presented in Table 4-5, which gives the mean number of responses by each group in phase 3. By standard statistical measures, Groups I and II do not differ, and both made significantly more responses than Groups III and IV, which also do not differ. This is exactly the predicted pattern of results.

Wickens and Briggs's study is an indirect test of the common response hypothesis of sensory preconditioning, indirect in that it employs an overt behavioral analog of the processes postulated to occur covertly in sensory preconditioning. Therefore, while Wickens and Briggs's results are fully consistent with the common response hypothesis, they do not enable the strong conclusion that the transfer ob-

Table 4-5. Mean number of responses during the transfer test of Wickens and Briggs's experiment

Group	Response made to	Mean
I	Tone and light together	7.6
II	Tone and light separately	7.1
III	Light only	2.1
IV	Tone only	0.2

served in the usual sensory preconditioning experiment is due to the operation of a common r-s element. However, their experiment greatly helps to substantiate the response-reinforcement interpretation by showing that a common mediating response is responsible for transfer within a close behavioral parallel of the sensory preconditioning experiment.

The Classical Conditioning Model

The second approach to assessing the response-reinforcement interpretation of sensory preconditioning drew upon the data of classical conditioning. The logic here is that since the covert conditioned responses assumed to operate in the preconditioning phase must follow the laws of conditioning, it should be possible to predict the effects of various alterations in preconditioning procedures by considering the effects those conditions would have in conventional conditioning.

Silver and Meyer (1954) were among the first to pursue this line of inquiry. The rationale of their experiment can be illustrated by reference to Figure 4-3, where the light stimulus can be seen to function as a UCS and the tone as a CS in the conditioning of the light-attending response to the tone. It is this CR that is held to mediate transfer from phase 2 to phase 3, as shown in Figures 4-4 and 4-5. The impetus to Silver and Meyer's experiment was the consideration that classical conditioning of overt CR's is highly dependent upon the temporal relations between CS and UCS. Specifically, classical CR's are commonly found to be established more readily when the CS precedes the UCS by a brief interval than when the two are presented simultaneously, and conditioning is poor when UCS onset precedes the CS. On this basis Silver and Meyer predicted that, with reference to Figures 4-3 to 4-5, there would be greater transfer in phase 3 if the tone preceded the light during preconditioning than if the two were presented simultaneously or in the reverse order.

Silver and Meyer tested this proposition in relation to transfer of a shock-elicited running response in rats. Three experimental groups each received preconditioning involving 3,000 presentations of a one-second light and a one-second buzzer. For one group the light and buzz were presented simultaneously. For a second group, the light and buzz were separated by a ½-second interval with the stimulus that was to function as CS for the response mediator presented first. In the third group, the stimuli were also separated by ½ second but the stimulus that was to function as CS for the mediator was presented second. To illustrate, a training sequence for an animal in the second (CS-UCS) experimental group would consist of the following phases: preconditioning with light followed by buzz, conditioning of the running response to buzz, test for transfer of running to light. A training sequence for an animal in the third (UCS-CS) group would be as follows: preconditioning with light followed by buzz, conditioning of the running response to light, test for transfer of running to buzz. Within these experimental groups, one half of the animals received the buzz and light as CS and UCS, respectively, and these assignments were reversed for the other half. Three control preconditioning groups were also employed, one given 3,000 buzzer presentations, one given 3,000 light presentations, and one without any form of pretraining. Conditioning of the running response and tests for transfer proceeded in exactly the same way for all experimental and control groups. In the conditioning phase, the subjects were placed in an apparatus equipped with a grid floor and given repeated presentations of a one-second light or buzz followed ½ second later by shock through the grid floor. A conditioned response was defined as a move of at least 6 inches prior to onset of the shock. Conditioning trials continued until the subjects made seven CR's in a block of ten trials. At that point, the subject received the transfer test, which consisted of 100 additional trials with the stimulus not used during the conditioning phase.

The results are shown in Figure 4-6, which depicts the mean number of CR's by each group in the transfer phase in blocks of ten trials. Statistical analyses confirmed that, as predicted, the CS-UCS group made significantly more CR's than the UCS-CS and "simultaneous" groups, which did not differ from one another. All of the experimental groups, however, were superior to the three control groups, which did not differ among themselves. In general, these data are consistent with the proposition that sensory preconditioning should vary in ac-

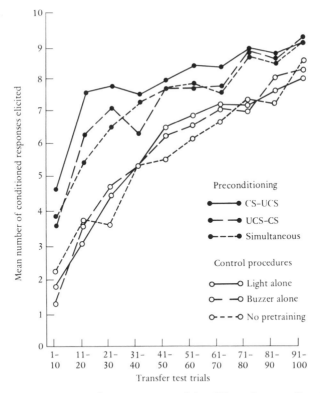

Figure 4-6. Learning curves for acquisition of the CR to the transfer stimulus. (After Silver and Meyer, 1954, p. 58. Copyright 1954 by the American Psychological Association. Reprinted by permission.)

cordance with the laws of temporal relations in classical conditioning. However, it is surprising that the UCS-CS group did not perform more poorly. A subsequent experiment by Coppock (1958), discussed next, sheds light on this outcome.

Coppock's study provides further analysis of sensory preconditioning from the viewpoint of classical conditioning. His experiment employed conditioning of the galvanic skin response (GSR) in human adults. The GSR, it will be recalled, is a change in the electrical resistance of the skin, elicited in this case by a brief shock to the hand. The experimental design encompassed five different training sequences, as summarized in Table 4-6. One group underwent the usual sensory preconditioning procedure. These subjects, labeled Preconditioning Group in the table, first experienced ten presentations at about 30-second intervals of a three-second tone and a two-second light. Tone onset preceded light onset by one second, and thus from the response-reinforcement perspective the tone would be assumed to

Table 4-6. Summary of Coppock's experiment

Group	Preconditioning		Preextinction	
	Stimulus presentation	Assumed S-R change	Stimulus presentation	Assumed S-R change
Pre-conditioning	Tone-light	Tone $\dashrightarrow$r(L)-s(L)		
Control	Tone, light unpaired	None		
Preextinction	Tone-light	Tone$\dashrightarrow$r(L)-s(L)	Tone	Weakening of tone$\dashrightarrow$r(L)-s(L)
Inverted preextinction	Tone-light	Tone$\dashrightarrow$r(L)-s(L)	Light-tone	Weakening of tone$\dashrightarrow$r(L)-s(L) Acquisition of light$\dashrightarrow$r(T)-s(T)
Inverted pre-conditioning	Light-tone	Light$\dashrightarrow$r(T)-s(T)		

Note. Solid arrows indicate unconditioned connections. Broken arrows indicate conditioned connections.

function as a CS in this preconditioning phase with respect to a UCS function of the light. The next phase of training for the Preconditioning Group involved GSR conditioning, which was accomplished by ten pairings of the light signal with a one-second electric shock, the light preceding shock by one second. The pairings were separated by intervals of about one minute. The final phase for the Preconditioning Group consisted of five transfer test trials in which the tone alone was presented. The Control Group had ten preconditioning presentations of both the light and the tone, but unpaired and in a staggered order. The GSR conditioning and transfer test procedures for the Controls, and for the remaining groups in the experiment, were identical to those of the Preconditioning Group. Table 4-6 indicates the critical behavior changes and outcomes assumed by the response-re-

Conditioning		Test	
Stimulus presentation	Assumed S-R change	Stimulus presentation	Predicted behavior
Light-shock	Light ──→ r(L)-s(L) ---→ GSR	Tone	Strong GSR via tone-elicited r(L)-s(L) mediator
Light-shock	Light ──→ r(L)-s(L) ---→ GSR	Tone	No GSR since tone elicits only its un-conditioned r(T)-s(T)
Light-shock	Light ──→ r(L)-s(L) ---→ GSR	Tone	Weak GSR since tone elicits a weakened r(L)-s(L) mediator
Light-shock	Light ⟨ r(L)- s(L) / r(T)-s(T) ⟩ GSR	Tone	Strong GSR since tone elicits both an r(T)-s(T) mediator and a weakened r(L)-s(L) mediator
Light-shock	Light ⟨ r(L)- s(L) / r(T)-s(T) ⟩ GSR	Tone	Strong GSR via tone-elicited r(T)-s(T) mediator

inforcement position for the Preconditioning and Control Groups. The analysis of preconditioning proceeds as previously described in Figures 4-3 to 4-5.

The three other groups were concerned with tests of the hypothesis that CR's are formed during the preconditioning phase. Coppock reasoned that if these hypothetical CR's are like all other CR's, as the response-reinforcement position claims, then they should be subject to the principle of extinction. Therefore, if following the preconditioning phase the assumed CS of that phase is repeatedly presented alone, the mediator CR associated with that stimulus should weaken, with consequent reduction in observed transfer on the final test trials. This prediction was tested by a Preextinction Group, which received preconditioning identical to that of the Preconditioning Group but followed by ten presentations of the tone signal unaccompanied by the light. While it can be questioned whether ten tone presentations would eliminate a conditioned $r(L)$-$s(L)$ reaction, there should certainly be substantial weakening of such a reaction, and hence it can be safely predicted from the response-reinforcement viewpoint that, as outlined in Table 4-6, the Preextinction Group will show a relatively weak GSR in the test phase.

Another group tested a different means of extinguishing the assumed response mediator. Following the preconditioning treatment, these subjects received ten additional presentations of the stimulus pair but with the presentation order inverted relative to that of the preconditioning phase, i.e., light-tone instead of tone-light. This treatment, labeled Inverted Preextinction in Table 4-6, should weaken an $r(L)$-$s(L)$ reaction to the tone, since tone is no longer followed by the light. But the inverted preextinction treatment has another interesting implication. Since tone follows light in these presentations, the tone-attending response should now become conditioned to the light. Consequently, when the light is paired with shock in the GSR conditioning phase, some portion of the attending response to tone will be evoked and stimulation arising from that response will thus become part of the CS conditioned to the GSR. This $r(T)$-$s(T)$ will be elicited in similar form by the tone itself in the final test phase, thereby yielding transfer of the GSR (see Table 4-6). On this basis Coppock predicted that the Inverted Preextinction Group would show better transfer than the Preconditioning Group.

The final group was treated as the Preconditioning Group was but with the inverse order of stimulus presentation in the preconditioning phase, i.e., stimulus pairing was in the order light-tone rather than tone-light. Coppock termed this group the Inverted Preconditioning Group. The consequence of inverting the order of stimulus pairing while keeping all other aspects of procedure like those of the Preconditioning Group is that the CS of the preconditioning phase then appears as the stimulus paired with shock in GSR conditioning rather than as the transfer test stimulus. The behavioral implications of this change are sketched in Table 4-6. For the Inverted Preconditioning Group, the preconditioning pairings condition the tone-attending response to the light. Then, in the GSR conditioning phase the light elicits the tone-attending response and stimulation arising from this response becomes part of the stimulus compound conditioned to the GSR. Finally, when this attending response is elicited by the tone as the test stimulus in the final phase, it mediates elicitation of the GSR. Therefore, a strong GSR is predicted for the Inverted Preconditioning

Group in the transfer phase. Essentially, this group tests the strong deduction from response-reinforcement theory that good transfer from preconditioning can be expected regardless of which of the paired stimuli is used in the overt conditioning and test phases of the procedure.

The Inverted Preconditioning Group should also be considered in relation to Silver and Meyer's UCS-CS treatment, which, it will be recalled, resulted in greater transfer than expected on the basis of conventional classical conditioning with UCS preceding CS. In the light of Coppock's analysis, the relatively good showing of Silver and Meyer's UCS-CS group may have been due to the conditioning of a response mediator in the manner outlined for the inverted preconditioning treatment.

Before proceeding to the results of Coppock's experiment, the reader should master the analysis and predictions outlined in Table 4-6. Coppock's study is a significant theoretical as well as empirical venture, and the generation of clear, testable predictions of the outcome and ordering of five experimental manipulations bearing on the sensory preconditioning phenomenon is a significant accomplishment, quite apart from the outcomes themselves. It requires some patience to work through such analyses of hypothetical response processes in relation to the postulates of the theory, but the return for so doing is appreciation of the strength and resilience of the response-reinforcement position.

The results of Coppock's study are summarized in Figure 4-7, which shows the median GSR response for each group on each of the five test trials. It should first be noted that the transfer scores of the Preconditioning Group were well above those of the Control Group on each test trial, a performance difference that was statistically highly significant. This outcome confirms at the human level the basic phenomenon of sensory preconditioning as seen in experiments with animal subjects, and does so under conditions that are directly comparable in all essential respects.

The ordering of transfer scores in the five groups was, with one exception, in good agreement with the response-reinforcement predictions. Transfer of the GSR response in the Preextinction Group was generally less than that in the Preconditioning Group throughout the test trials, as would be expected if preextinction weakened the response mediator. At the same time, GSR transfer in the Preextinction Group over the five test trials was significantly greater than that in the Control Group; hence by implication the preextinction treatment still left the response mediator at considerable strength, as could

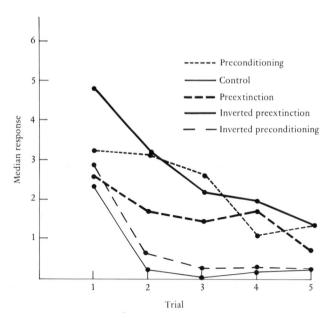

Figure 4-7. Median percentage GSR on the five test-phase trials of Coppock's study. (After Coppock, 1958, p. 216. Copyright 1958 by the American Psychological Association. Reprinted by permission.)

be expected under the response-reinforcement assumptions. Finally, the overall transfer scores of the Inverted Preextinction Group were significantly greater than those of the Preextinction Group, an outcome consistent with the response-reinforcement argument that the inverted preextinction treatment would condition a second mediator which would offset the weakening effect of extinction of the original mediator.

The only notable exception to the response-reinforcement predictions is the relatively poor transfer performance of the Inverted Preconditioning Group. This group had been expected to exhibit as much transfer as the Preconditioning Group but in fact did not differ significantly from the Control Group. In discussing this outcome, Coppock noted the common finding in conventional classical conditioning that the presentation of a novel or relatively intense stimulus about the time the CS occurs is likely to disrupt or inhibit the CR, a phenomenon known as external inhibition. Coppock suggested that during the conditioning phase of the Inverted Preconditioning treatment, the r(T)-s(T) reaction to the light CS (see Table 4-6) might have been inhibited by presentation of the shock stimulus and thus did not become part of the stimulus complex conditioned to the GSR. Such external inhibition is a possibility, but it is not a convincing explanation of the weak transfer in the Inverted Preconditioning Group in view of the strong transfer exhibited by the Inverted Preextinction Group, which is assumed to have the same conditioned mediator operating during the conditioning with shock. The performance of the Inverted Preconditioning Group remains a problem for the response-reinforcement interpretation.

Further studies on response mediation in sensory preconditioning could be described, but the foregoing are sufficient to characterize the response-reinforcement treatment of the phenomenon. That treatment closely resembles those accorded latent learning and latent extinction. First, possible implicit responses and subtle reinforcers are pointed out in the experimental situation. Then these hypothetical processes are linked to principles of overt behavior, and logical deduction is made both of the specific behavior at issue *and* of previously unknown aspects of the phenomenon. For example, the Silver and Meyer study and the Coppock study contain five predictions concerning the outcomes of previously untested conditions of training, predictions derived from the same postulate set used to account for sensory preconditioning itself. Confirmation of these predictions, and most were confirmed, both buttressed the response-reinforcement interpretation and extended empirical knowledge of the phenomenon in a systematic way.

In this way, response-reinforcement theory can be said to have weathered the challenge of sensory preconditioning, much as it weathered those of latent learning and latent extinction. But more important than the question of the adequacy of the opposing theories in each of these research issues was the cumulative impact these controversies were having on the viability of both theoretical positions. Before considering that impact we shall examine one further set of findings that bear on the central assumptions of cognitive and response-reinforcement theory.

PLACE VERSUS RESPONSE LEARNING

If asked what an animal learns in the course of mastering a maze, the response-reinforcement theorist replies, "The movement sequence leading to reward with least delay," and the cognitive theorist replies, "An expectation of food at the goal." But if one inquires further of the cognitive psychologist just how we are to know the animal has this expectation, the reply would be, "Because when hungry the an-

imal engages in the movement sequence leading directly to the goal."
The point is that, in application to the usual case of learning, the
cognitive and response-reinforcement conceptions of what is learned
are not distinguishable in any pragmatic way.

While acknowledging that the two theories often point to the same
behaviors to operationally define their conceptions of what is learned,
Tolman maintained that acquiring an expectation means considerably
more than merely engaging in the appropriate movement sequence.
The two theoretical conceptions of what is learned can be differentiated, he argued, by their predictions of how an animal will behave
when an already-learned path to the goal is suddenly blocked. If the
animal has acquired a knowledge of the location of the reward, it
should take the most direct available route to that place; but if the
animal has learned a specific set of movements, it should engage in
the most similar possible movements.

In a test of these views, Tolman, Ritchie, and Kalish (1946a) gave
rats five food-rewarded runs from starting point A to goal G in the
elevated maze shown in Figure 4-8A. A lamp, which provided the
only illumination in the room, was located at H. Following the rewarded runs, a block was placed in path C, sections D through G
were removed, and 18 new paths were placed around the table top as
shown in Figure 4-8B. Each animal then received a single test trial
with the new maze arrangement, and the critical question was whether
the subject would choose the paths most closely approximating the
formerly correct *movement* sequence (paths 9 and 10) or the path
pointing most directly to the location of food (path 6).

The percentages of animals choosing each path are shown in Figure 4-8B. Path 6 elicited by far the largest proportion of choices (.36),
whereas very small percentages of the subjects chose paths 9 and 10,
bordering on the original route. These data are consistent with the
view that the subjects had learned to expect food at a specific place
and adjusted their behavior in accordance with that expectation. Noting the small number of training trials in this experiment, the authors
argued that if additional training had been given, even more animals
would have chosen the shortcut.

Now consider a very similar study by Kendler and Gasser (1948),
investigators within the response-reinforcement tradition. During the
training phase of this experiment rats were run through only the T-
maze portion of the alley maze apparatus shown in Figure 4-9. This
was accomplished by placing movable partitions, indicated by broken
lines, in the semicircular area at the base of the T-maze, which is
indicated by cross-hatching in the figure. The only light in the room

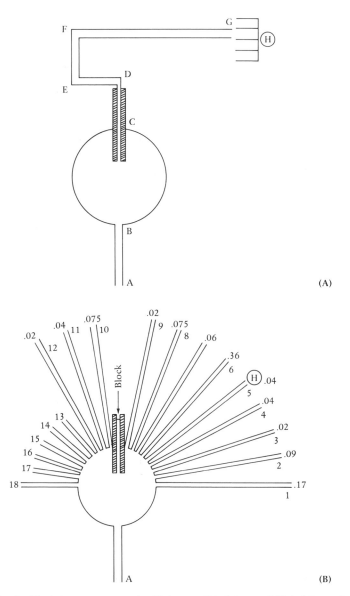

Figure 4-8. A. Training apparatus in Tolman, Ritchie, and Kalish's study: (A–B) starting alley; (C) alley bounded by 18-inch-high plywood walls; (G) goal boxes; (H) light. B. The test apparatus showing original pathway blocked and radiating alternate pathways provided. Numbers given are percentages of animals choosing various pathways.

(From Tolman, Ritchie, and Kalish, 1946a, pp. 16, 17. Copyright 1946 by the American Psychological Association. Reprinted by permission.)

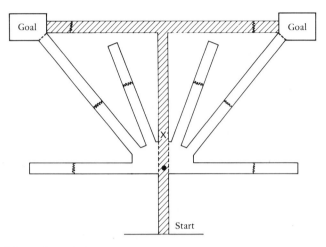

Figure 4-9. Floor plan of the apparatus used in the Kendler and Gasser study. (After Kendler and Gasser, 1948, p. 180. Copyright 1948 by the American Psychological Association. Reprinted by permission.)

was suspended above the choice point of the T. Different groups of animals were given free-choice trials in the T until they had obtained either 0, 5, 20, or 100 food reinforcements for choosing a designated arm. After the subjects had achieved the specific number of reinforced runs, a block was placed at the entrance to the T-maze stem at point X; the partitions in the semicircular area were removed, allowing access to the three alleys on either side of the T-stem; and each subject was given 20 test trials involving free choice of the six radiating alleys. Black curtains (wavy lines in the figure) suspended midway in the alleys prevented the subject from seeing the termini of the alleys from the choice point.

Table 4-7 gives the results of the first test trial, the measure most comparable to that in the Tolman, Ritchie, and Kalish study. The numbers assigned to the six alleys have as their point of reference the goal box that contained food during the training series. If a subject had been rewarded for approaching the left goal, the alleys were numbered 1 to 6 from left to right; if reinforcement had been in the right goal, the alleys were numbered 1 to 6 from right to left. For each animal, then, alley 2 represented the path leading most directly to the rewarded goal, and alley 1 required the response most similar to the behavior rewarded during maze training, a 90-degree turn.

It is apparent that the bulk of the animals chose alley 1 and that almost no animals chose alley 2, the "shortcut" path. These data indicate that Kendler and Gasser's rats learned specific responses dur-

Table 4-7. Percentage of subjects in each training condition choosing each of the six alleys on first test trial

Number of reinforcements in training	Alley number					
	1	2	3	4	5	6
0	50.0	0.0	8.3	0.0	8.3	33.3
5	36.4	9.1	18.2	0.0	9.1	27.3
20	41.7	8.3	8.3	16.7	0.0	25.0
100	53.9	0.0	23.1	0.0	7.7	15.4

ing maze training, not place of reward—an outcome precisely the reverse of that of Tolman, Ritchie, and Kalish! The data from the remaining test trials were generally consistent with those of trial 1, although choice of alley 2 was somewhat more likely, particularly following an intermediate number of reinforcements.

How can such diametrically opposing outcomes be accounted for? And should we conclude that animals are place learners or response learners? Despite the close similarity of method in these two studies, there is one seemingly trivial but actually quite significant difference— a difference that distinguished the earliest studies of maze learning by Small and Watson (pp. 46–47). Like Small, Tolman and his co-workers used an open (wall-less) maze; like Watson, Kendler and Gasser used an alley (walled) maze. The open elevated maze exposes the animal to stimulation from all areas around the maze, and in the Tolman, Ritchie, and Kalish study would enable the subject to locate the reward in relation to a salient cue, such as the nearby light. Kendler and Gasser's maze, on the other hand, sharply reduces stimulation by extramaze cues, thereby favoring response learning as the likely solution mode. It is interesting to note that cognitively oriented investigators typically used open mazes and response-reinforcement investigators tended to rely upon the alley maze. Here is a prime instance of a dependence of finding on technique and, even more basically, of the dependence of finding on theoretical orientation since, as noted in our discussion of Small's and Watson's work, this difference in technique is not a matter of chance but arises from differing assumptions about the capabilities of the organism and about appropriate strategies of research.

Whether an animal will exhibit place learning or response learning, then, depends significantly upon the particular experimental arrangement. With this idea in mind, let us consider another experi-

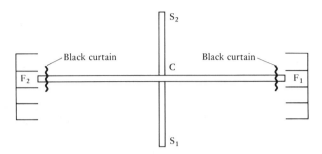

Figure 4-10. The cross maze; S_1 and S_2 are starting points, F_1 and F_2 are food boxes, and C is the center (choice) point.
(From Tolman, Ritchie, and Kalish, 1946b, p. 223. Copyright 1946 by the American Psychological Association. Reprinted by permission.)

ment by Tolman, Ritchie, and Kalish (1946b, 1947) which sought to determine the relative priority of place and response learning in the rat. They devised a clever "cross maze" (see Figure 4-10) which allowed them to put place and response learning tendencies in direct competition. As usual, the maze was an elevated affair and several prominent extramaze cues (e.g., a cage rack) were in the experimental room. Subjects in a place-learning group were required to go to the same goal on each trial to find food but were started on half of their trials from S_1 and on the other half from S_2, in a scrambled order. This meant that to solve the place problem these animals had to make opposite turning responses over the training trials. A response-learning group was trained to make the same turning response on each trial, also being started equally often from S_1 and S_2. To solve the response problem these animals had to go to opposite places on different trials. Since solution of the response problem required the subject to disregard place-going tendencies and solution of the place problem required overcoming specific response-making tendencies, the idea was that the relative speed of learning by the two groups would indicate whether place or response is the more basic mode of learning.

Tolman et al. found that the place-learning group proved far superior to the response group. Only three of the eight subjects in the response group reached the criterion of learning (ten successive errorless runs) within the 72 training trials alloted, whereas all eight subjects in the place-learning group achieved the criterion in eight trials or less. The authors concluded that while the rat may exhibit either place learning or response learning, place learning is the simpler and more basic form of learning in situations where there are marked extramaze cues.

The last phrase of this conclusion is at least implicit acknowledgment that the relative priority of place and response learning as observed in Tolman, Ritchie, and Kalish's study might be reversed under conditions which reduce extramaze stimulation. As one might guess from our review of Kendler and Gasser's study, evidence in that direction was not long in coming. It was supplied by Blodgett and McCutchan (1947) in the form of a comparison of response learning under ordinary laboratory conditions with response learning under conditions designed to eliminate extramaze cues. The experimenters were painstaking in setting up the latter conditions. The training was conducted in a room so dimly illuminated that even after becoming dark-adapted the experimenter could barely discern the subject's movement through the maze. Then, to eliminate whatever sparse patterned light might reach the subject from extramaze sources, the maze (an elevated T) was virtually enclosed within a homogeneous, light-diffusing dome. The dome also served to preclude possible cues derived from sound drift within the room, or from temperature, or other gradients. Finally, the placement and axial direction of the maze were changed for every trial. Under these conditions one group of rats received 48 free-choice trials in the T with food reward for making a particular turning response. Another group received the response training in the same maze but under conditions comparable to those used for the response-learning group of Tolman, Ritchie, and Kalish. Specifically, for the latter group the open T without dome was placed in a fixed position within an illuminated room containing the spatial landmarks ordinarily found in laboratory rooms, e.g., doors, windows, heat sources.

The experimenters found, first, that response learning by their "room" group proceeded at about the same low rate that Tolman, Ritchie, and Kalish had observed in their response group. For example, in each of these groups the same number of animals (five out of eight) failed to solve the problem within the training period. Second, Blodgett and McCutchan found that response learning by their "dome" group was significantly superior to that by their room group. In fact, the dome group was performing in essentially errorless fashion by the end of the training trials. These comparisons permit the conclusion that there is nothing absolute about the slow response learning observed by Tolman, Ritchie, and Kalish. Rather than being inherently slow, response learning may proceed rapidly when interference from spatial cues is eliminated.

In sum, the experiments we have considered indicate that neither place learning nor response learning is uniformly dominant. This conclusion is the major outcome of the very sizable number of studies that were devoted to this issue (see Restle, 1957, for a review). Whether an animal learns places or responses and the relative ease of place versus response learning depend primarily upon the stimulus situation presented to the animal. Since neither place nor response learning can be said to be the rule, neither cognitive nor response-reinforcement theory could claim to be supported in any telling manner by this body of research.

Nevertheless, this line of research did provide some clear instances of both place and response learning, and these instances had to be accounted for by the opposing theoretical orientation. For example, how did the response-reinforcement theorist account for the fact that Tolman, Ritchie, and Kalish's rats tended to select the most appropriate path to the goal upon finding the original path blocked (Figure 4-8)? Such outcomes in the place-learning research are related to a variety of observations scattered throughout the literature on animal learning. For example, Lashley found perfect retention of a learned maze route by rats that had been subjected to surgical lesions in the central nervous system, lesions that destroyed the animals' capacity to produce the movements that constituted the original maze learning (Lashley and Ball, 1929; Lashley and McCarthy, 1926). Although some of the lesioned animals could do little more than flounder through the maze, slipping and falling continually, they nevertheless entered only correct pathways. In another study of the role of movement patterns in maze learning, MacFarlane (1930) immersed a complex maze in water and required one group of rats to learn the maze by swimming to the goal and another group to learn by running on a floor placed just below the water surface. When each group was suddenly shifted to the training condition of the other, and hence to the distinctive movement patterns of the other, there was generally little or no disturbance of correct choice behavior. Finally, Zener (1937) showed that even classical conditioning may result in behaviors strikingly different from those observed during training. Using typical Pavlovian procedures, Zener gave CS-UCS pairings to dogs restrained in a harness upon a table. The unconditioned stimulus was a food pellet delivered by a chute to a pan placed just beneath the dog's nose; the conditioned stimulus was movement of a small card mounted nearby. After a number of conditioning trials, Zener placed his subjects without restraining straps on a different but empty table in the same room and presented the CS to the animals in that position. While the subjects were content to remain on the new table in the absence of the CS, as soon as the card began moving they exhibited a behavior pattern far more complex than simple salivation to the visual cue. Without hesitation they jumped from the second table, crossed directly to the top of the training table, and approached and looked into the food pan.

In each of the foregoing studies, as in the Tolman, Ritchie, and Kalish experiment, the particular responses associated with a reward appear to be subordinate to a knowledge of the whereabouts of the reward. In each case, when the usually rewarded movements were prevented, there appeared new, functionally equivalent movements. These studies are all instances that fit the cognitive theorists' dictum, "Given knowledge, behavior will take care of itself." In the face of such observations, the response-reinforcement theorists drew upon two long-established concepts in their armamentarium—the habit family hierarchy and the r_g.

Application of the habit family concept proceeds in the manner discussed in our presentation of Hull's theory (pp. 94–96). Hull argued that as a consequence of general locomotory experience, the animal acquires different ways of moving between any two points in space and that these alternative motor sequences become organized into families by means of the r_g arising from the terminal point and conditioned throughout each motor sequence leading to that point. Thus, if the r_g derived from terminal point B is evoked at start point A, there will tend to be evoked at A, via s_g, all the motor habits bringing about movement to B. In relation to the experiments at issue, the response-reinforcement theorist could invoke the habit family hierarchy concept to argue that once the r_g defining "being at the goal place" is conditioned back to the start point, the animal has available its family of previously learned ways of moving to the goal place. Since the family is ordered by strength of its members, and since the strongest members will as a rule be the shortest motor sequences (because they are more strongly reinforced), we can expect the animal to choose the most direct available means to the goal when previously rewarded behaviors are no longer possible. While this analysis is applicable to each of the experiments described above, it is not fully satisfactory, since Hull left rather vague just how the r_g reaction defines the place of reward.

It is also possible for the response-reinforcement theorist to explain place learning and related phenomena in terms of an r_g mechanism alone. As we know, it follows from Hullian theory that fractions of the goal reaction will become conditioned to cues at and near the goal region and will be evoked with increasing magnitude as the animal approaches the goal. If it is further assumed that increase in the magnitude of reward-derived r_g-s_g's facilitates any movement that brings about that increase (and we saw that certain of the latent learning experiments forced such an assumption on the response-reinforcement theorist), then an animal that oriented or initiated a movement toward the goal region would be drawn to the goal simply by an ever-increasing r_g-s_g reaction. To illustrate, consider Zener's experiment discussed above. In the conditioning phase, the moving card becomes conditioned to evoke salivation and other anticipatory consummatory responses. When the animal later orients to the moving card from a more distant point in the room, the consequent r_g-s_g reaction elicits further movement toward the CS, which in turn further increases the magnitude of the r_g-s_g reaction, which in turn elicits further approach, etc. Similarly, in the Tolman, Ritchie, and Kalish experiment (Figure 4-8) the prominent light behind the goal box would, on the test trial, evoke continuous approach to the goal by the same r_g-s_g steering mechanism. Of course, the assumption that increased r_g-s_g's directly elicit approach behavior is a significant modification of the original r_g-s_g mechanism, which operated solely by means of conditioned links between r_g-s_g and overt responses.

Cognitive theorists, for their part, were similarly strained to account for some instances of response learning which appeared in the course of this research controversy. In regard to response learning in experiments like that of Kendler and Gasser (Figure 4-9), the usual argument was that the experimental situation severely restricted opportunities for place learning. By using an alley maze and deliberately avoiding salient extramaze cues, Kendler and Gasser made it difficult for their subjects to locate the reward in relation to the choice area. In such situations, the argument went, there was little the animal could do but master the task by learning the specific response sequence leading to reward. To use Tolman's vocabulary, enclosed mazes lead to the formation of narrow, striplike cognitive maps rather than broad, detailed cognitive maps.

Such an explanation is not unreasonable in relation to experimental procedures like Kendler and Gasser's, but it is inadequate in application to some other studies showing response learning. Consider, for example, the following experiment by Gentry, Brown, and Lee (1948). Rats were first trained in the elevated multiple-T-maze shown in Figure 4-11. The apparatus was placed in a room containing such extramaze cues as windows and a door. A lamp located at H was the only source of illumination in the room, and it was placed so that its

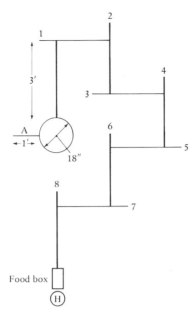

Figure 4-11. Maze used in the learning phase of the Gentry, Brown, and Lee experiment.
(After Gentry, Brown, and Lee, 1948, p. 313. Copyright 1948 by the American Psychological Association. Reprinted by permission.)

light would shine directly down pathway 8. On each training trial the subjects were placed at starting section A, from which they moved to the circular table top and thence through the eight-choice-point maze to the goal box, where they received food reward. After 11 such trials the animals had attained practically errorless performance on the maze. At this point, the multiple T was removed, the path leading from the table during original learning was blocked, and open pathways were arrayed around the table as shown in Figure 4-12. The animals were then placed in starting section A and allowed a single free choice of the new pathways. Would the subjects choose pathway 8, the pathway pointing directly to the goal location? Or would they choose pathway 2 and 3, those most adjacent to the original learning pathway and involving movements most similar to those of the initial choices during learning?

The percentages of animals choosing each path are shown in Figure 4-12. Some 52.5 percent chose paths 2 and 3, but only 5 percent chose pathway 8. There is no evidence in these data that the animals had developed or could use a cognitive map of the learning situation. Note that the experimental setup had all the ingredients that appear

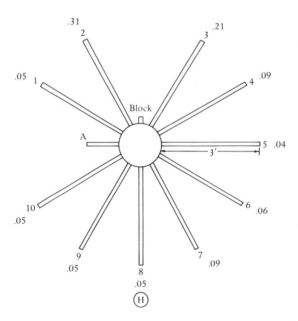

Figure 4-12. Apparatus used in the test phase of the Gentry, Brown, and Lee experiment.
(After Gentry, Brown, and Lee, 1948, p. 314. Copyright 1948 by the American Psychological Association. Reprinted by permission.)

to favor place learning—an open maze, extramaze cues, and a salient cue highlighting the goal location. And yet the subjects not only failed to manifest place learning on the test but chose paths that were at once most similar to those in the original learning and most directionally opposed to the goal location. Faced with such outcomes (this was not the only study showing response learning under such conditions), the cognitive theorist is forced to appeal to the possible effects of poorly understood conditions of training. For example, perhaps the subjects had been overtrained in the initial learning phase to the point that they ceased attending to the extramaze stimulus relationships, or perhaps it is unusually difficult to learn the location of a goal that is in directional opposition to the initial movements leading to it. While there may be validity in such arguments, they do not derive in any compelling way from the postulates of cognitive theory.

In summary, both cognitive and response-reinforcement theory received partial support from the research on place and response learning, but neither could lay convincing claim to a majority of the experiments in this area and each had its difficulties in accounting for some outcomes.

In the context of this general conclusion, it will be useful to consider one further paper which, without contributing an additional experiment, can be said to have resolved the place versus response learning controversy. The paper is by Restle (1957), and while it properly belongs to the next stage of an historical treatment, it will be instructive to consider the paper at this point as a transitional work as we close out our review of these classical theoretical controversies.

Restle's Analysis

The cognitive versus response-reinforcement controversy naturally disposed investigators to analyze place-response experiments in an "either-or" manner. Each side assumed that place and response learning were fundamentally different forms of learning, and each side sought to establish that one form was paramount and the other of little or no importance. The essence of Restle's contribution was to break this either-or mentality by showing that place and response learning can be seen as manifestations of a common process.

Restle's analysis drew considerably upon a very early line of research on maze learning that had sought to determine the stimulus basis of learning. The question that motivated these early investigators was, what type of stimulus input constitutes the basis of maze learning? Do rats use primarily visual cues? Olfactory cues? Proprioceptive cues? One of the strategies in this early research, devised by Watson (1907), was to study the effects of eliminating or severely restricting a sense modality, usually by surgical means. The idea was that if the stimuli conveyed by that modality were important in learning, then maze performance should be significantly hindered. In a series of studies Watson *separately* restricted the visual, olfactory, auditory, and tactual modalities and, finding no deleterious effects on performance, concluded that the important stimuli in maze learning must therefore be the proprioceptive stimuli, the only type of input he had not restricted. Proprioceptive stimuli were thought to provide a basis for learning by serving as cues to successive responses in the true-pathway sequence. If each correct response becomes associated with the stimulation arising from the previous correct response, the true-pathway movement sequence becomes a self-sustained chain independent of external sensory input.

This analysis proved vulnerable on several grounds. First, Watson had eliminated a single sense modality at a time, and so it was possible that the null effects he observed reflected not the unimportance of the stimulus class eliminated but rather an increased reliance by the animal upon sensory data from other channels. That the latter is the case was indicated by Lindley (1930), who showed that rats made blind *and* anosmic (deprived of the sense of smell) had far greater difficulty in maze learning than merely anosmic *or* blind rats, whose learning approximated that of normals. Similarly, Honzik (1936) found that blind-deaf-anosmic rats had greater difficulty learning an elevated maze than animals restricted in only two of these modalities, who in turn had greater difficulty than animals restricted in just one of these modalities. Moreover, Honzik found that his blind rats had significantly greater difficulty in learning an *elevated maze* than normals, while Watson and Lindley had found no effects of blinding upon performance in an *alley maze*. These several observations indicate that the rat does not generally rely upon any single type of input but instead may use input from any modality and can shift among these inputs as need or advantage dictates.

A final blow to the view that proprioception is paramount is that Honzik's blind-deaf-anosmic animals were actually unable to learn the maze despite considerable training. Since these animals had proprioceptive stimuli to fall back upon, their failure to learn indicates that proprioception is not a reliable basis for acquisition of maze habits, at least in complex multiple-unit mazes such as employed by Honzik. Actually, this outcome is suggested by purely logical considerations as well. If a given turning response is to become associated with stimuli arising from preceding correct movements, the sequence must occur with some regularity in order to be strengthened by the terminal reinforcement. But in complex mazes, which afford many possible locomotory sequences, it is unlikely that regular proprioceptive stimuli-response relationships obtain until after the maze has been learned on some other basis. Of course, in simple mazes such as a single T, which afford only a few locomotory sequences, learning on the basis of proprioception should, and in fact does, proceed much more rapidly.

Another facet of the early research on the stimulus basis of maze learning concerned the relative importance of intramaze and extramaze cues. The basic research strategy here was to scramble one type of cue during learning, thereby forcing the animal to rely on the other type. Extramaze cues, such as visual or auditory stimuli from the room containing the maze, could be scrambled in their relation to choice responses simply by rotating the entire maze in random fashion from trial to trial. In general, studies of this nature found that maze rotation seriously retards learning, particularly if the maze is elevated and the room contains salient cues. However, the rotated-maze problem was usually solvable so long as the *intramaze* cues were not disturbed. Intramaze cues, e.g., visual, olfactory, and other stimuli arising from within the maze itself (including proprioceptive cues), could be scrambled by interchanging the maze segments from trial to trial and in this way destroying any consistent relation between correct response and stimuli within the immediately preceding maze segment. Again, while learning was generally found to be retarded by this procedure, rats could solve the problem as long as the goal was kept in a fixed position relative to *extramaze* stimuli. The scrambled-cue research, then, showed that both intramaze cues and extramaze cues contribute to learning, and that each can support learning in the absence of the other. Finally, this research indicated that in elevated mazes, extramaze cues usually are more important to learning than intramaze cues, whereas if the maze is an alley maze, or an elevated maze within a homogeneous surround, intramaze cues become more important than extramaze cues.

An important part of Restle's analysis consisted in pointing out that these early findings on the stimulus basis of maze learning fully predict the outcome of the later research on place versus response learning. From the perspective of the early research, comparisons of place versus response learning are actually comparisons of learning on the basis of extramaze cues versus learning on the basis of proprioceptive cues. As noted above, rats are capable of using input from *any* source, and the early research made clear that the relative ease of learning on the basis of extramaze versus proprioceptive cues depends sharply on the type of maze employed (elevated vs. alley, complex vs. simple) and upon its surround (whether rich or barren in stimulation). Just as learning on the basis of proprioceptive cues was shown to be difficult or impossible unless the maze was simple or its surround homogeneous, so response learning was found to predominate only in such situations. And just as learning on the basis of extramaze cues was shown to be ascendant over learning on the basis of intramaze cues in elevated mazes with rich surrounds, so place learning was found to predominate in these situations. In this fashion, Restle showed that the outcomes of all tests of place versus response learning could be predicted from the empirical knowledge supplied by the early investigators, and could be interpreted not as evidence of the primacy of one of two forms of learning, but rather simply in terms of whether the stimulus situation presented to the animal favored the use of extramaze cues or the use of proprioceptive cues to accomplish the task. Most important, Restle's analysis showed that place learning and response learning can be seen as manifestations of the *same process*, namely, the animal's discriminating out which cues are consistently associated with reward, regardless of their source.

It is significant that, having advanced a persuasive argument that place learning and response learning are fundamentally the same thing, Restle did not then attempt to fit the overall data in this area within the framework of either cognitive theory or response-reinforcement theory. Instead Restle chose to describe the data within a narrower theoretical formulation which he had developed to handle phenomena of discrimination learning, the process whereby organisms come to respond differently to different aspects of stimulation. In Restle's formulation the stimulus situation facing the learner is thought of as a set of cues (i.e., different aspects of stimulation), each of which can be categorized as either *relevant* or *irrelevant*. Relevant cues are those that can be used to predict the receipt of reward in the given learning task, and irrelevant cues are those that bear no consistent relation to receipt of reward. For example, if a subject is always rewarded for approaching a black card that is randomly placed in either of two positions from trial to trial, then the cues from the black card are relevant and the position cues are irrelevant. Solving any discrimination problem is a matter of the subject's learning to relate its choice responses to the relevant cues while at the same time learning to ignore the irrelevant cues. Two further assumptions of the theory allowed Restle to make quantitative description of the course of learning. These assumptions were, first, that on each trial a constant proportion, θ, of relevant cues and of irrelevant cues become conditioned and ignored, respectively, and, second, that θ is the proportion of relevant cues in the problem. Within these assumptions, the value of θ for a given problem can be estimated by mathematical treatment of the subject's learning curve on that problem (for equations and methods of computation, see Restle 1955).

As applied to place and response learning, the theory says that in either type of learning the subject is simply discriminating out any and all cues that predict reward within the problem. An important deduction from the theory is that the set of cues relevant in a problem that consistently rewards the animal for making the same response sequence leading to a single place will be the sum of the separate place and response cues. Restle termed this deduction the *hypothesis of cue additivity*, and it enables a quantitative test of his interpretation of the place-response data. Consider the three problems diagrammed in Figure 4-13. In problem A the animal learns to go to the same

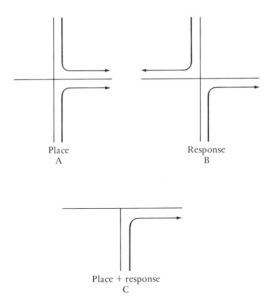

Figure 4-13. Runs rewarded in place, response, and place + response learning.

place for reward from alternate starting positions and hence makes different responses from trial to trial. In problem B the animal learns to make the same response for reward but must go to different places from trial to trial. In problem C the same place and response are rewarded on every trial. Now if problem A (place) and problem B (response) are, as the theory maintains, just different discriminations that proceed by the same process, and if that process is validly described by Restle's model, then the set of cues relevant in problem C should be simply the sum of the place and response cues in problems A and B, respectively, and θ_{A+B} will equal $\theta_A + \theta_B$. Recall that the model permits estimation of the value of θ for a given problem on the basis of a performance curve on that problem. Therefore, once θ's for problems A and B have been determined in this way, these can be used to *predict* the θ and performance curve for problem C and these predictions can be checked against actual performance on that problem. Surveying the place-response literature, Restle found several experiments that had trained groups of animals under the three conditions of Figure 4-13, and thus he was able to use available data to test the cue additivity principle. When Restle analyzed the data from these experiments, using such performance measures as the number of errors to solution, he generally found that actual performance on the place plus response problem agreed closely with performance as predicted from θ's based on the learning of the separate place and response problems.

To put the analysis more generally and simply, a strong prediction of the theory is that place plus response learning will always be faster than place or response learning alone because, if subjects do use whatever relevant cues are available, the combined problem will necessarily result in a larger effective relevant cue set and hence a higher learning rate. There has now developed a good deal of evidence for cue additivity in the discrimination learning literature (although the research also shows that degree of additivity is significantly affected by the relative salience of the cues involved). But in the present context, the important thing to note is that evidence for cue additivity in place and response learning is not only supportive of Restle's interpretation but logically inconsistent with the either-or interpretation of place and response learning as pursued by the cognitive and response-reinforcement theorists.

Considering the place-response literature as a whole, and bearing in mind that it stemmed from the opposition of the cognitive and response-reinforcement theories, we can see some of the factors responsible for the decline of interest in such major theories of learning. In retrospect, the place-response research appears to highlight several negative aspects of the attempt to establish general theories of learning. Not only was neither cognitive nor response-reinforcement theory able to derive much support from the very extensive research efforts expended in this vein, but it could even be argued that the theoretical controversy obscured perception of an already-existing, sound empirical basis for predicting the outcomes of the various tests. And most significant, Restle's analysis showed that a satisfactory resolution of the controversy and a more precise prediction of data in this area could be obtained by working from a small set of theoretical principles closely tied to the paradigm at issue. This was to be increasingly the form that theoretical work took in the decade ahead—the development of theoretical models intended to provide relatively precise, and frequently quantitative, accounts of a limited range of phenomena, as opposed to the development of broad theories intended to account for all learning phenomena at a more general level.

SUMMARY AND OVERVIEW OF THE CONTROVERSIES

Cutting across the separate controversies were the two most fundamental questions in learning theory—whether learning can occur
without reinforcement and whether learning can occur without response. What can now be said about the status of these issues? Considering first the role of reinforcement in learning, the research paradigms that sought to resolve this issue were those of latent learning
and sensory preconditioning. When all the data had settled on these
controversies, response-reinforcement theory could still maintain that
some form of reinforcement is essential for learning, but this position
was maintained only at a price of drastically altered and questionably
tenable conceptions of the nature of reinforcement. At the outset of
the latent learning research, reinforcement was generally thought of
in terms of the action of such conspicuous need-reducing stimuli as
food and water to a deprived organism, as evidenced by the serious
consideration which response-reinforcement theorists gave to Blodgett's claim that he had demonstrated learning without reward. As the
latent learning research progressed it became necessary for the reinforcement theorist to argue that rewards might take such forms as
novel stimuli or activity per se. And the sensory preconditioning research left the reinforcement theorist no alternative but to argue that
offset of the paired preconditioning stimuli provides a rewarding reduction in drive stimulation. Such arguments are a legitimate recourse of the theory, and they were sufficiently plausible, primarily
on logical rather than empirical grounds, to enable preservation of
the reinforcement postulate. However, it did devolve upon the theorist to put the hypothesized rewarding stimuli and the needs they were
presumed to serve on a firm empirical basis if their usage was to have
meaning in the long run. For example, since all learning involves
activity to at least some degree, invoking activity as a reinforcement
for learning becomes meaningless unless accompanied by specification of independent conditions defining the operation and reduction
of an activity drive. Thus, while the latent learning and sensory preconditioning experiments could not resolve the role of reward in
learning because of the impossibility of ruling out all theoretically
possible forms of reward, they did expose the glaring gap in the response-reinforcement theorist's knowledge of the drive and stimulus
incentive conditions that could define a rewarding state in the Hullian sense. Efforts to develop an adequate theoretical and empirical
knowledge of drives, incentives, and reinforcers became a major con-

cern of workers in the response-reinforcement tradition. The outcome of these efforts belongs to another part of our story; the point to be noted here is that the latent learning and sensory preconditioning research made clear that a need-reduction theory of learning could not get far with reliance on obvious, autonomically defined need states, such as hunger, thirst, and pain.

The other fundamental question in learning theory, whether learning can occur without response, was addressed in each of the controversies. The latent extinction and sensory preconditioning paradigms dealt directly with this issue, since there was no observable response basis for the learning in these experiments. The latent learning and place learning paradigms did not eliminate overt responding, but these experiments can be said to have seriously questioned the importance of *specific movement patterns* in learning. In the final analysis, none of these experimental attacks were sufficient to dislodge the proposition that learning always involves a peripheral response process, but, as in the case of the reinforcement postulate, the controversies forced the response-reinforcement theorists to significantly alter their conception of the form that the process might take. Consistent with Watsonian psychology, response-reinforcement theorists preferred to explain learning whenever possible solely in terms of direct associations between observed behaviors and the environment. While mediating response processes in the form of anticipatory reactions were an early feature of Hull's thinking, the initial conceptions stressed peripheral and potentially measurable, if indeed not directly observable, components of strong consumatory reactions. During the course of research on the various controversies, the response-reinforcement theorists were pushed progressively farther upstream, so to speak, in the brain-to-behavior sequence with regard to the locus of the response processes underlying learning. Finally, as exemplified in the response-reinforcement accounts of Strain's research and of sensory preconditioning, even perceptions, dutifully defined in terms of peripheral orienting and receptor reactions, were held to qualify as conditionable processes that could mediate observed behavior changes. At this point, the distinction between the cognitive and response-reinforcement positions was not so much a matter of where the critical mediating processes were taking place, since for both camps the mediators were well beyond any possible measurement operation, but rather a matter of how these processes were to be thought about. Should the mediators be conceived of as *ultimately* peripheral in source and therefore as following the laws of conditioned overt behaviors? Or should they be seen as perceptual or central in origin and perhaps following different laws? This issue became a persistent legacy of the major controversies and reappears as a central theme of later theoretical developments.

In addition to altering conceptions of what is learned and of the necessary conditions for learning, the controversies had important effects on the structure and development of the two major theories. Much of the research inspired by cognitive theory had a "let's see how you get out of *that* one" quality, whereas the response-reinforcement theorists were generally concerned to show that their theory *could* handle "that one." In this way, the various controversies stimulated development within the response-reinforcement framework of specific behavioral mechanisms to account for the acquisition of knowledge about the environment and for translation of that knowledge into performance. Cognitive theory, on the other hand, primarily concerned as it was with efforts to demonstrate the independence of learning from response and reinforcement, did little to elucidate the perceptual and demand mechanisms determining acquisition of cognitions or their relation to performance. During this period, for example, cognitive theory did not advance beyond the rather simple view that given cognition and demand, appropriate behavior will follow. In contrast, the response-reinforcement theory elaborated the r_g mechanism, based on facts of conditioning, to detail the links between perception, learning, and performance. As we have seen, r_g's were employed in this regard in two ways. The theory assumed, first, that r_g-s_g's might control performance via a direct conditioned link to overt behavior and, second, that r_g-s_g's conditioned to some aspect of the learning situation could either facilitate or inhibit ongoing behaviors depending on the nature of the original goal reaction. In the decades ahead, each of these conceptions became the basis for theoretical analysis of a number of important learning phenomena. There were no analogous legacies from the cognitive orientation. Thus, although the postulates of cognitive theory underwent little change as a result of the controversies, it was nevertheless the response-reinforcement framework, which had been consistently on the defensive, that proved to have the greater utility for later work on learning.

Viewing the several controversies as a whole, we can see a clear pattern in the course of research on these classic issues. First, a phenomenon is demonstrated that appears interpretable only within the cognitive framework. The demonstration takes the form of an apparent case of learning without response or without reinforcement. Then closer examination of the situation suggests a possible basis for the operation of responses and reinforcement, usually in covert or as yet unmeasured form. By recourse to such hypothetical processes, the response-reinforcement theorists explain the phenomenon in a fashion consistent with the postulates of their theory. The recourse to such processes is a setback for response-reinforcement theorists, since they must abandon the strong reliance upon peripheral responses and patent biological need reducers that characterized the initial conceptualization of the theory. In these respects the theory must move closer to the cognitive orientation. On the other hand, by linking the postulated hypothetical processes to principles of overt behaviors, response-reinforcement theorists were often able to predict new aspects of the phenomenon; such predictions at once lent credence to the basic postulates of the theory and indicated the value of the theory in extending empirical knowledge about the phenomenon in a systematic way. For their part, cognitive theorists could be credited, either directly or indirectly, with compelling evidence for the control of learning by cognitive processes, evidence in the form of successful demonstrations of latent learning, latent extinction, etc. On the other hand, lacking formal links to explicit behavioral mechanisms such as characterized the response-reinforcement position, cognitive theorists usually did not progress beyond the initial demonstration to further knowledge of the phenomenon. Latent learning, latent extinction, sensory preconditioning, and place learning all fitted the broad postulates of cognitive theory and forced significant modifications of the response-reinforcement position, but cognitive theory frequently could not match the eventually more detailed behavioral predictions and analyses of the modified response-reinforcement position. The overall result of the several controversies was, then, a kind of standoff in which neither side "won."

The fact that the validity of neither theory could be convincingly assessed through these extended research controversies was a clear signal that the effort to construct testable general theories of learning was premature. Indeed, as we have noted, the question of whether learning is at base a matter of s-s or of $S-r_g-s_g-R$ had assumed a form that was clearly beyond the resolving power of available techniques. Consequently, the 1950's and 1960's saw a general decline of interest in testing the merits of these broad theories, and investigators increasingly turned to construction of theoretical models intended for more limited application, as in Restle's discrimination learning model. However, the broad concepts of the cognitive and response-reinforcement frameworks remain strongly influential to this date.

Although none of the controversies produced an *experimentum crucis* with respect to the relative merits of cognitive versus response-reinforcement theory, they worked significant changes in conceptions of the nature and conditions of learning, as we have reviewed in this section, and they had an important cumulative impact on the course of learning theory and research in general. Essentially, the controversies made clear the need to distinguish between learning and performance. It was driven home that under a much wider range of stimulus conditions than heretofore credited, organism-environment interactions could result in unobserved changes in the organism's behavior tendencies, changes that are reflected in performance only under *other* conditions. There could scarcely be a more important lesson for a behavioral psychology. In this way, the controversies destroyed forever any simplisitic notion of learning as primarily a matter of *direct* associations between overt behavior and the environment, and they seriously challenged the view that learning can be broadly explained solely in terms of directly observed stimulus-response relations. It was now clear to many S-R psychologists that an S-R analysis would have to be critically concerned with processes mediating between observed stimuli and final behavioral output, however those processes were to be conceptualized. And indeed mediational theories of learning constituted a substantial part of the theoretical work of the 1950's and 1960's.

However, we turn now to consider two other major theoretical approaches that had their inception and development largely in the 1930's to 1950's—the conceptual treatments of Edwin R. Guthrie and B. F. Skinner. Both of these theorists developed systems that skirted the theoretical controversies, and both firmly maintained that it is possible to achieve comprehensive explanation of behavior without reliance upon mediating processes.

5

The Conceptual Treatment
of Edwin R. Guthrie

Edwin R. Guthrie approached learning in a manner quite distinct from that of Hull and Tolman yet solidly within the S-R framework. In particular, the system developed by Guthrie did not lead to extensive theory-testing programs of research such as those engendered by the theories of Hull and Tolman. Nevertheless, Guthrie's theory became an important reference point in the study of learning and must be ranked along with those of Hull and Tolman as a preeminent S-R system.

As in the case of our treatment of Tolman's and Hull's theories, we shall begin by considering Guthrie's broad orientation to the subject.

GENERAL ORIENTATION

All that the most sophisticated man can do in any situation is to contract his muscles in some order and pattern.

This quotation from Guthrie (1942, p. 24) captures the essence of his orientation. This rather stark expression of behaviorism makes the point that however subtle or complex the human capability, it is inevitably at base a matter of muscle movements. Of course, all behaviorists accepted Watson's dictum that any aspect of psychological activity, including thought and emotion, can be given response definition, but Guthrie was the most insistent in pointing out that all responses, and hence all psychological activity by the behavioral definition, must reduce to particular muscles acting in particular ways. It is this conception of muscle action as the ultimate vehicle of psychological reality that led Guthrie to give primacy to the organism's actual movements, i.e., specific patterns of muscle and gland action, as the appropriate level of analysis for the psychologist. The organism's movements, in turn, were seen by Guthrie as linked to equally specific patterns of sensory input. Under this view the typically smooth flow of our behavior is the result of the combination of many elemental action patterns, or movements, linked to changes in sensory input. Guthrie's unit of behavioral analysis is the most molecular in S-R theory.

Also implicit in the foregoing quotation is the idea that behavior is directly controlled by the environment. Note that the "sophisticated man's" response is limited to the order and pattern of muscle action elicited *by the situation*. There is no room here for intervening central processes in the direction of behavior, or for a conception such as Tolman's in which the particulars of behavior conform to the organism's purposes and cognitions. In Guthrie's conception, the organism's specific movements are not the servants of higher processes—they *are* the higher processes, and they are situationally controlled. Organisms are the sum of their observable stimulus-movement relations.

In his general view, then, Guthrie was faithful to the letter as well as the spirit of behaviorism. Animals are machines and, in principle at least, simple machines at that. All human and animal activity is analyzable into particular movements that are directly linked to immediate environmental stimuli. Psychological explanation consists of determining (a) the movements that define the phenomenon in question, (b) the stimuli controlling those movements, and (c) the rules governing the development of that control. This is behaviorism with a capital B.

Probably the first question that arises in considering application of this explanatory system concerns the adequacy of its descriptive units. Is it really possible to have a comprehensive description of animal activity in terms of such atomistic descriptive units? And is not the natural activity of higher animals characterized by more extended and meaningful units of behavior? An important part of Guthrie's position on these questions lies in a distinction he drew between *movements* and *acts*. Whereas movements refer to patterns of muscle action, acts in Guthrie's lexicon refer to a group of alternative movements that produce some common end. For example, opening a door is an act in that it refers to a behavioral event that can be accomplished by many different movements. In general, acts are behaviors defined in terms of their end results, or what the behavior accomplishes in the world. "Acts are defined by consequences but executed by movements" (1952, p. 196).

Ordinarily we tend to describe behavior in terms of acts. For example, we might speak of a behaving individual as writing in a notebook, walking to the door, greeting a companion, and so forth. These are acts, not movements, in that writing can be done in a number of ways, a variety of movements will bring the individual to the door, and the companion may be greeted in many different ways. Such common sense use of acts as descriptors is understandable since we are usually concerned with the ends of behavior rather than with its movement composition. But Guthrie argued that whatever the utility of acts in everyday communication, they are not sufficient to the task of the learning psychologist. The learning psychologist must be concerned with movements, asserted Guthrie, because that is what the nervous system, the physical basis of response, is concerned with. In Guthrie's words, "the [individual's] nerves connect his sense organs with his muscles, not with his notebook, the door, or his companion," (1942, p. 22). The actions of the nervous system define movements, and hence these must be the natural units of learning. Acts themselves are not learned in any direct sense but rather emerge as the product of the learning of many stimulus-movement relations that produce a common outcome.

> The difference between act and movement is of vital importance in learning theory because our accepted notion of the mechanism of response is that nerve impulses actuate muscular contraction, and if association or conditioning is to be related to changes in the nervous system it is specific movement patterns which must be dealt with (Guthrie, 1952, p. 28)

In taking this position Guthrie did not deny that acts have a place in a scientific description of behavior. He saw two quite distinct problems of learning. One is the problem of *what we learn;* the other is *how we learn.* The language of acts was held to be appropriate to the first of these problems, and the language of movements to the second. By "what we learn" Guthrie meant what humans and animals are capable of accomplishing through learning. Obviously this is a question that requires consideration of goals, purposes, and behavioral consequences, and it naturally disposes one to a molar level of description. Guthrie felt that determination of what is learned in this sense is a valid and significant, but nevertheless distinctly secondary, question for the learning psychologist. (He felt that Tolman was too concerned with this question.) The truly fundamental issue in learning theory is how we learn, and in Guthrie's view the answer to this question is to be found only in careful study of what the organism actually *does* in the learning situation. To the degree that we allow our descriptions of learning per se to be influenced by our perception of the ends, or consequences, of learning, to that degree do we distort the description of precisely how the organism learns to achieve these ends.

While Guthrie's distinction between movements and acts makes clear his position on the appropriate role of these units in behavioral analyses, it leaves unanswered the question of whether it is feasible to attempt a comprehensive description of learning in terms of movements. It would surely be impossible to reduce all the basic phenomena of learning to their constituent stimulus-movement elements. For that matter it would be extremely difficult to specify the scores of movements that doubtless occur in the course of acquisition of even simple habits, such as the bar press response. The measurement task could be made more manageable by studying only those movements that occur at presumably critical points in the learning situation, e.g., those movements that occur just prior to reward. Guthrie's own studies of the role of movements in learning (Guthrie and Horton, 1946) employed just such a selective measurement technique. But even with a restricted population of movements there remain the knotty questions of how best to classify and record the movements observed.

Given the formidable problems involved in movement analyses, it is not surprising that the actual measurement of movements in Guthrie's program was more a matter of ideal aim than of realization. The Guthrie and Horton study of movements (referred to above) came late in the development of the theory, and most of the observations upon which Guthrie based his theory were in fact drawn from the experiments of others, experiments which recorded behavior at relatively molar levels. But if Guthrie seldom dealt with measured movements, he surely dealt with behavior at a considerably finer level of analysis than Tolman or Hull. His belief in the primacy of movements gave him a deep respect for the importance of the subject's moment-to-moment behavior as a determiner of learning and led him to scrutinize the details of behavior and stimulation in any learning situation. Taking this perspective, Guthrie was usually able to point out possible, if not actually observable, stimulus-movement relationships that could be argued to play a critical role in the learning observed.

Practically speaking, then, Guthrie's concern with movements manifested itself not so much in measurement as in the way he analyzed learning situations, and in his hands at least, the idea that movements are the units of behavior proved a valuable analytical tool.

It was noted above that Guthrie's theory was based largely on experimental data supplied by others. In this respect, Guthrie's preferred mode of explanation differed considerably from that of Tolman and Hull. For the latter theorists, explanation took the form of theoretical predictions that could be subjected to direct experimental test, and both men were prolific experimenters in this regard. In contrast, Guthrie published but two experimental assessments of his position (Guthrie, 1933; Guthrie and Horton, 1946). Generally, explanation for Guthrie took the form of showing that a wide variety of existing behavioral observations were consistent with the principles he espoused. He was, as we shall see, very adept at finding such consistencies.

Another significant feature of Guthrie's approach was his concern with real life behavior. He was a pioneer in extending data and concepts from the study of animal learning to analysis of human behavior in natural life settings, and at least half of his writings are concerned with practical applications of learning principles. In particular, he made important contributions to the areas of education, personality and social psychology, and clinical psychology.

Guthrie's strong concern with real life behavior may come as a surprise in the light of his spare, mechanistic conception of human behavior. But it must be remembered that Guthrie began his work during the full flush of behaviorism's early promise. Like other behaviorists, he accepted learning as the primary determiner of human behavior, and he had absolute faith in the power of an S-R analysis to provide a valid account of the learning process at all levels of complexity. The following quotation will help to convey Guthrie's outlook as he undertook the explanation of human activity in S-R terms and will serve to summarize several of the points we have made about Guthrie's general orientation. The passages quoted occur in the context of a discussion of the defining feature of "mind." After considering and rejecting several criteria that might define the presence of "mind" in any life form, Guthrie continues as follows:

Inanimate objects respond to the forces about them, to blows, to heat, but their responses are not adaptive. Living creatures respond to a limited range of forces with adaptive changes, changes like the protective contraction of the sea anemone [when it is prodded]. Those living creatures that have minds do more than this: they improve their adaptation with experience. This is something that never happens in the anemone. No matter how often we prod it, it does not improve its device for meeting this disturbance. It may grow fatigued and fail to respond. This is not learning, because as soon as rest has restored it we will find its response what it was before. The anemone and its descendants will continue to have only one answer to being prodded. The anemone has no capacity for changing its answer to our attack, and therein consists its mindlessness. It is, like the radish, a living being but a witless being.

The mental character of a response does not lie in the adaptive nature of the response because all living beings are capable of adaptive response. Responses are mental when they are subject to modification. This modification lies in the capacity for attaching responses to new stimuli. Organisms which can learn to give their adaptive reactions to new cues have at least the rudiments of mind. . . .

To endow a sea anemone with mind, several additions would be necessary. One of these would be a new receptor, a new sensitive area. Another would be a connection through nerve fibers with a central ganglion (brain) in which impulses from the new sensitive area could be routed to the muscle effectors. If the added sensitive area were a pigmented spot conveniently placed to receive the shadow of an approaching enemy and this were sensitive to light, and, further, if the central ganglion were so constructed that activity in the pathway leading to the muscle led to the establishing of connections between the nerves from the pigment spot and the nerves to the muscle, the conditions for learning would be satisfied. The anemone would not at first contract when the shadow fell upon it, but if the shadow accompanied the touch and the contraction which follows the touch, the next occasion would find the animal prepared. The shadow would be a cue for the defensive contraction and the touch would be avoided.

The anemone of our illustration would be, of course, capable of learning only one thing, whereas most men have astonishing possibilities of learning. Instead of one response to begin with and one possible new cue for that response, human beings have a large number of responding organs and an almost infinite number of possible new cues for these, with a practically infinite variety of combinations of both cues and responses. (Guthrie, 1938, pp. 11, 15–16).

These passages incorporate several significant assumptions. First, it is clear that Guthrie effectively equates learning and mind. Capacity for change in stimulus-movement linkage is taken to be the defining feature of mind, or intelligence, at all levels of organismic complexity. Second, differences between species in learning or mental abilities are viewed as matters of degree rather than of kind. Note that the difference between the learning ability of the "modified" sea anemone and that of humans is rendered solely in terms of the number and variety of possible stimulus-response associations. Third, the nervous system is conceived of as merely a switchboard for routing incoming sensory signals directly to effectors; and in organisms that have mind, the brain is simply a means of establishing new, but no less direct, routings of nervous impulses between sensory organs and effectors. In sum, mind, or intelligence, is the capacity for learning, learning is the formation of stimulus-movement linkages, and the principles and physiology of their formation are fundamentally the same in all species. Hence, when we study the attachment of a movement to a stimulus, even in a simple organism, we are studying the process that lies at the heart of the most complex mental phenomena. This philosophical outlook is the basis of Guthrie's willingness (and that of other behaviorists) to apply S-R principles fully and freely in explanation of human activity.

To recapitulate, the hallmark of Guthrie's approach was simplicity. Of the S-R theorists, he provided the most direct, elemental, and uncomplicated expression of behaviorism. His units of analysis (movements), his conception of the determinants of activity (direct elicitation of movements by stimuli), his virtual exclusion of central intervening determinants, his switchboard model of the central nervous system, and his confidence in the universality of S-R principles all represent behaviorism in a pure and simple form. Moreover, Guthrie's insistence on movements as the proper data inclined him to root explanation in the concrete details of behavior and environment. While full measurement of movements was not feasible, it was nevertheless the case that, for Guthrie, what the organism actually does would always be more important than what the organism might be perceiving, thinking, or remembering. Also, it may be speculated that the difficulty of measuring movements, coupled with Guthrie's strong interest in accounting for all behavior including real life activities in terms of stimulus-movement associations, led him to a mode of explanation that relied less on experimental test and more on a search for consistency between existing behavioral observations and his explanatory principles.

Finally, it is worthwhile to briefly contrast Guthrie's approach with that of Tolman in order to underscore the importance of the theorist's decision on unit of analysis. In selecting behavior acts as the "real" units, Tolman was led to ignore the particulars of behavior and to concentrate upon the purpose and cognition inherent in behavior acts. For Guthrie, the particulars of behavior were all-important, and acts and their immanent higher-order properties immediately became secondary and derivative.

GUTHRIE'S THEORY OF LEARNING

There is often some discrepancy, usually unavoidable, between the position a theory takes in the abstract and its position in relation to actual situations. In interpreting Guthrie's theory, it is particularly important to keep this dual aspect in mind. The theory in its purest form can be stated in its entirety as follows: *Whenever a stimulus pattern is contiguous with a movement, it becomes maximally associated with that movement.* This principle gives Guthrie's position on the two fundamental questions in learning theory—what is learned? and under what conditions does learning take place? The principle states that learning consists of the formation of stimulus pattern-movement associations, that the formation of such associations requires nothing more than temporal contiguity of stimulus and movement, and that the association goes to full strength in a single pairing or trial. Guthrie held firm to these beliefs throughout his writings. However, in analyzing actual learning situations Guthrie generally employed a somewhat modified form of the above principle, as follows: *A stimulus which has accompanied a response will on its recurrence tend to be followed by that response.* This statement simply redefines the abstract theoretical principle in the light of the difficulty of measuring the total effective stimulus pattern at any moment. Essentially, it represents Guthrie's acknowledgement that a stimulus judged by an observer to be a recurring pattern may in fact contain some undetected variation that weighs against repetition of the original response to that pattern. Hence, recurring stimuli are said to *tend* to be accompanied by the same response.

Our analysis will generally follow Guthrie's practice of making reference to the modified form of the principle, but the reader should keep firmly in mind Guthrie's absolute conception of the nature and conditions of learning.

Clearly, the theory is remarkable for its brevity. The single postulate discussed above is the assumptive base from which all of Guthrie's explanations were generated. Theorizing for Guthrie was largely a matter of showing how this principle might broadly account for all the phenomena of learning. At first blush, the postulate that recurring stimuli tend to be accompanied by the same response seems rather sparse basis for a comprehensive explanation of learning. But one of the enjoyments for the new reader of Guthrie is to see just how far Guthrie *was* able to progress in reducing the variety and complexity of learning to this single principle.

Our review of Guthrie's theory will first consider its ability to handle the phenomena of conditioning. The theory will then be applied to research on the role of reward and punishment in learning. Later sections will be addressed to experimental assessments of the theory and to a consideration of Guthrie's ideas in applied form.

Guthrie's Theory and the Phenomena of Conditioning

Guthrie's first paper on learning, published in 1930, was an attempt to account for the major established facts of conditioning in terms of his theory. Both the facts and Guthrie's interpretations have remained remarkably constant over the years, and hence this paper still has an air of currency. The phenomena discussed are those of classical conditioning, but in most cases the interpretation is easily generalized to instrumental conditioning.

Acquisition of CR's. The first fact is that of conditioning itself—the fact that pairing of CS and UCS leads to a response to the CS resembling that to the UCS. This outcome can be seen as an illustration of the principle that stimuli accompanying a response (here, the UCR) tend on their recurrence to be followed by that response. This is precisely how Guthrie did view conditioning; that is, he saw conditioning not as a model or primary form of learning but simply as an instance of the all-encompassing principle of association by temporal contiguity.

The fact that the classically conditioned CR tends to resemble rather than duplicate the UCR is often cited as an apparent exception to a simple contiguity theory of learning. But from the outset Guthrie argued that in light of the realities of the conditioning situation we must expect some difference between CR and UCR. He pointed out that even in well-controlled settings the subject is continually beset by internal and external stimuli, some of which are likely to be attached to responses incompatible with the conditioned response. The result of such conflicting behavior tendencies will be what Guthrie termed a "compromise response," a behavior pattern that includes components of the conflicting responses. For example, a child who, accompanied by its mother, encounters a fierce dog that earlier had frightened the child into running away will probably not flee again. The mother's presence alters the situation and evokes approach responses incompatible with running. The result is that the child is likely to cling to the mother or get behind her, with these behaviors being accompanied by some emotional signs of fear. Depending, then, upon the degree of stimulus variation, particularly upon the difference between the stimulus patterns operating at the point of CS and UCS onset, one can expect wide variation in the comparability of CR and UCR, although some significant part of the UCR should always be present in the CR.

Effects of Repetition. The fact that acquisition of a conditioned response is typically a gradual affair requiring a number of pairings of CS and UCS suggests that the strength, or lasting quality, of the underlying change increases with repetition of the CS-UCS pairings. Thus, an assumption common to many theorists is that repetition of the environmental sequence involved in Pavlovian conditioning, or in any form of learning, works a "wearing in" of connections in the neural pathways that represent that sequence. In contrast, Guthrie viewed learning as more analogous to the setting of a switch than to the wearing of a path. A switch in this view is an all-at-once connection between a stimulus pattern and a contiguous movement, and the typically gradual appearance of the conditioned response results from increase in the *number* of stimulus pattern-movement connections rather than from increase in the *strength* of any individual connection. Consider, for example, a Pavlovian salivary conditioning experiment involving simultaneous presentation of CS and UCS. In Guthrie's analysis the experimenter-manipulated CS, e.g., a tone, is but part of the stimulus pattern connected to the salivary response on each trial. Other important aspects are external stimuli other than the tone and the proprioceptive stimuli arising from the animal's orienting movements to the tone. These additional components can be expected to differ somewhat from trial to trial because of such unavoidable factors as differences in the animal's receptor activity and posture at the onset of the tone. On any trial the particular pattern (tone plus other exteroceptive stimuli and proprioceptive stimuli) operating on that trial is conditioned to the contiguous behavior. Prominent in that behavior, of course, will be the UCS-elicited salivary reaction, but this will be embedded in an overall movement pattern that will also show some trial-to-trial variation. Repetition of the CS-UCS pairings serves to establish an increasing number of such stimulus pattern–salivary reaction connections but does not increase the strength of any previously established connection. With an increasing number of CS-UCS pairings, it becomes increasingly likely that the stimulus pattern present on each tone presentation will have been previously connected to a salivary reaction. The result, then, is the gradually sloped learning curve that characterizes many conditioning and learning experiments.[1]

In support of this analysis, Guthrie was fond of citing Pavlov's experience that better control of extraneous stimuli in his new soundproof laboratory reduced the number of CS-UCS pairings required for stable conditioning from about 50 in his earlier studies to about ten. Guthrie would argue that if it were possible to ensure a truly constant stimulus situation (e.g., the subject in exactly the same posture throughout the experiment), there would be practical certainty of conditioning with one CS-UCS presentation.

In Guthrie's analysis of the effect of practice, the pertinence of the distinction between movements and acts is obvious. The relatively permanent change we call learning takes place at the level of movements and occurs in all-at-once fashion. But acquisition of stable performance takes place at the level of acts and occurs gradually through the learning of many functionally equivalent stimulus-movement associations.

The analysis we have illustrated here in relation to Pavlovian conditioning was applied by Guthrie to all forms of learning. Thus, in instrumental conditioning, as exemplified by acquisition of bar pressing, the response becomes more and more probable as more and more of the stimulus patterns afforded by the experimental situation occur contiguously with, and hence become connected with, the bar press response.

Remote Conditioning. A fact of conditioning that appears to challenge Guthrie's theory is that a conditioned response can be established despite intervals of many seconds between onset of the CS and UCS. In such *remote conditioning* the CR tends to appear just prior to onset of the UCS whether or not the CS persists through the CS-UCS interval. For example, if onset of a five-second tone CS is repeatedly followed 30 seconds later by presentation of meat powder to a hungry dog, a stable conditioned salivary reaction will appear some 20 or so seconds after tone onset. How could the CS come to control a response so widely separated in time if, as Guthrie's theory claims, strict contiguity is necessary for formation of stimulus-response associations? The answer, said Guthrie, is that the salivary response is in fact conditioned to stimuli that *are* contiguous to it. The argument is as follows: When the tone sounds, the dog responds by "listening," which involves a pattern of orienting responses that generally persist for some time—postural changes, turning of the head, autonomic changes such as acceleration of heart rate, and so forth. When the salivary glands begin to secrete in response to the UCS, the contiguous stimuli will be those arising from the orienting responses to the tone, responses that are routinely given on each occurrence of the tone. In this way, the tone serves to instate internal stimuli which bear a constant relation to the salivary reaction. These internal stimuli are the true conditioned stimuli, and they *are* contiguous with the response they control.

This explanation would also account for the fact that remotely conditioned responses are highly subject to disruption by the introduction of new, extraneous stimuli during the CS-UCS interval. This would result from alteration by the new stimuli of the regular series of movements and movement-produced stimuli initiated by the CS and filling the CS-UCS interval and the consequent destruction of the true conditioned stimuli for the reaction.

It should be noted that remote conditioning is not established as readily as conditioning under short CS-UCS intervals. In fact, several experiments have found an optimum CS-UCS interval on the order of a half-second. This observation, too, can be nicely accommodated within Guthrie's analysis if it is assumed that a half-second interval produces the most consistent internal stimuli in conjunction with the UCR. It does not seem unreasonable to suppose, as Guthrie did, that as a rule a half-second interval is just sufficient to allow for perceptual response to the CS and the cessation of ongoing activities, whereas longer intervals afford increasing opportunities for distraction and for the introduction of interfering behaviors.

Generalization. Another phenomenon of conditioning that would appear to pose a problem for Guthrie's formulation is primary stimulus generalization, the fact that stimuli physically similar to the conditioned stimulus will evoke the conditioned response. If appearance of the CR requires recurrence of a previously connected stimulus pattern, it is not obvious why stimuli merely similar to the CS should also evoke the CR. Guthrie's explanation is fully consistent with his basic theoretical principle and proceeds from the observation that the experimenter-manipulated CS is seldom, if ever, the stimulus most contiguous with the to-be-conditioned response. In the usual conditioning procedure the CS precedes the UCS by a short interval. This means that stimuli arising from the animal's orienting responses to the CS will be more contiguous with the UCR than onset of the CS itself. Even when CS and UCS are presented simultaneously, some overlap can be expected between the stimuli arising from the subject's response to the CS and the occurrence of the unconditioned response. Given these considerations, generalization can be accounted for if it is assumed that physically similar stimuli produce the same orienting response pattern, and hence the same pattern of movement-produced stimuli. Thus, an animal conditioned to salivate to, say, a 500-cps tone will also salivate to a 10,000-cps tone because the two evoke common listening reactions, and it is the stimuli from these reactions that are the true conditioned stimuli for the salivary response. As the generalization stimulus becomes increasingly dissimilar to the CS, evocation of the CR becomes less likely although still possible by the same mechanism. For example, one might observe some generalization of a conditioned response from a tone CS to the sound of a voice because these stimuli could be expected to evoke some common attentional responses even if they do not produce identical listening reactions. In Guthrie's view, then, all generalization is mediated generalization, where the mediation is provided by common movements in the responses to the training and generalization stimuli.

Extinction. Extinction, the weakening and elimination of conditioned responses by withdrawal of the reinforcing stimulus, is another basic aspect of conditioning that must be accounted for by any theory of learning. It will be recalled that Hull devoted two major postulates to extinction, each containing several subassumptions. In contrast, Guthrie interpreted extinction as another straightforward instance of his basic principle.

The question that extinction poses for a strict contiguity theory, of course, is why conditioned responses ever disappear if their occurrence depends solely upon recurrence of previously conditioned stimuli. The answer is that the previously conditioned stimuli (or *conditioners*, as Guthrie was apt to call them) do not invariably recur; even after considerable training there will likely be some environmental variations that bring unconditioned stimulus elements into combination with CS presentation. In Guthrie's words,

> With the unconditioned stimulus withdrawn only occasional combinations of conditioners elicit the response, for it should be remembered that the animal is, in spite of soundproof room and uniform lighting, in constant motion and subject to a continuously changing pattern of stimulation. At times there are more conditioners present, and at other times fewer. When the response fails, or is diminished because relatively few conditioners are present, these and other stimuli present become inhibitors, or, what is the same thing, conditioners of other responses. (Guthrie, 1930, p. 423)

In other words, extinction consists of the attachment of responses other than the conditioned response to the stimuli of the learning situation and the consequent displacement of the conditioned response. The foregoing quotation stresses one reason to expect such new learning. That is, unavoidable variations in the stimulus situation evoke new responses that become connected to an increasing portion of the cues originally attached to the conditioned response. But another reason should also be stressed. During extinction, the subject typically emits a long run of unreinforced CR's at the outset of the session, thereby fatiguing the response system. At this point, the conditioned stimuli are likely to be followed by other responses, e.g., sitting passively, grooming, etc., that then become attached to former conditioners of the CR by virtue of their contiguous occurrence with those stimuli.

In sum, extinction is just another case of learning and will be facilitated by any condition that promotes responses that interfere with the previously learned response.

A related behavior requiring explanation is spontaneous recovery, the tendency for extinguished CR's to reappear when the subject is reintroduced to the experimental situation. Guthrie argued that an extinction session, particularly the initial session, is likely to have stimulus details that are specific to that session, e.g., stimuli arising from frustration reactions to the discontinuation of reinforcement. Removing the animal from the situation terminates such stimulus aspects and allows some original conditioners to prevail upon reintroduction to the situation, and hence there is some recovery of the response. But after several extinction sessions, virtually all of the initial conditioners should become attached to responses other than the CR, and spontaneous recovery would no longer be observed. The data show that spontaneous recovery does indeed disappear with repeated extinction treatments.

Guthrie applied his theory to other details of conditioning that will not be considered here. The analyses described above should be sufficient to indicate the way he elaborated his basic principle to account for the facts of conditioning. It can be seen that application of the principle to data requires the spelling out of a number of assumptions that might be regarded as concealed in the principle. However, these assumptions have a definitely subsidiary character in that they either are implicit in the proposition itself, e.g., the movement-act distinction which follows from the units of analysis, or are generally accepted within the S-R framework, e.g., the assumption that movement-produced stimuli may control behavior. In the handling of conditioning, then, the parsimony and explanatory power of the theory should be regarded as real achievements.

We turn now to consider Guthrie's position on another matter of central importance to any theory of learning—the role of reward and punishment in the control of behavior. Guthrie was also able to treat the findings in this area under his basic principle.

Guthrie and the Law of Effect

At a strictly empirical level there is obviously a relationship between behavior and its aftereffects. General observation tells us that organisms do tend to repeat behaviors that are followed by satisfying consequences and tend not to repeat behaviors that are followed by aversive consequences. In common sense terms, rewarded actions are likely to persist or become habitual and punished actions are likely to be suppressed or eliminated. As we have seen, how learning theorists react to this *empirical law of effect* is a critical point. Thorndike and Hull both formulated a *theoretical law of effect* for learning, in which stimulus-response associations are assumed to be strengthened *only* if followed by consequences of a certain nature. Tolman, on the other hand, rejected a law of effect for learning but accepted rewards and punishments as powerful determiners of performance. Guthrie accepted neither of these approaches.

Our prior statement of Guthrie's theory makes clear that he fully rejected a theoretical law of effect for learning. Attaching a response to a stimulus requires contiguity. Period. Similarly, Guthrie had no need to appeal to aftereffects as a performance principle since in his theory learning and performance are one. That is, the only requirement for performance of a learned response is recurrence of the stimulus pattern connected to that response. How, then, did Guthrie handle the empirical law of effect, the obvious and undeniable correlation between behavior and its consequences? The answer is typical Guthrie—parsimonious, provocative, and plain. Rewards and punishments, he argued, are simply stimuli and like any stimulus have two direct effects—they alter the situation within which they occur and they evoke responses. These effects, Guthrie maintained, account for all of the action of reward and punishment.

Let us first apply this proposition to the reward paradigm, where receipt of a rewarding, or positively reinforcing, stimulus is made contingent upon emission of the desired response. When reward occurs and alters the existing stimulus situation, there is of necessity a change in the stimulus pattern that has just elicited the desired response, thereby leaving that response connected to that stimulus pattern. The result is that the stimulus-response association just prior to the onset of reward is *preserved*. Consider, for example, an animal who must learn to manipulate an escape latch in a Thorndike-type problem box in order to gain access to food placed outside the box. Thorndike argued that the animal gradually learns the correct response over successive trials because repeated receipt of the food reward strengthens, or "stamps in," the association between the box and the appropriate response to the latch. But Guthrie's interpretation is as follows:

What I am . . . urging is that the food reward does not intensify the latch opening. That is the erroneous assumption made by Thorndike in his argument for a law of effect. *What encountering the food does is not to intensify a previous item of behavior but to protect that item from being un-learned.* The whole situation and action of the animal is so changed by the food that the pre-food situation is shielded from new associations. These new associations cannot be established in the absence of the box interior, and in the absence of the behavior of search that preceded latch-opening. The unlearning of a response to a stimulus situation requires the presence of the situation and the forming of new responses in that situation. (Guthrie, 1940, p. 144)

Another example, and one that does not involve spatial displacement of the subject, is acquisition of bar pressing behavior in the Skinner box. In training this response, the usual procedure is to first remove the bar from the box while the subject is trained to eat from the food dish at the sound of a pellet dropping into it. After this behavior is well established, the bar is inserted, the food delivery mechanism is set to deliver a pellet whenever the bar is depressed, and the subject is left on its own to learn the response. The first bar press response, which typically occurs inadvertently in the course of general movement, is followed immediately by delivery of a food pellet to the dish. Food delivery at once alters the stimulus situation and evokes the responses of approaching the dish and seizing and ingesting the pellet, thereby preserving the association between "sight of bar" and "pressing." Therefore, when the food is consumed and the animal again approaches the bar, it will make the last response it made to that stimulus, namely, pressing. This in turn again produces the sound of pellet delivery, which again evokes approach to the food dish, and so on. In this way, there rapidly develops a stable chain of alternating pressing and eating behaviors, the details of which are quite stereotyped. A strong feature of this analysis is that it predicts not only the effect of reward but also the repetitive stereotypy that food-rewarded bar pressing, and other similar behaviors, do in fact display.

In considering Guthrie's proposition in relation to the punishment paradigm, where delivery of an aversive stimulus is made contingent upon emission of an undesired behavior, the response-eliciting property of the punishing stimulus is of critical importance. To illustrate, suppose that an animal is first trained to traverse a runway for food and is then punished for engaging in this behavior by being given a brief shock through the floor of the runway a short distance from the goal box. From Guthrie's viewpoint, the outcome will depend entirely upon what the punishing stimulus makes the subject do. If the response to the shock is to run forward to the goal box, the punishment will be totally ineffective in eliminating approach behavior since the original associations between alleyway cues and forward locomotion are left intact. If the response to punishment is to withdraw toward the start area of the alley, the punishment will be effective in eliminating approach behavior since withdrawal responses will then become attached to the cues in the start region and thus displace the approach responses to those cues. If the response to punishment is to stop in place at the point of punishment, the punishment will have little immediate effect on approach to that point. On the next exposure to the apparatus the subject should traverse the runway and again encounter shock, since the punishment has not evoked any responses incompatible with approach to the shock point. However, if the alley were of similar appearance throughout, one could expect that the stopping response would eventually generalize to the early portion of the alley, since cues associated with the punishment-elicited stopping would also be present in the start region. Guthrie's analysis thus has provision for an anticipatory response mechanism much like Hull's which enables organisms to avoid aversive events (and to be directed toward appetitive events).

A critical question in the foregoing analysis is, how are we to know just what the punishment stimulus will make the animal do? This will depend in part upon parameters of the stimulus, such as its intensity and duration. But it will also depend upon a factor that is difficult if not impossible to specify in advance, namely the organism's precise behavior at the onset of punishment. For example, an animal moving slowly is likely to react quite differently to a brief shock than an animal moving rapidly. Therefore, as Guthrie emphasized, one must carefully observe the details of behavior in a punishment situation in order to properly understand the outcome. This point is nicely illustrated by an experimental exchange between Brogden, Lipman, and Culler (1938) and Sheffield (1948). In one condition of the Brogden et al. experiment, guinea pigs were placed in an activity wheel and given pairings of a 2-second tone (CS) and a .1-second shock (UCS) administered through the floor of the wheel, the tone and shock terminating at the same time. The shock was observed to elicit a running response, and over the early CS-UCS pairings a conditioned running response began to develop during the tone and in advance of shock onset. However, with continued tone-shock pairings the running response to the tone became unstable and, despite extended training, never appeared to the CS on more than 50 percent of the trials. This outcome can be taken to refute the contiguity principle, since running is presumably being consistently elicited in contiguity with the tone CS and hence should become strongly attached to it. Sheffield, who had studied with Guthrie, repeated the Brogden et al. experiment but with more sensitive measures of the subject's locomotion upon CS and UCS presentation. Specifically, by taking polygraphic records of detailed motions of the wheel, Sheffield was able to measure such aspects as the latency, speed, duration, and direction of locomotion. Sheffield found that the probability of unconditioned running to shock (as reflected in the wheel motion measures) was in fact significantly *lower* when shock came on while the subject was running than when the subject was not running. Moreover, he observed that shock during conditioned running often evoked behaviors incompatible with running, behaviors such as hopping, turning, or crouching. Given these observations, it can be argued that as conditioned running develops to the tone, shock onset becomes increasingly likely to evoke responses incompatible with running to the tone, and therefore stable conditioned running to the tone cannot be established. This is precisely the outcome that Sheffield, like Brogden et al., observed. Sheffield's findings, then, indicate that the Brogden et al. data are actually consistent with the contiguity principle and that the animals learned to respond to the CS in the way they responded to the shock. But this conclusion would not have been possible without careful observation of the subject's response to the aversive stimulus.

To recapitulate, Guthrie held that the action of rewards and punishments can be understood solely in terms of their functions as stimuli. No new principles or mechanisms are required. Like any stimulus, reward and punishment change the situation and occasion responses, and as relatively strong stimuli they can be expected to exert these effects to a greater degree than most other stimuli. In accounting for the effects of reward, Guthrie generally emphasized the stimulus-changing function of rewards, which acts to terminate the stimulus pattern attached to the rewarded behavior and so preserve that association. In accounting for the effects of punishment, Guthrie generally emphasized the response evocation properties of punishment which serve to determine just what response the subject will make in the presence of the cues for the punished behavior and hence what behavior will be given to those cues when they recur.

GUTHRIE'S IDEAS IN APPLIED FORM

An important feature of Guthrie's writings is the skill with which he extended laboratory data and theoretical ideas to analysis of everyday life situations. Of the major learning theorists Guthrie is without peer in the use of example, analogy, and anecdote to fortify, enliven, and demonstrate the utility of the general behavioral formula. His selections in this regard had a particular aptness and "good sense" quality that had the effect of at once clarifying the theoretical point at issue and persuading the reader of the true generality of the principles. Applications of research and theory to natural life situations so pervaded Guthrie's writings that his accomplishment in this vein can be properly appreciated only by reading the original works (particularly *The Psychology of Learning* and *The Psychology of Human Conflict*). But no review of Guthrie's theory would be complete without some attempt to convey this dimension of his work. There follow a number of excerpts from Guthrie's writings selected primarily on the basis of their relation to the ideas discussed in the foregoing sections. While the major purpose here is to give a sense of Guthrie's style and skill as a popularizer, an effort has also been made to select capsule statements of his views on topics at issue in learning theory.

A *Guthrie Sampler*

What practical advice does Guthrie have for bringing about learning in everyday life situations? He proposes a simple rule applicable to any type of learning: *Using any means, get the new, desired behavior to occur in the presence of the stimuli we want to produce that behavior.* This is all that is necessary. This rule is at the basis of each of the examples in the following excerpt.

> If we wish to teach a dog to come when he is called, our method will be to get him to come to us by hook or crook. There are no rules for this except what we know of dogs in general. We may hold up a bone, start running away from the dog, or pull him toward us with a check line, or use any device which experience has suggested. While he is coming we speak the dog's name. If we take care not to speak the name on any occasion when we foresee that he will not come, when he is, for instance, chasing a cat or gnawing a bone (when we believe an unwanted response is dominant), we can readily establish a stable conditioned response. We say the dog "knows" his name. If we are so misguided as to try to call him back from the pursuit of a passing car before we have insured the effectiveness of calling, we have reconditioned the dog and made his name a signal for chasing cars, not for coming to us.

The skilled trainer uses his dog's name only when the prompt response is highly certain. If the response fails, he does not repeat the name, but uses his practical knowledge to remove the cause of failure or waits until the cause is removed. The dog may have been occupied in looking at another dog or watching a passerby. The trainer waits until he has the dog's attention before he repeats the name. Otherwise the name tends to become a cue for looking at the passerby or noticing other dogs.

It is on exactly the same ground that the student officer is cautioned never to give a command that he is not confident will be obeyed. If the command is followed by acts other than those commanded the command becomes merely a cue for disobedience and the officer loses his authority.

If we are ourselves concerned to learn to address a new acquaintance by name we can achieve that result by following a simple rule: Speak his name while looking at him. Social convention will prevent our using a method which would insure remembering. This method would consist in shouting his name at the top of our voice while looking at him.

A memory for names consists in very simple habits of using the rule of conditioning. The person who excels at it has a settled habit of using names in conversation while looking at his victims, or of rehearsing them subvocally under the same circumstances. The reason most of us fail is that we do not look at the man while naming him. The person with the memory for names takes occasion to do this as often as he can manage: "Yes, Mr. Walker. . . . No, Mr. Walker. . . . Don't you think so, Mr. Walker?" (1935, pp. 45–47)

The reader can surely guess the identity of the psychologist referred to in the following excerpt, if only on the basis of the advice given.

The mother of a ten-year-old girl complained to a psychologist that for two years her daughter had annoyed her by a habit of tossing coat and hat on the floor as she entered the house. On a hundred occasions the mother had insisted that the girl pick up the clothing and hang it in its place. These wild ways were changed only after the mother, on advice, began to insist not that the girl pick up the fallen garments from the floor but that she put them on, return to the street, and re-enter the house, this time removing the coat and hanging it properly. (1935, p. 21)

The point of insisting upon re-entry while wearing the coat is, of course, to have the desired behavior occur in the presence of the stimuli the mother wants to elicit that behavior, i.e., the stimuli associated with entering the house.

Finally, an example of the rule in a lighter mode.

> Two small country boys who lived before the day of the rural use of motor cars had their Friday afternoons made dreary by the regular visit of their pastor, whose horse they were supposed to unharness, groom, feed and water and then harness again on the departure. Their gloom was lightened finally by a course of action which one of them conceived. They took to spending the afternoon of the visit re-training the horse. One of them stood behind the horse with a hay-fork and periodically shouted "Whoa" and followed this with a sharp jab with the fork. Unfortunately no exact records of this experiment were preserved save that the boys were quite satisfied with the results. (1935, p. 48)

Of course, Guthrie saw the influence of conditioning everywhere, as this following quotation indicates.

> Illustrations of . . . conditioning stimuli and their resulting facilitation are plentiful in everyday life. The literary man accustomed to writing while smoking a pipe finds it difficult to work without the pipe in his mouth. The clergyman is moved to greater eloquence when wearing his cassock, and would find it difficult to preach a sermon on the street corner. The college instructor, because he frequently uses chalk during lectures, finds facilitation to his speech through holding a piece of chalk in his hand. Our familiar surroundings increase our personal efficiency, and this law gives a psychological justification to the so-called right of personal property.
>
> A man sleeps best in his own bed, not only because he is negatively adapted to the distracting stimuli of his neighborhood, but because he has gone to sleep many times in these surroundings and they have a facilitating effect in producing slumber. A child often refuses to sleep unless covered by a familiar blanket, or allowed to suck his thumb, or permitted to take a certain doll to bed with him. (Smith and Guthrie, 1921, p. 96)

The movement-act distinction and the effects of repetition on learning are both touched upon in the following analysis of a familiar habit that is "in reality made up of thousands of habits."

The reason that one occasion is not enough to rid us of an annoying habit, though the cue is present and we have succeeded in inhibiting the response, is that not all the cues were present, and not all the possible conditioners were alienated. On each successful inhibition some of these cues may be attached to the inhibiting response and eventually we may have enlisted such a proportion of cues for our inhibiting behavior that the annoying habit will appear only occasionally, when some one of the more rare stimulus patterns which have not been reconditioned happens to be present. The habit of smoking is in reality made up of thousands of habits. The sight of tobacco, the smell of it, the mention of it, the finishing of a meal, finishing an office task, looking at the clock, and innumerable other situations all have become signals for smoking. We resolve to stop smoking. We substitute for a few of our conditioned responses inhibitory responses, a grim closing of the mouth, a tendency to push away the pipe, groping movements ending in substituting chewing gum, or nails, or pencil. We suddenly find ourselves smoking. Some cue which we had not alienated has taken us unawares and had its usual response. As a result we find it necessary to practice not-smoking on numerous occasions before our rejection is a settled habit. The apparent exceptions to the rule that the most recently practiced stimulus-response sequence will prevail, turn out not to be exceptions in fact. (1935, p. 102)

Here is Guthrie rationalizing the stimulus basis of so-called remote conditioning.

Such a movement as listening or looking is not over like a flash or an explosion. It takes time. The movement, once started, maintains itself by the stimuli it furnishes. When the telephone bell rings we rise and make our way to the instrument. Long before we have reached the telephone the sound has ceased to act as a stimulus. We are kept in action by stimuli from our own movements toward the telephone. One movement starts another, that a third, the third a fourth, and so on. Our movements form series, very often stereotyped in the form of habit. These movements and their movement-produced stimuli make possible a far-reaching extension of association or conditioning. They make possible remote association in which the remoteness is only limited by the length of such a regular series of movements. . . . Strictly speaking, we are answering the bell only for the first half second or so: after that we are answering our own actions. (1935, pp. 54–55)

There follow some examples of the mechanism of remote conditioning in everyday activities.

These [remote conditionings] are familiar in everyday experience. A good drill-master always gives his commands in the same tempo. After his preparatory command the command of execution must follow at just the practiced instant or there will be conditioned and premature movements of execution. Where traffic is changed by a double bell signal, a delay of the second signal will find many drivers starting without it. Keeping time with the world is so familiar that our illustrations are striking only when something has gone wrong. The sound of a footstep leads us to expect a visitor after an interval; the musician resumes playing after a musical rest without disturbing the rhythm; Mark Twain describes the tense anxiety produced by the failure of the retiring hotel guest in the room above to throw on the floor his second shoe; we have all waited breathlessly for the inexpert pianist next door to play the delayed chord. (1935, pp. 56–57)

Guthrie's interpretation of generalization, which we illustrated in the previous section by reference to a laboratory conditioned response, is here used to explain a child's generalized fear reaction to animals and animal-like stimuli.

A child frightened by an animal will show distress on being confronted with other animals of different shape, or even with bits of fur, or cloth with a long nap. These generalized stimuli are not the same that accompanied the first fright, but they have some common feature to which we have given an identical response, or a response identical in some respect. The movement-produced stimuli of the response serve as the conditioners for the generalized behavior. Out of many experiences with the family dog the child eventually has a perceptual response to the bark, the footsteps, the sight of the dog in many positions and distances because these have been present when the perceptual behavior was in progress. When this has been achieved, one unfortunate experience with the dog resulting in fright will be found to be generalized. The child is frightened at the sound of the footsteps, at the bark of the dog when it is out of sight, at any of

the visual patterns which have previously been associated with
the perception. Very few of these have been directly conditioned
or associated with the fright, but they are all able to evoke it.
They have the power to evoke fright because they evoke the perception of the dog, which has become a signal for fright. (1935,
pp. 91–92)

As we have seen, Guthrie viewed the extinction of an association
between a stimulus and a response as a matter of associating a new
and incompatible response to the stimulus. He applied this formula
to the problem of getting rid of any undesired behavior, as the following examples indicate.

Unmaking a bad habit is thus essentially the same process as
establishing a good habit. Bad habits are broken by substituting
for them good habits or innocuous habits. The rule for breaking
an undesired conditioned response becomes this: So control the
situation that the undesired response is absent and the cue which
has been responsible for it is present. This can be accomplished
by fatiguing the response, or by keeping the intensity of the cue
below the threshold or by stimulating behavior that inhibits the
undesired response. If the cue or signal is present, and other
behavior prevails the cue loses its attachment to the obnoxious
response and becomes an actual conditioner of the inhibiting
action.
   Many an adult who suffers from "cat fear" has been thus re-
trained by tolerating in the house a kitten so small and helpless
that the fear is not called out. The kitten's growth is so gradual
that habits of caring for it and tolerating it, petting it, persist even
when it has reached maturity, and the patient finds that cats no
longer call out panic. A psychoanalytic treatment is essentially
the same process. An effort is made to recall the circumstances
under which the original conditioning of panic occurred. The
circumstances recalled, or re-enacted, which is much the same
thing, include many of the conditioners for the panic which are
also called up by the actual presence of cats, and which are a
necessary condition for the production of the fear. To these, under the careful management of the analyst, the patient responds
without fear. As a result, the sight of a cat or mention of a cat
may become merely cues for talking about cats or talking about
the original incident. (1935, pp. 76–77)

The following excerpt humorously illustrates (at Guthrie's own expense) the problem of getting rid of habits that have been established to a multiplicity of stimulus situations.

The chief difficulty in the way of avoiding a bad habit is that the responsible cues are often hard to find, and that they are in many bad habit systems extremely numerous. Each rehearsal is responsible for a possible addition of one or more new cues which tend to set off the undesired action. Drinking or smoking after years of practice are action systems which can be started by thousands of reminders. . . .

The original wakening of the desire may be caused by any of the chance accompaniments of previous smoking, the smell of smoke, the sight of another person smoking, or of a cigar, the act of sitting back in the office chair, sitting down to a desk, finishing a meal, leaving the theatre, and a thousand other stimulus patterns. Most smokers can, while busily engaged in activities not associated with smoking, go for long periods with no craving. Others find that the craving is strictly associated with such things as the end of a meal, if it has been their practice to smoke at that time. I had once a caller to whom I was explaining that the apple I had just finished was a splendid device for avoiding a smoke. The caller pointed out that I was at that moment smoking. The habit of lighting a cigarette was so attached to the finish of eating that smoking had started automatically.

When the cues for a bad habit are as varied and as numerous as they are in the case of smoking, it is clear that a general unconditioning of all cues is a long and arduous process. A more successful method in dealing with such a habit is to "side-track" by attaching other responses to the initial movements of the habit itself. (1935, pp. 138–140)

In the following passage Guthrie shows how his response interference theory of extinction can be generalized to account for the forgetting of any type of learning. Note the consistency with which the central idea is applied across the forgetting of motor, cognitive, and emotional learning. Forgetting of any stimulus-response association is, like extinction, a function of the opportunities for new, interfering responses to become attached to the stimulus.

As time elapses, more and more of the conditioned cues of the original learning are alienated from their responses and their new responses tend to break up the continuity of the memorized series.

This is probably the reason why so many studies have shown forgetting to be very rapid at the start of the interval, and increasingly slower as time goes on. When the response occurs it probably enlists many thousands of conditioning cues from among the stimuli acting. More of these cues are eliminated at the beginning because there are more cues to eliminate. Accompanying postures and movements and tensions are, as time goes on, incidental accompaniments of other situations, and so lose their associative connection with the response we have in mind. Furthermore, the evidence from experiment that in nearly all cases some faint traces of learning are present even if months or years have elapsed is very plausibly explained by the supposition that there are generally some cues which have not occurred in the interval and so have not been alienated.

The reason why we can go from winter through spring, summer, and autumn and find the next winter that our skill in skating is almost what is was when we left off ten months before is that the postures and movements of skating simply do not find a place in our domestic routine, or in our summer sport. . . .

Popular tunes heard daily on the radio or in the restaurant lose their power to please. Played while we are conversing, while we are eating, or while we are casting up the amount of the check, they become conditioners of these activities and active inhibitors of the pleased attention they once caused. In the age of the radio, popular songs have a life of a few weeks as compared with the occasional record of over a century for songs which before the days of easy mechanical reproduction were heard only on occasions when they could be given undivided attention. It is not time that robs beauty of its charm, but preoccupation with other affairs in its presence.

Grief is recalled to a large extent by the objects and scenes which were associated with it. A change of residence and new scenes will not always do away with it because it has become conditioned on our own behavior. But the demands of practical affairs will be effective even if we remain in the same surroundings. Those who are at leisure to enjoy their grief may preserve it indefinitely. Those who are compelled to an active part in living find that forgetting has taken place. The reminders which conditioned the emotion have become conditioners of other activity. (1935, pp. 121–123)

We discussed Guthrie's interpretation of the effects of reward in relation to laboratory situations. Here are some applications of Guthrie's stimulus-change hypothesis to considerably more complex problems in human behavior.

When a man is in a state of continued anxiety, when some inner conflict keeps him active and discontented and when under these circumstances he has, with the pressure of companions, taken in a sufficient amount of alcohol to relax his tensions and bring him peace, he is in a fair way to fix this as his mode of escape from worry just as the cat [in the problem box] has fixed whatever act it was that let it out of the box. The reason that drink tends under these circumstances to become a habit is that it brings relief, forgetfulness. And this makes so radical a change in his stimulus situation that no forgetting can occur. The next time he is in a state of jitters, the obvious association will be alcohol. No new associations are present because the jitters have not been present and *without being present cannot be associated with any new response.*

To the person who thinks in terms of pleasure and pain the habit of getting drunk remains a perpetual mystery. Why shouldn't the next morning's nausea teach him a lesson? The answer is that the next morning is so different from the state of jitters in which he began to drink that there is no strong association. To "teach him a lesson" would require that the association between jitters and drinking be replaced by associating jitters with some other response. This can not be done the next morning.

Morphine acts in much the same way. No matter what the drive or annoyance, no matter what the persistent source of excitement, whether pain or a bad conscience or a conflict of interests, morphine acts as a consummation. It relieves, not by removing the persistent stimuli, but by acting directly on the response mechanism. With restlessness gone, the morphine-taker is in the same state as an animal that has found food or any other consummation of a drive. The morphine relaxes, and the next time the state of excitement recurs, the obvious association is morphine. This was the last association with the excited state. (1938, pp. 99–100)

As was pointed out earlier, Guthrie emphasized the stimulus-change function of rewards in his account of the empirical law of effect. The

following selection shows that the response-evocation properties of rewards must also be taken into consideration.

> I would not hold that all satisfiers tend to fix the associative con-
> nection that has just preceded them. When a satisfying situation
> involves breaking up the action in progress it will destroy con-
> nections as readily as punishment. In teaching a dog to sit up,
> tossing his rewarding morsel to a distant part of the room will
> prove a very ineffective method. There is no doubt of the satis-
> fying character of the meat. The dog certainly "does nothing to
> avoid, often doing such things as attain and preserve," not, of
> course, the meat, but the eating of it. But the effect of the re-
> ward will be that the dog instead of sitting up stands ready for
> another dash across the room. (1935, pp. 153–154)

The final excerpt below presents Guthrie's interpretation of the effects of punishment. It seems likely that Guthrie is the "certain psychologist" referred to in the passage, and the incident thus suggests
that the source of much of Guthrie's supporting material was his own
shrewd observation of natural life situations. This excerpt is a fitting
one with which to close out our Guthrie sampler in that the opening
eight words of the passage might well serve as Guthrie's primary guide
to understanding learning in natural life situations.

> What is learned will be what is done—and what is done in in-
> tense feeling is usually something different from what was being
> done. Sitting on tacks does not discourage learning. It encour-
> ages one in learning to do something else than sit. It is not the
> feeling caused by punishment, but the specific action caused by
> punishment that determines what will be learned. To train a dog
> to jump through a hoop, the effectiveness of punishment de-
> pends on where it is applied, front or rear. It is what the punish-
> ment makes the dog do that counts or what it makes a man do,
> not what it makes him feel. The mistaken notion that it is the
> feeling that determines learning derives from the fact that often
> we do not care what is done as a result of punishment, just so
> that what is done breaks up or inhibits the unwanted habit.
>
> My own view of the way in which unpleasant or unsatisfactory
> consequences of action affect learning might be further illus-
> trated by a minor incident in the routine of a certain psycholo-
> gist. He rented an apartment for the summer with a garage which
> had a large swinging door. From the top of the door hung a

heavy chain. Opening the door hurriedly the first morning the chain swung about slowly and struck a blow on the side of the subject's head, a distinctly painful and "unsatisfactory" event. But this continued to happen each morning for some two weeks. Why the long delay in learning to stand aside?

The answer, I believe, is that the act of opening the door was performed while looking at the exterior of the door. The chain struck after the door had opened and the scene changed. Dodging was not conditioned on the sight of the door because a sight of the door had not accompanied flinching from the blow. The flinching movement which occurred as the rear of the car came into view was too late. Only after the bruised ear became a chronic reminder and the incident had been talked about and finally had been told to a visitor on the way to the garage, did caution show itself in time.

The whole incident is not to be explained in terms of pain or annoyance, but in terms of the action and its cue. It is not the annoyance, but what the annoyed person does that determines what will be learned. (1935, pp. 158–159)

A prime function of theory is to reduce variety to uniformity—to show seemingly disparate phenomena as manifestations of a single underlying principle. When the foregoing extensions of Guthrie's theory are considered in conjunction with the conditioning data handled by the theory, it is clear that Guthrie's formulation serves such simplifying and unifying functions. Whatever questions one may have about Guthrie's interpretation of particular phenomena, one cannot help but be impressed by a theoretical effort that is at once so parsimonious and so comprehensive. Another function frequently asked of psychological theory, although seldom fulfilled, is that of assisting analysis of human behavior in natural life situations. Even those who disagreed with Guthrie's theoretical position were apt to find in his writings many useful suggestions, as exemplified above, for dealing with practical problems of behavior.

EXPERIMENTAL ASSESSMENTS OF THE THEORY

As previously noted, Guthrie's theory did not generate extensive testing of its propositions. Moreover, Guthrie largely ignored the research controversies that so embroiled the advocates of cognitive and re-

sponse-reinforcement theory. (An interesting exercise for the reader
would be to consider how Guthrie might handle latent learning, latent extinction, sensory preconditioning, and place-response learning.) Guthrie's avoidance of these classical controversies probably stems
from the fact that they were largely based on phenomena of maze
learning. Guthrie viewed maze learning, with its emphasis upon
overall trial, error, and speed scores, as inappropriate to the study of
how learning proceeds at the level of movements. In particular, he
felt that maze learning obscures observation of the possible all-or-
none nature of associations.

There were, however, a small number of studies designed to bear
directly upon Guthrie's assumptions, and these did produce some valuable and intriguing data. In this section we shall consider several
representative pieces of Guthrian research on animal learning. While
the outcomes of these studies were generally consistent with Guthrie's
ideas, they are more notable in the final analysis for their limitations
as critical tests of the theory.

The Guthrie-Horton Study

This was Guthrie's only published study of animal learning, and it
stands as a beautiful example of the interaction of theory and method.
The study was designed to remedy what Guthrie felt to be the deficiencies of mazes and the various conditioned response apparatus employed in existing research. These devices, he felt, captured behavior
at altogether too abstract a level to provide information useful to the
question of *how* behavior change comes about. If learning occurs at
the level of movements, what is needed is a record of the movements
by which "correct" performance is attained. Guthrie and Horton
(1946) sought such a record.

It is interesting that Guthrie and Horton chose for study the very
experimental situation that had led to formulation of the theoretical
law of effect—the behavior of cats in escaping from a puzzle box. But
Guthrie and Horton's puzzle box differed from Thorndike's in several
significant respects. Their release mechanism was a floor-mounted
pole set on a rocking base so that a slight movement of the pole in
any direction activated the exit door. The door-opening mechanism
operated with virtually any response that displaced the pole, e.g.,
pushing, nosing, clawing, biting, running or rolling against, or backing into. The pole was spring-mounted and so returned to its original
position following each displacement. (This was not a property of

Thorndike's release mechanism.) Guthrie and Horton's situation therefore allowed a great variety of effective release behaviors but at the same time made possible repetition of any previously effective movement.

Another critical feature of the Guthrie-Horton apparatus was that the entire front of the box was made of glass, thereby permitting full observation of the animal throughout the escape episodes. The movement that activated the release mechanism simultaneously activated a camera which took a picture of the animal through the glass front and thus recorded the subject's precise posture at the moment of "success." Also, motion picture records were made of the complete behavior of some subjects throughout a number of trials.

On each trial the subject entered the puzzle box through a narrow tunnel attached to the rear of the box and upon exiting was allowed access to a bit of salmon placed just outside the box. Following three preliminary escape trials on which the exit door was left ajar, the subject received a lengthy series of trials which required escape by its own movements to the pole.

Guthrie and Horton's procedure sacrificed some degree of experimental control in that the glass front exposed the subject to considerable extraneous stimulation. But the experimenters felt this a small price to pay for the greater observation of behavior afforded by their procedure. In their words,

> In general, the practice of guarding against secondary cues in animal experimentation . . . has tended to lessen . . . the collection of a great deal of data that is important in the study of learning. . . . The puzzle box *allows the experimenter to see what the animal is doing* in the situation. This knowledge of what the animal is doing is more important than any record of time or errors. (1946, p. 9)

Here again we see surfacing the issue that separated Small and Watson. Guthrie and Horton were clearly on the side of qualitative and wholistic observation of behavior at the expense of quantitative measurement and rigorous control by isolation.

Guthrie and Horton's results literally present a picture of learning, and that picture is strikingly consistent with expectations from Guthrie's theory. The photographs of escape postures indicated that escape movements were learned in all-at-once fashion and then repeated in virtually identical form over many trials. These features are illustrated in Figure 5-1, which shows the complete record of escape postures of

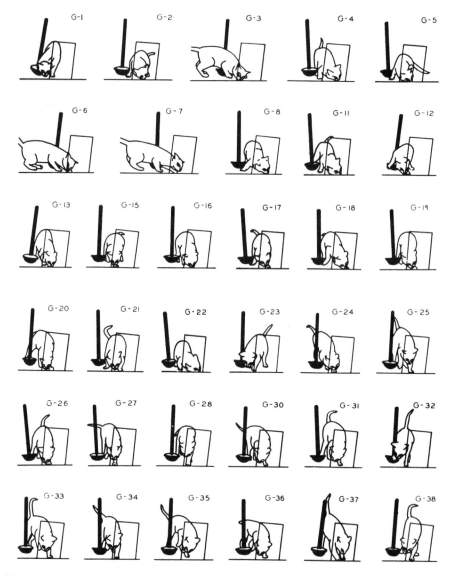

Figure 5-1. Complete record of escape positions for cat G. Trials for which pictures are missing involved camera failure. On trial 41 the pole was moved to a new position in the box. Because of an apparatus failure on trial 41, the subject's first successful escape by pole manipulation did not occur until trial 42.

(From Guthrie and Horton, 1946, pp. 50–52. Copyright 1946 by Holt, Rinehart and Winston. Reprinted by permission.)

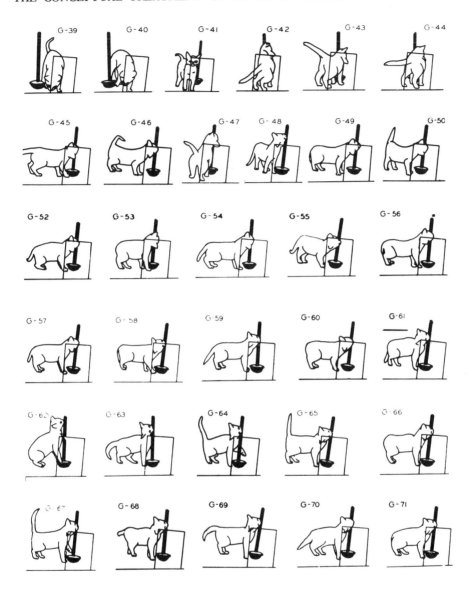

Cat G over a training series involving an experimenter-manipulated change in the stimulus situation. The escape postures are depicted in the form of tracings (from the full camera record) of the outlines of cat, pole, and exit door. It can be seen that on trials 1 through 40 the cat is almost always in the same position at the moment of success, nose down toward the edge of the exit door and the right side of the body brushing or nearly brushing the pole. Note that this posture had also obtained on the subject's *very first escape*. On trial 41 the pole was moved to the right rear of the box, as shown in Figure 5-1, and thereby a new stimulus situation was created for the subject. After repeatedly making the movement that had brought release on the first 40 trials, a movement that was ineffectual with the pole in its new position, the cat eventually made its first pole-mediated escape by turning and biting the pole, as shown in trial 42. This movement was then repeated on each of the additional 30 training trials given this subject, as seen in Figure 5-1.

The degree of stereotyping seen in these photographs is all the more impressive when it is realized, first, that each escape episode typically involved a good deal of other behavior prior to the final escape movement and, second, that very slight differences in the subject's spatial position (e.g., distance from the pole) during the execution of two identical movement series would cause the camera to be tripped at somewhat different points in the series and therefore distort the true identity of the escape postures.

Many other subjects showed similar all-at-once learning and stereotyping of the final escape movement. One of the more bizarre instances is depicted in Figure 5-2, which shows the "rolling" escape movement of Cat T.

In addition to stereotyping of the final escape movement, the experimenters reported that the movements immediately preceding the final escape movement were also highly routinized and that stereotyping was often manifested in earlier portions of the escape episodes as well. Some subjects exhibited stereotyping of long series of movements about the box, series that might occupy a minute or more.

These observations follow nicely from Guthrie's theory. Consider, for example, the pole-biting response of Cat G. Under Guthrie's assumptions, this behavior becomes fully conditioned to its contiguous stimulus, e.g., "sight of pole when approached from right," on the occasion of its very first contiguous occurrence with that stimulus (trial number 42). Since pole biting removes the cat from the situation, no new responses can become attached to stimuli in the puzzle box. Therefore when the cat again encounters "pole approached from right" on the second and successive trials, biting will again be elicited on each occasion. Note that this account treats escape from the box as Guthrie treated any reward, namely, as an event that alters the stimulus situation and so preserves the preceding S-R associations.

The same interpretation applies to the tendency toward stereotyping in earlier segments of the escape sequences. Each movement in the first escape episode becomes fully associated to its contiguous stimuli and, upon the cat's escape from the box, is protected from unlearn-

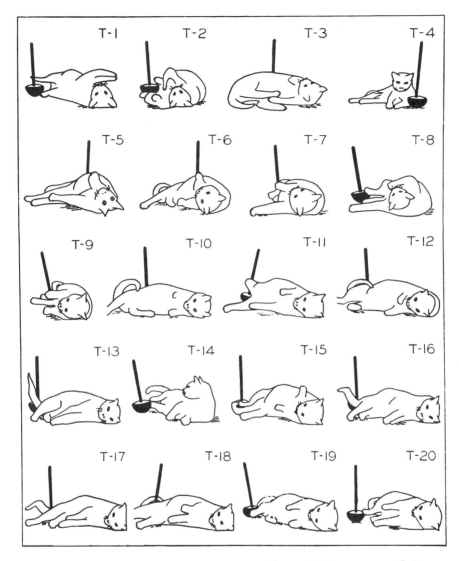

Figure 5-2. Complete record of escape postures for cat T. Escape in each instance was attained through lying down and then rolling in such a fashion as to operate the pole release.
(From Guthrie and Horton, 1946, p. 60. Copyright 1946 by Holt, Rinehart and Winston. Reprinted by permission.)

ing. To the degree that the conditioned stimuli are reinstated on the following introduction to the box—and the narrow entrance tunnel helps to ensure a constant starting stimulus from trial to trial—the same movements should be elicited in series. As Guthrie and Horton state, *"The cat learns to escape in one trial* and will repeat the specific movements of its first escape except in so far as new trials by accidental variations of situation cause new associative connections to be established" (1946, p. 41).

In general, Guthrie and Horton's observations support the prediction from Guthrie's theory that what an organism will do in any situation is what it last did in that situation. This predictive principle was termed the *principle of postremity* by Voeks (1948).

Whatever one's theoretical persuasion, it is hard not to be impressed by the behavioral stereotypy found by Guthrie and Horton. Their data picture the organism as a strictly mechanical object that moves in particular ways when particular signals occur. Other behavioral theorists, of course, also assumed environmental determination of behavior, but preferred to work at levels of observation that accept a good deal of variability in the movements that compose behavior. The Guthrie and Horton study, on the other hand, pursued the proposition that if movements are the items associated in learning, then the details of the movements in learning should be predictable. Their observations are surely consistent with this line of reasoning. Moreover, their data are consistent with the idea that learning consists of *nothing more* than movement changes. There is nothing in the behavior of Cat T, for example, which suggests that it had learned anything other than a movement that, in relation to its environmental consequence, was awkward and effortful. Moreover, despite many repetitions this movement showed no trend toward an efficient form as might be expected if movements are merely the servants of a higher organization in learning. It is true that with repeated escape episodes, most subjects took less time to escape. But Guthrie and Horton's observations indicate that this was due to the dropping out of movements prior to the terminal escape series. The escape movement itself, or the several alternative escape movements displayed by some subjects, tended to occur in many consecutive escapes following their first appearance and were sustained throughout training.

The Guthrie and Horton study, then, gave some force to Guthrie's ideas, which until this study had been based primarily on analyses of behavioral observations at relatively molar levels.[2]

Other Studies

Most of the animal learning studies that relate closely to Guthrie's theory were concerned with the stimulus-change hypothesis of reward. In this area Seward (1942) contributed yet another of his procedurally simple yet pointed tests of theory. Seward reasoned that if, as Guthrie claimed, rewards act only to preserve preceding stimulus-response associations, then simply removing the animal from the learning situation should constitute an effective reward for any preceding response. Since the response preceding removal would always be the last response in the situation, it would remain attached to situational cues and be evoked upon the next exposure to those cues. Each response evocation would condition the response to further variations of the situational cues, and each removal would protect that further conditioning from unlearning. As a result the response should become increasingly likely upon successive introductions to the situation. Moreover, Seward reasoned, if stimulus change is indeed the sole function of rewards, then removing the animal from the situation should strengthen preceding behavior fully as much as a conventional reward such as food.

Seward tested these ideas by comparing acquisition of bar pressing in a Skinner box by animals given the usual food reward with that by animals given "removal reward." In the Food Group each rat received a pellet of food upon pressing the bar and was then removed from the box. Each subject in the Removal Group was simply lifted out of the box as soon as it had pressed the bar. All animals received one daily trial, i.e., one bar press—reward sequence, over a several-week period, followed by several daily extinction sessions. Animals in both groups were fed one pellet of food just *before* each trial in order to minimize between-group differences in internal stimulation resulting from ingestion of the reward pellet. Also, because immediate return to the home cage might constitute a reward for bar pressing, all animals were placed in a small detention cage upon removal from the Skinner box for at least 15 minutes before being returned to their living cage. Finally, a Control Group that received daily placements in the Skinner box without any form of reward provided a measure of the rat's spontaneous, or unreinforced, behavior to the bar.

Two features stand out clearly in Seward's data. First, both the Food and Removal Groups showed systematic reduction in bar press response latency with successive trials and gave significantly more pressing responses in extinction than the Controls. Thus, as Guthrie would predict, removal as well as food reward strengthened the bar pressing behavior. Second, on both measures the Food Group was far and away superior to the Removal Group, an outcome that questions the view that conventional rewards act solely by preserving preceding stimulus-response associations.

What are we to make of these data? On the one hand, the fact that the Removal Group did learn supports Guthrie's theory and poses difficulty for effect-type theories such as Hull's. On the other hand, to account for the marked superiority of the Food Group, a Guthrian must argue that receipt of food reward preserves the associations between bar pressing and situational cues more effectively than does actual removal from the presence of those cues. But perhaps this argument is not as implausible as seems at first blush. A food pellet to a hungry animal doubtless results in strong new internal stimuli from emotional-motivational reactions as well as marked changes in external stimuli as a consequence of the restriction of orienting responses to the food. And with regard to animals in the Removal Group, it is not unreasonable to suppose that during the few seconds that were doubtless required to effect full removal of the animal after a bar press, defensive reactions and other responses incompatible with bar pressing behavior may have become conditioned to the critical cues. Of course, to determine whether it is the food reward or the removal that most effectively preserves preceding S-R associations would require *full* measurement of the stimuli and movements surrounding receipt of each form of reward. Such measurement is simply not possible. Seward's data, then, can be interpreted in a fashion consistent with Guthrie's theory, but the interpretation must remain essentially speculative in nature.

The suggestive but inconclusive nature of Seward's results characterizes other experiments in this area. Zeaman and Radner (1953) found that rats readily learned to cross from one side of a lighted apparatus to another when crossing was followed by sudden darkening of the apparatus, but did not learn when crossing occurred in the dark and was followed by light onset. The fact that a stimulus change from light to dark can produce learning fits Guthrie's theory nicely, but the theory is then embarrassed by the failure to observe learning under the presumably equal stimulus change from dark to light. Zeaman and Radner suggested that Guthrie's theory could handle the data if it were assumed that the sudden onset of light evoked minute "freezing" responses which then became anticipatory in the dark and thus interfered with crossing behavior. But they also noted that a conclusive evaluation of this interpretation would require a microanalysis of the animal's movements during both dark and light conditions, an analysis that is beyond the capability of any present measurement technique.

Finally, in a now classic experiment, Sheffield and Roby (1950) showed that saccharine has the properties of a primary reward in instrumental learning. The significance of this observation lies in the fact that saccharine is a nonnutritive substance that goes through the body without chemical change and without effect on the organism's food intake. Ingestion of saccharine thus amounts to sham feeding that produces stimulus and response changes similar to those in true feeding (e.g., restriction of orienting behavior) but without any apparent need or drive reduction. The fact that hungry rats will learn a T-maze for a saccharine reward, as Sheffield and Roby showed, is therefore congenial to Guthrie's stimulus-change theory of reward action and is presumptive evidence that food and other conventional rewards act through a stimulus-change mechanism. But before this interpretation can be granted the status of a conclusion it is necessary to ensure that the S-R changes associated with saccharine ingestion *are* comparable to those of true feeding (and that possible subtle forms of drive reduction are ruled out). But once again such assessment would exceed existing measurement capabilities.

In sum, Guthrie's theory prompted a number of intriguing findings that sensitized theorists to the reinforcing possibilities of stimulus change per se. But the data were at best consistent with the theory, rather than uniquely supportive of it. These several experiments highlight the primary limitation of Guthrie's theory, a limitation inherent in its molecular units of analysis. Explanations that rested on such molecular units of analysis could not be subjected to unambiguous experimental test.

OVERVIEW OF GUTHRIE'S WORK

Guthrie's work can be roughly characterized as a 30-year exploration of the explanatory possibilities of the principle of association by contiguity. In general, he sought to show that there was nothing in either laboratory or real life learning that (as he was likely to put it) went contrary to or in violation of that principle. The concern in Guthrie's writings to establish the simple congruence of data to principle is striking. From the outset he saw explanation as a matter of showing a logical consistency between observations and theoretical principles, not as a matter of deciding among alternative theoretical principles by experimental test. For Guthrie, a good theory was one that organized existing knowledge so as to facilitate its communication and practical application, and the best organization is that which covers the most data with the fewest principles. In short, he gave primacy to parsimonious organizing power over experimentally assessed validity.

Guthrie's theory is indeed parsimonious, at least relative to other major S-R theories. It is true that any application of the principle of one-trial learning by contiguity is likely to involve a number of additional assumptions regarding the nature of the stimulus and response, the role of movement-produced stimuli, the distinction between movements and acts, and the like. But it is nonetheless true that such assumptions are distinctly secondary in nature and occur as elaborations of the basic principle. The parsimony of the theory is real in the sense that at base it has but a single explanatory principle that is applicable to all instances of learning. We should emphasize, too, one other point of theoretical economy in Guthrie's system. By defining drives as stimuli that constitute part of the stimulus patterns conditioned to movements, Guthrie did away with the learning-performance distinction and with the need for a separate principle of motivation to account for the conversion of learning to action. In Guthrie's system, learning and performance are one: behavior occurs when its unconditioned or conditioned signals occur and does not occur when its signals are absent.

We have stressed that Guthrie's theory represents behaviorism in relatively pure form. Guthrie's behavioral orthodoxy was particularly evident in his choice of specific movements as the units of behavioral analysis. There is strong face validity in the argument that the response must be defined in terms of muscle movements because these are what the nervous system, the physical mechanism, actually produces. But in fact there was no reason to suppose that higher (cortical) centers might not activate a *movement class or system* (e.g., locomote toward stimulus X) while leaving the particular movements in that action class to be determined by moment-to-moment input from feedback systems. On the basis of what we know about the motor system, one could equally well hypothesize either specific movements or a class of movements as the unit of conditioning. Guthrie chose muscle movements not so much because it was good neurology as because it was good behaviorism. Movements have an apparent material form, a reality that is hard to demonstrate for central neural action systems. Here again we see in retrospect the strong influence of the reaction against structuralism and mentalism.

But the question we wish to pursue here is whether Guthrie's determined peripheralism was of particular advantage in implementing the behavioral formula. Like other behavioral theorists Guthrie preferred to base his explanations solely upon overt S's and R's but, like other behavioral theorists, was required to postulate processes within the organism in order to account for various phenomena of learning. As we have seen, Guthrie conceived of these intervening controlling processes as stimuli arising from explicit or implicit movements that were triggered ultimately by external stimuli. It is clear that Guthrie regarded movement-produced stimuli as behaviorally respectable variables in that the movements in question, he claimed, were observable, whether they be the receptor and postural adjustments held to underlie delayed conditioning and generalization or the implicit movements of the vocal cords held to account for thinking. But in fact, given the molecular level of analysis Guthrie pursued, his movement units defied measurement and had a status more logical than empirical. Thus, while Guthrie's behavioral analysis had a greater air of physical specificity about it than did that of Hull or Tolman, Guthrie's molecular peripheralism was no more adequate to handle the phenomena of learning without recourse to unmeasured intermediary processes.

The molecularity of Guthrie's stimulus and response units is the basis of what is usually the chief complaint against the theory, namely, the difficulty of arranging a clear test of its postulates. As we saw in reviewing several experiments on Guthrie's stimulus-change hypothesis of reward action, seeming departures from theoretical predictions could always be rationalized in terms of postulated, but unmeasured, changes in the stimulus situation, including those stimuli arising from the organism's own movements. Even in the Guthrie and Horton study, where a major effort was made to secure a detailed record of the critical stimuli and movements, it was still not possible to pin down the theory. For example, as inspection of Figure 5-1 indicates, while stereotypy of escape movement was usually observed, it was not invariably observed. Do these departures from stereotypy in the presence of a seemingly constant stimulus pattern constitute critical evidence against the theory? Not at all, since they can be ascribed to unobserved accidental variations in the stimulus chain. Clearly, as Osgood (1953) points out, it is difficult to back this theory into a tight empirical corner.

> The very simplicity and flexibility which renders this theory so readily applicable to all conceivable situations makes it hard to evaluate scientifically. *This does not mean that the theory is wrong.* No evidence disproving it has been presented, and none

is likely until we are in a position to measure the stimulus situation, including internal ones, with much greater accuracy and detail. (Osgood, 1953, p. 372)

While testability is of course a critical aspect of any theory, we have pointed out that Guthrie himself gave priority to other valued features of theory—organizing power, parsimony, communication, and practical utility. Surely Guthrie's theory rates high on these attributes. In regard to substantive ideas, his theoretical efforts gave new force to the proposition that contiguity may be the necessary and sufficient condition of learning, and more than any other theorist, he made clear the possibility that learning at base and in the individual is frequently, if not always, sudden rather than gradual in nature. In addition, the fruits of his careful attention to actual movements led investigators to have increased respect for the details of behavior. In his applied work, he contributed many valuable insights into real life behavior problems and gave psychologists a view of the exciting possibilities for analysis of complex human activity in terms of learning principles. Finally, while the theory itself did not rank high in testability, its assumptions lent themselves to development of some important quantitative models and tests of learning in the decades ahead.

All in all, no small achievement.

6

The Behavior System
of B. F. Skinner

GENERAL ORIENTATION

We have made the point that the theorist's unit of analysis is fundamental to the explanatory system developed, and B. F. Skinner is no
exception to this dictum. In fact, of the several major theorists we
have considered, only Skinner can be said to have developed a system
based upon a truly thorough examination of the unit of analysis.
Skinner's ideas in this regard were expressed in an early (1931) paper,
"The Concept of the Reflex in the Description of Behavior," in which
he reviewed various usages of the reflex concept through history. By
"the concept of the reflex" Skinner referred to the idea of strict stimulus determination of behavior, particularly as evidenced in patently
involuntary, direct stimulus-movement linkages, e.g., the scratch reflex, the salivary reflex. Skinner's historical review sought to isolate
the invariant features of the reflex concept as it had been applied
throughout analyses of behavior-environment relationships. The logic
was that these features should define the appropriate analytical unit
for an S-R psychology since they had in fact consistently characterized *actual* scientific conceptions of the relation between organism
and environment. A proper understanding of Skinner's approach requires our consideration of the ideas in this paper.

Major Steps in the Historical Development of the Reflex Concept

Skinner attributed to the philosopher Descartes (1596–1650) the first
conception of a strictly determined, automatic relation between as-

pects of the environment and aspects of behavior. Descartes's conception took the form of a mechanical model of the living organism, which he offered primarily as an instrument of persuasion toward the philosophical position of determinism. Essentially, the body was seen as a system of stored energy which might be released to specific muscles upon occurrence of particular stimuli. In mechanical terms, the organ of sense was thought to be set in motion by external forces and, in consequence, to pull upon a thread which in turn opened a valve at a central (brain) reservoir, letting fluid flow outward along pipe lines to specific muscles which were then activated. By today's knowledge, of course, Descartes's model seems a farfetched picture of the nervous system. But the truly significant achievement in Descartes's conception, from the viewpoint of a scientific analysis of human behavior, is the notion of *stimulus-controlled behavior,* as opposed to metaphysical conceptions of the causal agents of action, e.g., the soul, the mind, or the will. However, in advancing the possibility of a mechanical, stimulus determination of action, Descartes did not eliminate metaphysical conceptions from his description of man, for he conceived of the mechanical principles as subordinate to a soul, which might suspend them at any time. In Descartes's system, then, certain human actions might at times be determined by a physical causal chain initiated by external forces and at other times result from the higher-order action of spiritual or mental agents. This kind of dual thinking about the control of behavior was characterized by Skinner as an effort "to resolve, by compromise, the conflict between an *observed necessity and preconceptions of freedom* in the behavior of organisms" (p. 431). It was compromise of this nature, argued Skinner, repeated in the history of the reflex, that resulted in confusions in usage of the reflex concept and, more generally, in the way we conceive of the relations between environment and behavior.

The concept of the reflex also developed quite independently during the 17th and 18th centuries through studies on excised nerve and muscle tissue in animals. Skinner traced this development through several stages: (1) the discovery of spontaneous action of excised muscle, which showed the inherent contractility of muscles apart from any possible remote source of control; (2) demonstrations of the responsiveness of muscles to direct stimulation, which showed the inherent property of irritability of nervous tissue; (3) demonstrations that muscles will also contract upon stimulation of tissue connected to them from near or remote points in the nervous system, i.e., that the sensitive (irritable) and responsive (contractile) parts of the system may be spatially separate; and finally (4) demonstration that the spinal cord was a necessary part of the conducting path between specialized irritable tissue and its associated muscle. These investigations went far toward establishing the idea that movement, rather than being a manifestation of the activity of the soul, might be studied as a strictly organic process. But these studies also left the matter in a compromised state, for while the beginning (stimulation) and end point (muscle action) of the chain were seen as palpable enough events, the mediating conducting force was conceived of by these early investigators as a psychical, nonphysical process. Thus, the notion of a strictly determined stimulus-response mechanism was left "in the position partly of a description of observed fact and partly of superfluous interpretation" (p. 434).

The logical next step in the development of the reflex concept was taken in the middle of the 19th century by Marshall Hall, who argued the physical nature of the "conducting force" itself. The reflex arc, Hall proposed, is to be understood as a train of purely physical events initiated by the application of appropriate stimuli to a sensory surface which in turn starts a conduction process inward to the spinal cord and thence outward to particular muscles, the conduction process itself being an intrinsic property of the intervening nerve tissue. So conceived, the reflex mechanism could now be used to provide a complete physical account of any aspect of the organism's behavior that reliably exhibited an immediate relation to the occurrence of an environmental stimulus. Here, then, was an appropriate neural model of Descartes's logical possibility.

There were two aspects of Hall's doctrine that doubtless aided acceptability of the reflex as a purely physical and wholly involuntary mechanism. First, Hall restricted reflexes to the spinal level of the nervous system. That is, he assumed that the neural events underlying reflexive behavior were organized at the level of the spinal cord and did not involve action of the cerebrum or higher brain centers. Second, Hall drew a sharp distinction between reflexes and *volition*. Under volition Hall included any movement of the organism that is spontaneous in the sense of not being tied to presentation of particular environmental stimuli. By this criterion, of course, volition constitutes by far the predominate portion of the behavior of men and animals, since relatively few behaviors are invariably predictable from the mere presence of particular stimuli. Further, Hall argued that voluntary behaviors are controlled by the cerebrum, in contrast to the spinal control of reflexive behaviors. In general, reflexes were held to involve simple, unconscious, unlearned behaviors, whereas in contrast volition involved the broader spectrum of complex, consciously undertaken, intuitively voluntary activities.

From the perspective of the early history of the reflex, it can be seen that Hall did not really rid behavior of metaphysical concepts but rather simply excluded them from the realm of the reflex. Essentially, his doctrine identified the reflex with scientific necessity and volition with unpredictability and in this way, as Skinner pointed out, continued the earlier compromise between a strict environmental determination of behavior on the one hand and a principle of freedom in organismic activity on the other.

The compromise between the principles of determinism and freedom was finally broken in the next, and last, step in the development of the reflex concept, which established the hypothesis that the total behavior of the organism might be described in terms of the reflex. The fundamental contribution to this end was made by Pavlov in the form of the *conditioned reflex*.

It is sometimes forgotten that Pavlov's work on conditioned reflexes, despite its pivotal importance to psychology, is properly seen as a continuation of the earlier physiological investigations of the reflex. Pavlov's initial aim was to understand the physiology of the salivary reflex in dogs, but his work toward that end was frustrated by the persistent tendencies of his subjects to salivate not only to the natural stimulus of food in the mouth but also to stimuli that regularly preceded food, stimuli such as the mere sight of the trainer. Pavlov referred to the latter responses as *psychic secretions*, since they suggested that the dogs were thinking about the food. The essence of Pavlov's contribution lies in the fact that he brought these "psychic secretions" under experimental control and analysis.

As we detailed earlier, Pavlov devised procedures that enabled measurement and analysis of the full course of acquired associations between environmental stimuli and the salivary reaction. He termed such learned stimulus-response relations *conditional reflexes* to denote that they exhibited the automatic character of the original reflexes but were conditional upon pairing of initially neutral and initially eliciting stimuli. Although Pavlov typically measured only the salivary reaction, since this could be nicely quantified, he noted that the conditioned salivary reflex also regularly included movements that fell into the category of volition as defined by Hall, movements such as anticipatory eating movements and efforts to locomote toward the source of the food. In further contradiction of Hall's doctrine, the associative nature of the conditioned reflex clearly pointed to involvement of higher brain centers, and Pavlov himself interpreted all of the phenomena of conditioning in terms of underlying cortical processes. Finally, Pavlov's findings with the salivary reflex were extended by himself and other workers to a variety of response systems, including behaviors of a more patently voluntary nature.

By providing evidence that both reflexive and voluntary movements of the intact organism can become linked to virtually any environmental stimulus to form a new reflex, Pavlov and his followers established that the concept of a strictly determined, mechanical relation between the environment and behavior could be extended, at least in theory, to the description and analysis of all behavior. Any complex or "voluntary" activity could now be seen as a conditioned reflex, or a collection of conditioned reflexes, traceable ultimately to the pairing of initially neutral and eliciting stimuli. In short, there was now clear empirical basis to expect stimulus determination and hence lawful S-R relations throughout all organismic activity.

The foregoing summary of Skinner's history of the reflex omits many details of the original paper, but should enable the reader to follow the arguments that Skinner proceeded to draw from his historical review.

The Nature of the Reflex

As stated earlier, Skinner's aim was to define the *invariant* properties of the concept of the reflex in its various applications. Historical usage, argued Skinner, makes clear that the concept cannot be tied to particular properties of the constituent stimuli and responses or of their relationship. As we have seen, the reflex concept has been applied to both stimulus-elicited and so-called voluntary behaviors and has included responses ranging from the twitch of excised muscle to combinations of response systems in the intact organism. The stimulus aspects have ranged from mechanical stimulation of nerve tissue to remote and complex environmental events. Various restrictive properties attributed to reflexes, e.g., "simple," "unconscious," "unlearned," "involuntary," "spinal," proved irrelevant to later applications of the concept and, in historical perspective, are seen to have proceeded from unscientific presuppositions concerning the behavior of organisms.

But if concepts of the reflex based on structural, physiological, or psychological properties of the stimulus-response relations have not

endured, how *are* we to define the reflex? The important point revealed by the history of the reflex, Skinner argued, is that the reflex
should be conceived of as nothing more than an *observed correlation*
between some aspect of behavior and some aspect of stimulation, for
this in fact has been the single enduring defining feature of the reflex
through all its usages. By *observed correlation* Skinner meant simply
that given presentation of stimulus X we regularly observe appearance
of response X. The possibility of observed stimulus-response correlations or dependencies was at the heart of Descartes's thinking. Observed stimulus-response correlations were the basis of all subsequent
physiological investigations of the reflex. Observed stimulus-response
correlations define all the phenomena of conditioned reflexes. And
observed stimulus-response correlations are the single *invariant* in all
of this history. Therefore, concluded Skinner, if we take reference to
actual usage of the concept, we find that a reflex can legitimately be
defined solely as a stimulus-response correlation, a correlation that
may be demonstrated in any aspect of the organism's activity. All else
is short-lived interpretation.

But what of the long and continuing tradition of physiological investigation of the reflex—the effort to define the anatomical and physiological substrate of observed S-R relations? Has this work not contributed significant additional dimensions to our conception of the
reflex? Not at all, said Skinner, because physiological conceptions of
the reflex are perforce derivative of observed stimulus-response correlations; that is, there can be nothing relevant to behavior in the physiological conception that is not first in the behavioral conception. To
illustrate this argument, Skinner cited Sherrington's (1906) classical
work on reflex physiology, which featured an account of reflex action
in terms of central physiological states.

Sherrington based his analysis on characteristics of reflex action
that violate a simple one-to-one conception of the relation between
input and output phases of the physiological mechanism. Sherrington's list included the findings that the strength of stimulus just sufficient to elicit a reflexive response is variable, that repeated application of an initially ineffective stimulus may result in response
evocation, and that repetition of an initially effective stimulus evokes
progressively weaker responses. These and other characteristics of reflexes listed by Sherrington have in common the fact that the output,
or response, side is not in invariable correspondence with the input,
or stimulus, condition; instead, these characteristics point to an important contribution of the physiological system itself.

Drawing upon a variety of observations, notably upon studies of the effects on reflex function of surgically ablating or otherwise excluding parts of the nervous structure, Sherrington established that the foregoing characteristics could not be accounted for by the action of end organs or of the peripheral inward or outward conducting nerves. He therefore assumed that the characteristics arise from the operation of a special structure, which he termed the *synapse*, hypothetically located at the junction between sensory and motor neurons in the central nervous system. From this notion, it was but a short step to theories of synaptic structure and function, proposed by Sherrington and others, designed to account for the phenomena of reflex action. For example, the variability of the threshold stimulus in reflex action might be due to transient changes in the chemical state of the synapse due to activity in neighboring neurons; the eventual effectiveness of a repeatedly applied but initially subthreshold stimulus could be due to the buildup of an excitatory substance at the synapse which facilitates transmission of neural impulses across the synapse; the progressive weakening of a response upon repetition of a stimulus could reflect the buildup of an inhibitory substance which blocks neural transmission, and so on.

Note that in this line of analysis the synapse and the events taking place there assume the status of physical realities that exist prior to the behavioral phenomena they are held to give rise to. As such, accounts of synaptic functioning are readily seen as a more fundamental and inclusive level of explanation than description at the behavioral level only. But in fact, Skinner maintained, Sherrington's synapse does not add anything beyond the behavioral description of the reflex because the information upon which Sherrington based the synapse is derived essentially from observed stimulus-response correlations. In particular, the properties ascribed to synaptic function are dictated by the properties exhibited by the overt stimulus-response correlations themselves. Thus, for example, the idea that the central connection (synapse) between incoming and outgoing pathways may become more excitatory, i.e., more likely to transmit neural impulses as a consequence of subthreshold stimulation, is literally demanded by the behavioral observation that repeated applications of a subthreshold stimulus may eventually result in response evocation. Seen in this light, Sherrington's concept of the synapse is simply the logical requirement of a set of S-R relationships that evidence the role of prior experience and organismic state in reflex function. In Skinner's terms, "The synapse . . . *is the conceptual expression for the conditions of correlation of a stimulus and response, where the incidental conditions imposed by a particular stimulus and a particular response have been eliminated"* (p. 443). Sherrington's synapse, Skinner maintained, is at base more a matter of a conceptual nervous system than a central nervous system.[1]

There is nothing in the physiological investigation of the reflex, then, which calls into question the definition of the reflex as an observed stimulus-response correlation, because "there is nothing to be found there that has any significance beyond a description of the conditions of a correlation" (p. 443).

The Concept of the Reflex and Descriptive Behaviorism

Skinner's historical review can be said to have persuaded him of several points: (1) that the concept of the reflex, that is, the idea of strict environmental control of behavior, is applicable to all organismic activity; (2) that experimental reflexes, as instances of the environment-behavior relationship, are properly conceived of as being nothing more than observed correlations of stimulus and response; and (3) that conceptions of the environment-behavior relationship that go in any way beyond the level of observation of stimuli and responses add nothing to our understanding of behavior and are apt to detract from it. Given these conclusions, it follows that the general explanation of behavior is to be sought in terms of observed S-R relations. But how do we proceed from the study of particular S-R relations to the explanation of behavior in general?

Descriptive Behaviorism. All S-R theorists, of course, faced the problem of developing general explanations on the basis of limited samples of behavior.[2] Hull and Tolman, as we have seen, proceeded by using available data as a springboard for the construction of theories or "best guesses" as to the nature of the changes taking place inside the organism and the conditions responsible for those changes. Deductions were then made from those theories concerning behavior to be expected under particular conditions, and experimental test of those hypotheses served to assess the validity and generality of the theory. Skinner proceeded in a fundamentally different manner, but a manner entirely consistent with the points established through his historical review of the reflex. Theory construction, hypothesis testing, and speculation about events taking place inside the organism have no place in Skinner's approach. Instead, explanation is to be accomplished solely in terms of regularities detected through observation of stimulus-response relationships. These regularities will have the status of principles or laws but are no more than summary statements of repeatedly observed, and therefore reliable, relationships between stimuli and responses. For example, the principle of stimulus generalization is one such regularity. This principle states that conditioned responses will be evoked by stimuli other than those involved in the conditioning process and to a degree proportional to the similarity of the conditioned and new stimuli. So expressed, stimulus generalization is not a theory about an aspect of behavior or a guess as to a true statement. It is rather a concept that summarizes what has been actually observed in hundreds of conditioning experiments with different organisms. Stated as a principle, it simply expresses the belief that the observation has been made with sufficient frequency to warrant the assumption that it is a reliable characteristic of the conditioning process. The principle is explanatory in that it reduces uncertainty about behavior-environment relationships and enables improved prediction and control of behavior. For example, it enables improved prediction and control of the transfer of learned responses in everyday training situations.

Skinner maintained that careful observation of behavior in the laboratory will uncover a number of principles like that of stimulus generalization, principles that collectively will enable an account of all behavior. In arriving at these principles, one need never invoke concepts that go beyond the level of observation or be concerned about events taking place inside the organism. In essence, the explanation of behavior can be achieved simply by *describing* observed stimulus-response relations.

Descriptive Behaviorism and Positivism. The idea that there is no difference between explanation and description is the central point of the philosophy of *positivism*, which stemmed chiefly from the writings of the physicist and philosopher of science Ernst Mach (1836–1916). Mach argued that proper scientific explanation will always be reducible to statements about observations of the phenomena at issue, since propositions and concepts that are not reducible to observational statements cannot be checked against experience and therefore are metaphysical. Hence the scientist's task is simply to record observable relationships and describe them in abbreviated ways, that is, in summary principles and formulas. To employ propositions that go in any way beyond economical description of facts, Mach argued, is to invite metaphysical statements or false claims about the phenomena at issue. Note the congruence between these ideas and Skinner's conclusion that historical conceptions of the behavior-environment relation that went beyond statement of observed S-R relations were either metaphysical, as in the attribution of freedom to parts of behavior, or erroneous, as in the restriction of stimulus determination to unlearned reflexes, or reducible to statements about observed stimulus-response correlations, as in physiological conceptions of the reflex.

In introducing his behavior system, Skinner cited Mach as providing the appropriate model of scientific explanation and adopted the major features of his approach (Skinner, 1931). An important aspect of Mach's position (traceable ultimately to the philosopher Hume, 1711–1776) is the idea that to speak of causality between events is to go beyond observation. We never actually prove causality between events but rather simply demonstrate the concomitance of events, i.e., we only observe that event B follows event A, not that A causes B. Hence we are entitled to claim only a correlation of events A and B or, employing the neutral mathematical concept of relationship, we may describe event B as a *function* of event A. In Mach's system, the notion of function is substituted for that of causation, and the full explanation of any event is given by a description of its observed functional relationship with antecedent events. We explain a phenomenon by defining the conditions of which it is a function. Applying this language to psychology, Skinner held that the full explanation of behavior will consist of determining, through experiment, the functional relationships between behavior and the environment. Note that this approach has no concern at all with establishing *how* environment and behavior are linked in the organism; it is concerned only with establishing that given such and such a stimulus input, one will observe such and such a behavioral output.

Descriptive Behaviorism in Relation to Other S-R Theory and Research. In the descriptive behaviorism advocated by Skinner, theories appealing to events other than observable stimuli and responses are viewed not only as unnecessary but also as potentially harmful to the cause of behaviorism. Skinner's primary targets in this regard are explanations, such as Sherrington's, which seek to account for stimulus-response relations in terms of changes taking place in the nervous system. But Skinner also felt that the theories of Hull, Tolman, and Guthrie had recourse to events beyond the level of stimulus-response observation, either in the form of physiological conceptions of behavior or more generally in the form of variables postulated to intervene between overt stimuli and responses. While intervening variables such as "expectancy," "habit," "r_g," and "movement-produced stimuli" are all capable of reduction to overt stimulus-response definitions, they nevertheless ordinarily functioned in explanation without such reduction and hence could be argued to have a metabehavioral character. The argument against such theories is best put in Skinner's own words:

> A science of behavior must eventually deal with behavior in its relation to certain manipulable variables. Theories—whether neural, mental, or conceptual—talk about intervening steps in these relationships. But instead of prompting us to search for and explore relevant variables, they frequently have quite the opposite effect. When we attribute behavior to a neural or mental event, real or conceptual, we are likely to forget that we still have the task of accounting for the neural or mental event. When we assert that an animal acts in a given way because it expects to receive food, then what began as the task of accounting for learned behavior becomes the task of accounting for expectancy. The problem is at least equally complex and probably more difficult. We are likely to close our eyes to it and to use the theory to give us answers in place of the answers we might find through further study. It might be argued that the principal function of learning theory to date has been, not to suggest appropriate research, but to create a false sense of security, an unwarranted satisfaction with the status quo. (1950, p. 194)

Essentially, Skinner's position is that if the final goal is to relate behavior to observed stimulus variables, then why not proceed directly to that goal? The shortest, most certain route to understanding environment-behavior relations is to determine them directly. Theoretical guidance involving appeal to events taking place at levels other than the level of observable behavior is by definition an indirect attack on behavior and hence runs the risk of being in error; we cannot go amiss in understanding environment-behavior relations if we stay at the level of demonstrated relations.

Skinner's refusal to entertain conceptions of intervening processes that might fill in the causal continuities between overt stimuli and responses led to the charge that he dealt with "the empty organism." But his stance in this regard should be seen not as a denial of the contribution of central processes, but rather as a matter of scientific strategy. It stemmed from the conviction that whatever mental processes mediate organism-environment interactions, these processes are themselves the product of prior interactions with the environment and therefore are dealt with most meaningfully, from a scientific viewpoint, through study of overt stimulus-response relations. Implicit in this view, of course, is the assumption common to all S-R theorists that genetic and biological factors play a relatively minor role in behavior.

Skinner's desire to deal with behavior directly underlies another distinctive feature of his approach, namely, an emphasis upon the study of behavior in the individual organism. The bulk of S-R research on learning proceeded by experimental designs that compared the performances of groups of animals trained under different stimulus conditions. All of the research reviewed in Chapter 4 on controversies between cognitive and response-reinforcement positions was of this nature. In such designs, conclusions about the nature of the relation between stimulus conditions and behavior are based on statistical assessment of measures of group performance, e.g., group averages. But measures of group performance are, by definition, indirect measures of the organism-environment interaction and may not faithfully reflect the nature of the relation between stimulus conditions and behavior for individuals in the group. What is needed, argued Skinner, are procedures that enable one to see directly and continuously the course of behavior by an individual organism in relation to changes in environmental conditions, and a major part of his effort concerned the development of just such experimental procedures.

There is also a strong pragmatic strain in Skinner's individual-organism approach, as indeed in positivistic philosophy as a whole. In Skinner's eyes, the "bottom line" of any psychological theory or research is its utility in predicting and controlling behavior, and the most stringent as well as the most useful test to that end is control of the individual organism. In Skinner's words, "No one goes to the circus to see the average dog jump through a hoop significantly oftener than untrained dogs raised under the same circumstances" (1956, p. 228).

In sum, we explain any behavior by determining and describing the S-R relationships that enable us to control that behavior in the individual.

The Experimental Analysis of Behavior

These general ideas were translated in a direct fashion by Skinner into a distinctive research program that he and his followers have pursued for several decades now. The procedure is to select a particular reflex, or S-R correlation, as a working model of organism-environment relationships in general and then set about determining, in the individual organism, the conditions of which changes in the strength of that correlation are a function. The conditions are suggested simply by careful observation of laboratory and natural life behaviors, but in every case their relation to reflex strength must be tested and established through experiment. That is, each condition suspected of influencing behavior must be systematically manipulated in relation to measures of reflex strength while all other conditions are held constant. Only in this way can the nature of the functional relations between environment and behavior be unambiguously described. If the reflexes selected for study are true working models of behavior, the functional relations determined through patient, long-term experimental analysis will prove, in the aggregate, to be the general laws of behavior.

Skinner and his co-workers have termed this approach *the experimental analysis of behavior,* a phrase that might seem somewhat arrogant given that all S-R investigators, regardless of theoretical persuasion, regarded the experiment as the means of establishing facts about behavior. However, Skinner intended the phrase to convey a *purely* experimental analysis, an analysis in which theoretical interpretations of the Hull and Tolman variety would be bypassed, and in this light the phrase is not inappropriate.

The Unit of Experimental Analysis. There arises immediately the question of how to specify the particular S-R correlations that are to be intensively studied in the laboratory as models of behavior. If the reflex preparation is to be fruitful in the analysis of behavior, it must take into account "the natural lines of fracture along which behavior and environment actually break" (1935, p. 40). The basic problem here is that of defining a meaningful and reproducible unit of organism-environment interaction in the face of what Skinner termed the *generic nature* of the stimulus and response. The stimulus and response members of even the simplest reflex are not invariant units but rather are *classes* of similar stimuli and responses. Given this circumstance, the experimenter must demark the S-R observations that will be accepted as members of a single, "natural" organism-environment exchange. One procedure would be to specify in advance by some logical criterion the stimulus and response properties that must obtain to constitute an instance of the experimental reflex. For instance, in setting up a leg flexion reflex, the experimenter might specify the exact direction and extent of flexion that must be exhibited to count as an instance of the reflex and ignore all instances that do not exhibit these properties. But such specification is inevitably arbitrary in some degree and therefore likely to result in a unit of little value in the general analysis of behavior. The solution, said Skinner, is to determine, by observation, that specification of stimulus and response where smooth, orderly relations appear between measures of the reflex strength and the stimulus conditions manipulated in relation to the reflex, a specification that is, he argued, uniquely determined for any reflex. The experimental unit of analysis is to be determined, then, like everything else is determined in Skinnerian psychology—by observing what works. Does the experimental reflex show clear, lawful relations with the manipulated stimulus conditions? If so, we know that we have respected "the natural lines of fracture" and that these experimentally determined relations will of necessity mirror the general laws of the environment-behavior relationship.

Basis of the Program. Skinner's research approach assumes that all of the significant phenomena of animal and human activity are expressible in terms of relations between stimulus conditions and reflex strength. The test of this proposition lies in the long-run success or failure of the approach, and Skinner had scant evidence for the proposition at the outset of his program. He could point to the work of Pavlov and Thorndike, which clearly indicated that an incisive analysis of learning might be accomplished by relating particular stimulus operations to measures of reflex strength. But there were few data to suggest that motivation, emotion, cognition, and other complex phenomena could be understood through the experimental analysis of reflexes.

Skinner's faith in the experimental analysis of behavior rested at the outset not on any accomplishments but rather on the thesis developed through his historical and logical analysis of the concept of the reflex. Specifically, it rested on the convictions that all behavior is controlled by the environment; that the experimental reflex, as an observed S-R correlation, is both an instance of that control and, as such, the appropriate vehicle of its analysis; and that explanations of environment-behavior relationships that go beyond observed S-R correlations will inevitably prove superfluous. Given these convictions, it follows that scientific explanation of human and animal activity should be reducible to statements of reflex strength as a function of stimulating conditions and that we will not discover any aspects of the behavior of organisms that cannot be described in such terms. "From the point of view of scientific method, at least, the description of behavior is adequately embraced by the principle of the reflex" (1931, p. 454).

In Skinner's eyes, observed S-R correlations are all the behavioral scientist knows, and all he or she needs to know.

Overview

The consistency with which Skinner has adhered to observability as the touchstone of sound scientific practice is remarkable. It has probably not escaped the reader that Skinner's conclusion that the explanation of behavior is to be sought in terms of observable S-R relations was itself based on observation! That is, it was based on his review of working conceptions of the environment-behavior relationship. From that review flow directly or indirectly all the major features of his system—the bias against theory, the reluctance to utilize intervening constructs, the avoidance of physiological conceptions, the pursuit of a purely experimental and functional analysis of behavior, the desire to deal directly with behavior in the individual organism, the emphasis upon the practical control of behavior, and the empirical determination of the unit of analysis. The insistence on observability, which Skinner has now maintained for four decades, is the key feature and great strength of his system.

IMPLEMENTING THE PROGRAM

Marshall Hall, as we have seen, restricted the concept of the reflex to behaviors that are regularly elicited by specifiable stimuli, but he classified as "voluntary behaviors" all movements that are emitted without identifiable eliciting stimuli. While Skinner insisted that all behavior is lawfully related to environmental stimuli, he nevertheless retained Hall's distinction between stimulus-elicited (reflexive) and emitted (voluntary) behavior. This distinction parallels the common sense classification of behavior as being either highly automatic and concerned with body functioning or relatively spontaneous and directed toward the external environment. To avoid the multiple meanings attached to the terms *reflexive* and *voluntary*, Skinner substituted the terms *respondent*, to carry the sense of a relation to a prior event, and *operant*, to carry the sense of behavior that somehow operates upon the environment in service of organismic needs.

In the context of traditional research on learning, respondents correspond to the stimulus-elicited behaviors typically studied in classical, or Pavlovian, conditioning, and operants correspond to the behaviors studied in instrumental conditioning, including most notably the behaviors observed in problem boxes, runways, and mazes. For the purpose of describing the learning of respondents, Skinner was content to employ the procedures and principles of Pavlovian conditioning. However, he had little interest in furthering the analysis of respondent behavior, which he felt had received disproportionate emphasis in the study of learning. Operants constitute by far the major portion of activity, particularly at the human level, and hence operants would be the focus of the experimental analysis of behavior. But unlike the case for respondent behavior, existing laboratory procedures were deemed of little utility for the understanding of operant learning.

The major problem with puzzle boxes, mazes, runways, and the like, in Skinner's view, is that they fail to capture the essentially *emissive and probabilistic* character of operant activity. Unlike respondents, operants do not invariably follow particular stimuli but must instead be described in terms of their likelihood of emission within a general context. The appearance of a respondent can generally be guaranteed simply by presentation of an above-threshold value of an appropriate stimulus, but appearance of an operant may range from highly unlikely to virtually certain depending upon the stimulus context and learning history of the organism. The experimenter might arrange a test situation to make appearance of a given operant somewhat more or less likely, but fundamentally the likelihood of emission of an operant is a function of prior stimulus conditions, most of which remain to be defined. Therefore, Skinner concluded, *probability of occurrence* is the natural datum to be sought in describing operant behavior.

But how is response probability best measured? Mathematically, the probability of an event can be estimated by observing the number of times the event occurs with a span of opportunity; the more frequent the event per unit of opportunity, the more probable is a subsequent occurrence of the event. In relation to operant behavior, the estimate could take the form either of the number of response occurrences during an appreciable interval of time or the number of occurrences per number of possible occasions (trials). The occurrence per trial measure must be employed with mazes, puzzle boxes, and runways because of the spatial character of the behaviors involved. That is, in these situations the animal must learn to move to a particular place to secure reward or avoid punishment, and the learned behavior cannot recur until the experimenter reintroduces the subject to the starting position to begin a new, discrete trial of the behavior. Under this procedure, frequency of the "correct" response is obviously constrained by the number and timing of the trials set by the experimenter, and hence it can be argued that the resulting probability estimate is a relatively crude measure of the organism's readiness to engage in the behavior. A more appropriate learning paradigm, Skinner argued, would leave the subject free to emit the behavior at any time. This condition can be met by abandoning the spatial responses of conventional paradigms in favor of simple manipulative responses that can be rapidly executed and that leave the organism ready to respond again. For example, as is now generally well known, Skinner's early research relied heavily upon the lever pressing behavior of the rat. Such free operants—free in the sense that there are no procedural restrictions on response emission—are measured in terms of number of occurrences per time interval, i.e., in terms of *response rate*. The response rate of a free operant, Skinner maintained, is the desired measure of operant strength because it is appropriately sensitive to the naturally probabilistic nature of operant behavior.

The experimental paradigm Skinner developed for analysis of operant behavior consisted of a hungry rat placed in a small chamber containing a lightweight lever, a nearby food receptacle, and little else (see Figure 6-1). The lever protrudes from the wall in such fashion that a rat introduced into the chamber for the first time has some low probability of displacing it in the course of general movement about the chamber. Each displacement of the lever activates a food-delivery mechanism that discharges a small pellet of food to the receptacle inside the chamber. In this way, Skinner arranged a controlled environment incorporating a continuing contingency between a simple bit of operant behavior and a favorable consequence, and he could study change in the probability of that behavior as a function of any stimulus variation he chose to introduce to the situation. Other versions of the "Skinner box" provided for the study of punishment by substituting for food a negative consequence, such as a brief shock through the floor of the box. As Skinner's program developed, the paradigm was modified to include other organisms and response systems, e.g., disk or key pecking by pigeons and panel pressing by monkeys (see Figure 6-2). But the essential feature of the paradigm endured—a bit of operant behavior is isolated in relation to a motivationally relevant environmental consequence.

Experimental free operants are selected initially on the basis of

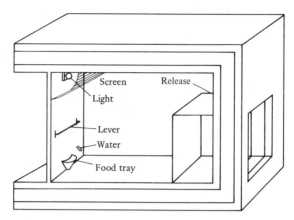

Figure 6-1a. An early version of an operant conditioning chamber. One side of the sound-resistant, lightproof chamber has been cut away to show the part occupied by the animal. The space behind the panel at the left contains the rest of the lever (a metal rod projecting about ⅜ inch inside the chamber) and the food delivery mechanism. The section labeled "release" could be activated from outside the chamber to allow the animal access to the bar area without further handling by the experimenter.
(From Skinner, 1938, p. 49. B. F. Skinner, *The Behavior of Organisms: an experimental analysis,* © 1938, renewed 1966. Reprinted by permission of Prentice-Hall, Inc., Englewood Cliffs, N.J.)

convenience. Thus, the rat's lever press is a relatively unambiguous response with a conveniently low frequency of occurrence before conditioning (a few responses per hour). Of course, as we detailed before, the criterion for *retention* of any free operant reflex as an experimental model of behavior is whether it is found to exhibit lawful relations with the stimulus variables manipulated by the experimenter.

In addition to mechanizing the behavior-environment contingency, Skinner also recorded response rate automatically by an ingenious modification of the standard kymograph. In the kymograph, a pen or stylus rests upon a moving strip of paper, yielding a continuous tracing. By mechanical or electrical means, the event to be recorded is made to produce a displacement of the pen and, in consequence, appears in the graphic record as a deviation from an otherwise straight-line tracing. In Skinner's application, each lever press operated an electrical device that produced a slight displacement of the pen to a new position on the paper, with successive displacements being *cumulative* in the same direction. For example, an animal that never responded during an experimental period would produce the horizontal straight-line record shown in Figure 6-3A, since the paper simply moved at a constant rate under a stationary pen throughout the pe-

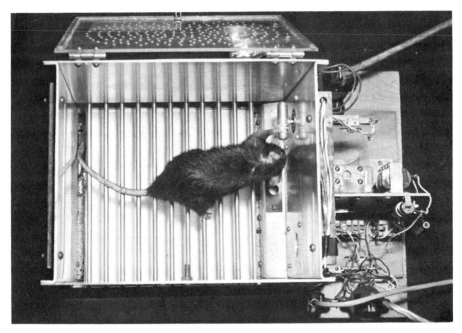

Figure 6-1b. This photograph shows a top view of the interior of a more contemporary version of an operant conditoning chamber with the subject in the act of pressing the lever. In the experimental arrangement shown, water is used as reinforcement rather than food. The water is delivered to the thirsty subject by means of the small dipper discernible immediately to the right of the subject's head. (Photograph courtesy of B. F. Skinner.)

riod. An animal who responded but once midway in the same period would show a record as in Figure 6-3B, the response appearing as a single vertical displacement. An animal that responded many times in quick succession during the first half of the period but infrequently in the last half would produce the record in 6-3C. It can be seen that with this device the animal draws its own curve, so to speak, a cumulative curve of response over time, the slope of which specifies rate of responding. If the animal responds rapidly, the slope is steep. If it responds slowly, the slope is gradual. By means of other pen displacements on the kymograph, a record can be made of any other event (e.g., food pellet delivery) the experimenter might wish to document in relation to bar pressing behavior. In this way, the cumulative recorder automatically provides a continuous picture of changes in response rate in the individual organism as a function of manipulated experimental conditions, a record that can be extended indefinitely at little cost to the experimenter.

Figure 6-2. The interior of an operant conditioning chamber for use with pigeons, showing a subject in the act of pecking a plastic key.
(Photograph courtesy of B. F. Skinner.)

As Skinner's research progressed he and his co-workers developed procedures that enabled full automation of experiments on operant behavior. By application of electronic timers and switching circuitry, it became possible to automatically regulate all experimentally significant conditions within the "operant space," e.g., presentation of discriminative stimuli, the availability and scheduling of reinforcing stimuli, or the temporal relationships between various stimulus events. Conducting an experiment became in essence a matter of presetting control equipment to administer the training program specified by the experimenter, placing the subject in the Skinner box, and then reading the cumulative record generated under that program. With these procedures, the subject need not even be seen during an experimental session, and in fact was commonly not observed by the experimenter. Observation of the organism-environment exchange was reduced to inspection of a cumulative curve, supplemented perhaps by counter readings.

Skinner's method, particularly in contrast to that of Guthrie, brings to mind once again the methodological differences that separated Watson and Small. Skinner's approach is clearly in agreement with Watson's emphasis upon precision and objectivity of measurement at the sacrifice of breadth of observation, and in fact the Skinner box can be regarded as the high point of the bias toward a restrictive methodology in S-R psychology. On the other hand, Guthrie's pictorial record of spatial escape responses from an "open" puzzle box stressed qualitative and wholistic observation of behavior at the expense of quantitative measurement and control by isolation. Skinner's methodology was to prove more in tune with the needs of S-R psychology than that of Guthrie. While learning investigators were not immediately receptive to Skinner's procedures, there was increasing disenchantment with the maze and related apparatus as standard learning paradigms. As described earlier, research on the issues separating the response-reinforcement and cognitive theories revealed that it is extremely difficult, if not impossible, to fully control extraneous stimulus influences in the maze situation. In contrast, the Skinner box successfully isolated the behavior processes at issue by the simple expedient of isolating the subject from the world. In addition, automatic experimentation greatly enhanced standardization of research procedures across laboratories while it minimized the role of experimenter bias and reduced the labor of experimentation. Ultimately, albeit slowly, Skinner box technology replaced the maze, the runway, and the puzzle box as the modal research procedure of the S-R learning psychologist. However, the reader can be assured that this tilt

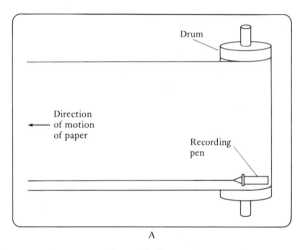

Figure 6-3. The cumulative recorder and illustrative cumulative records. (Photograph courtesy of Gerbrands Corporation.)

toward the "Watson end" of the methodological dimension was not a final stage. As with any fundamental, polar issue, there was the inevitable reverse swing of the pendulum. That swing would come with the realization that Skinnerian methodology allowed little or no play of biologically based dispositions in learning. But this realization, like any counter pendulum swing, took time to gather force and belongs to a later part of our history.

As reviewed here, Skinner's methodology appears a very direct and rational implementation of his broad program for the experimental

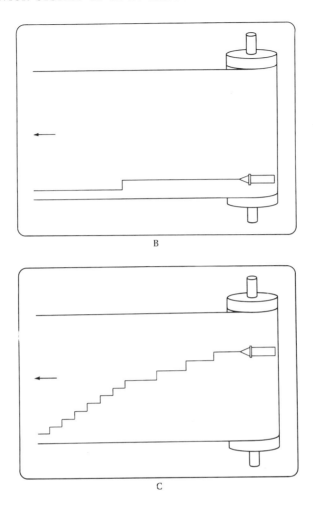

B

C

analysis of behavior. Actually, our brief account has ignored some significant steps in the evolution of the methodology, several of which were purely serendipitous, as Skinner humorously acknowledges in an autobiographical account of his early career (1956). But nevertheless the critical elements of Skinner's methods were present in his very first experiment (1930), which sought to describe the rate of certain eating reflexes in the rat. The Skinner box, the cumulative recorder, the technology of experimental control, and the free operant rationale were essentially elaborations of the procedures in this first study, published while Skinner was still a graduate student in psychology at Harvard.

THE DESCRIPTIVE PRINCIPLES

The descriptive principles that make up Skinner's explanatory system were derived largely through experimental analysis of bar pressing behavior of the rat. The assumption, of course, was that principles so derived would be found to hold in essential form for any operant activity in any organism. Our aim in this section is to describe the major descriptive principles and to contrast their character with the explanatory principles of other S-R theorists. In a following section we shall consider the application of Skinner's principles to complex human activity in natural life settings.

The Principle of Reinforcement

From the outset Skinner focused on the response-consequence relation as the major determiner of operant strength. In the Skinner box this relation is studied chiefly in terms of the programmed contingency between bar pressing and food pellet delivery. Setting up the contingency involves several steps: (1) a preliminary session in which a hungry rat is placed in the Skinner box with the bar absent and a supply of food in the receptacle until the animal moves and eats freely in the situation; (2) an additional session without the bar, during which the rat becomes accustomed to eating pellets discharged one at a time from the food dispenser by the experimenter; and (3) a final session with the bar in place for the first time and connected to the food magazine so that each press of the bar delivers a single pellet to the food dish. In the final session the rat typically approaches the food dish immediately, as trained in sessions 1 and 2, and, finding no food, begins general investigatory activity in the region of the dish and bar. In the course of this activity the rat is likely to depress the bar and, at the distinctive sound of dispenser operation, quickly moves to the food dish to seize the pellet, as trained in session 2.

Figure 6-4 shows typical cumulative records generated in the final session of such training. These functions are copied from actual records obtained in Skinner's experiments, and, as is typical in such records, the "response steps" of the recording pen are kept small to produce a fairly smooth curve of results. Note that after the first rewarded bar press or two the behavior immediately goes to a high and steady state of strength for the remainder of the session. On the basis of such observations, Skinner concluded that the conditioning of an operant occurs in all-at-once fashion *whenever the response-conse-*

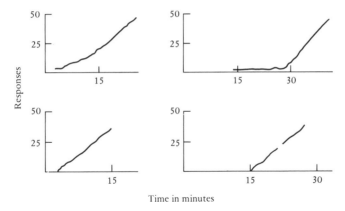

Figure 6-4. Typical cumulative records generated by hungry rats when first exposed to a contingency between bar pressing and receipt of a food pellet. Note that after the first reinforced bar press the behavior quickly goes to a high state of strength. (After Keller and Schoenfeld, 1950, p. 46. By permission.)

quence relation is immediate. In this regard, an important factor in the all-or-none character of the behavior changes in Figure 6-4 is the preliminary training, which had the effect of ensuring a close temporal succession of bar press and food reward. Consider the likely outcome of the final session for an animal *not* given prior training in approaching the food dish at the sound of the dispenser operation. Such a subject would not quickly look in the food dish at the sound of the dispenser and in fact would probably be frightened into immobility by this novel stimulus. There might be many minutes, and many intervening behaviors, between bar pressing and discovery of the pellet. Under such conditions operant conditioning of the bar press would be slow indeed. The three-step training sequence used to produce the results of Figure 6-4 illustrates an important principle of Skinnerian psychology, namely, that acquisition of an operant can often be greatly facilitated by the prior conditioning of responses that make likely the desired new behavior-consequence conjunction.

The curves of Figure 6-4 are instances of a behavior-outcome relationship repeatedly observed in operant conditioning experiments. Consistent with Skinnerian psychology, the reliability of this outcome warrants statement in a summary principle—*the principle of reinforcement:*

> *If the occurrence of an operant is followed by the presentation of a reinforcing stimulus, the strength is increased.* (Skinner, 1938, p. 21)

By *reinforcing stimulus* Skinner meant any stimulus *observed* to increase response strength. At first blush, this conception of reinforcement appears to make the principle circular and without predictive power—the strength of an operant is increased if followed by a stimulus observed to increase the strength of an operant. But noncircularity and predictive power *are* preserved by the consideration that in any application of the principle the reinforcing stimulus and behavior at issue need not have been previously observed in conjunction. That is, any stimulus observed to strengthen a response in one situation is predicted to strengthen other operant behaviors of that organism in other situations.

The reader may also be struck by the apparent resemblance of Skinner's reinforcement principle to Thorndike's law of effect ("responses . . . followed by satisfaction to the animal will . . . be more likely to recur") and to Hull's principle of reinforcement (effector activities followed by diminution of a need will be more likely to recur). But the conceptual gulf here is very wide. Note that unlike the principles of Thorndike and Hull, Skinner's principle says nothing whatsoever about the mechanism of reinforcement or its psychological concomitants. It does not seek to explain *why* the reinforcing stimulus strengthens behavior; it simply asserts that some stimuli are observed to strengthen preceding responses and that such stimuli will so affect any operant of the organism in question.

While Skinner was less concerned than other S-R theorists with defining the mechanism of reinforcement or the class of stimuli that will so serve, he was (for that very reason, he would argue) more inclined to explore the empirical domain of reinforcer-behavior relationships. His work in this vein forms the core of his experimental program and constitutes a major contribution to our knowledge of learning. We cannot do justice here to the tremendous volume of data amassed by Skinner and his followers on the behavior-reinforcer relation, but we can review some of the more important findings and principles from this research.

Superstitious Behavior. While this phenomenon is not the most important of Skinner's discoveries in the realm of reinforcement-behavior relations, it is useful to consider at the outset because it is particularly helpful in communicating the flavor of the Skinnerian analysis of reinforcement.

The principle of reinforcement leads us to expect that a reinforcing stimulus will strengthen *any* operant behavior that closely precedes it

in time regardless of the appropriateness of that behavior to the reinforcing stimulus, regardless of the organism's "intentions," and regardless of the relevance or irrelevance of the behavior to procurement of the reinforcing stimulus. Skinner (1948a) conducted a simple
experiment to test this expectation. A hungry pigeon was placed in
an experimental chamber with an attached food (grain) hopper that
could be swung into place so that the pigeon could eat from it. An
automatic timer held the hopper in place for five seconds at each
presentation, which allowed the bird to procure grain reinforcement
on each occasion. A clock was then arranged to present the food
hopper at regular intervals (for example, at 20-second intervals) *with
no reference whatsoever to the bird's behavior.* Several birds were treated
in this manner with the experimenter simply observing the general
behavior of each subject after a period of time.

What Skinner observed was the production of bizarre behaviors
that persisted at a high rate. The conditioned responses were so clear
that independent observers could readily agree on counting instances.
For example, one bird learned to "tour" counterclockwise around the
chamber, making two or three turns between reinforcements. Another subject acquired the habit of repeatedly thrusting its head into
one of the upper cage corners, and another developed a head tossing
response as if placing its head beneath an invisible bar and lifting it
repeatedly.

According to Skinner, the conditioning behind such behaviors is
usually obvious. The bird, of course, is always doing *something* when
the reinforcing stimulus is presented, e.g., it might be just completing
a movement around the chamber. Given the powerful effect of a
single reinforcement, the frequency of doing that something will immediately increase, and hence the act will be likely to be again occurring close to the moment of the next reinforcement. Only a few
such "contingencies" between any particular behavior and reinforcement are required to bring that behavior to a high state of strength
and to make even more probable further conjunctions between the
behavior and receipt of reinforcement. The result is the conditioning
of behavior that is totally irrelevant to the procurement of reinforcement. Nor should the reader feel that such habits are unstable or
short-lived. Skinner (1948a) reports that during extinction of one such
behavior (a side-to-side hopping response) more than 10,000 responses were emitted before the response returned to a low state of
strength.

Skinner's experiment might be said to demonstrate a sort of superstition. The bird behaves as if there were a causal relation between its behavior and presentation of the reinforcing stimulus, although such a relation is lacking. A few purely chance conjunctions of particular behaviors and a reinforcing state of affairs are sufficient to produce repetition of those behaviors, which then are maintained under further adventitious reinforcement.

Similar observations of superstitious behaviors, that is, behaviors arising from accidental correlations between behavior and reinforcement, are quite common in conditioning research. The writer has found that superstitious behaviors of this sort are readily established in classroom demonstrations. The procedure is much like that of Skinner's experiment. A hungry pigeon is placed in a large, glass-walled box equipped with a food magazine. The bird has been trained to approach the food hopper at the sound of its operation but has received no other training. The class first observes the bird for several minutes to establish that the subject has no bizarre or ritualistic behaviors at the outset of the demonstration. A member of the class is then given a hand switch to control presentation of the food hopper and is asked to sit facing away from the demonstration apparatus so that neither the apparatus nor the bird can be seen. Another student is asked to tap the first on the shoulder at, say, 30-second intervals, at which time the first student operates the hopper for a few seconds. The prediction, made at the beginning of the conditioning procedure, is that within ten minutes the subject *will be* emitting some form of ritualistic behavior at a high rate. In many such classroom demonstrations, the prediction has invariably been met. The exercise provides a striking demonstration of the automaticity of reinforcement and is highly effective in communicating Skinner's conception of reinforcement as a purely mechanical, behaviorally blind process.[3]

There are many analogies in human behavior. The sports world provides good examples. The ritualistic behavior of the baseball batter approaching the plate or addressing the pitcher, the twisting and turning of the bowler after releasing the ball down the alley, the dependence of the angler on specious natural "signs," and the stereotyped pre-serve behavior chains of the tennis player are all instances of learned behaviors that have no real influence on outcomes, just as the food would appear if the pigeon did nothing at all. Analogies abound, too, in more significant spheres of action, such as medical and religious practices, where time-honored treatments and rituals are sometimes later shown to have no causal relation to consequences. Such practices were presumably originally set up and maintained through purely accidental relationships to favorable outcomes.

The Method of Successive Approximations: The Shaping of Behavior. Like other S-R theorists, Skinner recognized the empirical principles of stimulus and response generalization, whereby reinforcement of any particular S-R conjunction is held to increase both the power of similar stimuli to evoke the behavior in question and the strength of similar behaviors. Given the conception of operants as behaviors without specifiable eliciting stimuli, Skinner was inclined to give greater weight than other S-R theorists to the principle of response generalization, which he termed *response induction.* Consistent with his notion of the reflex as a class of S-R correlations that behave as a lawful unit, Skinner maintained that reinforcement of any particular instance of a reflex automatically increases the probability of all specific responses making up the reflex, though in proportion to their similarity to the reinforced response.

An important practical implication of such response induction is that by judicious application of reinforcement it should be possible to shift the organism's behavior through overlapping response distributions to new, but initially improbable, forms of behavior. Consider a simple example. In conditioning the bar press, it sometimes happens that on first exposure to the bar the subject makes no responses whatsoever to it and after lengthy exploration around the chamber settles down to rest at some distance from the lever. The experimenter might then have a very long wait indeed before the subject depresses the bar. But in such circumstances the experimenter can greatly accelerate the learning of bar pressing by reinforcing successive behavioral approximations to that end. In the situation described the experimenter has little behavior to work with at the outset and would probably begin by operating the food magazine upon mere orientation of the subject's head toward the bar. At the sound of dispenser operation the animal will, as a consequence of its prior magazine training, move to the food dish and secure reinforcement. This single reinforcement will be sufficient to increase the frequency of a whole class of investigatory behaviors in the vicinity of the food dish, and the experimenter might now reinforce (by operating the food magazine) any investigatory response that involves raising the head in the direction of the bar. One or two reinforcements of head raising will immediately result in increased frequency of movements of this class, and now the experimenter will withhold reinforcement until there occurs a raising response that involves removal of the forepaws from the floor. Reinforcement of this response in turn shifts the distribution of raising responses, and the experimenter will withhold reinforcement until, within that distribution, the subject raises its paws in the vicinity of the bar. Reinforcement can then be withheld until the animal emits, within the class of paw raising movements, a movement that results in the animal's touching the bar with a forepaw. Finally, the experimenter will require depression of the bar for reinforcement, and with a few reinforced repetitions of bar pressing the response goes to full strength. The end result is that a behavior that might not ever have been emitted in acceptable form under the initial contingencies is induced to appear in a matter of minutes.

By applying the method of successive approximations Skinner and his colleagues have trained animals to perform many acts that common sense would hold to be impossible. Rats have been trained to operate and ride in mechanical conveyances, pigeons to play ping-pong, hens to play melodies on a keyboard, chimpanzees to type, and so on. The trick quality of these behaviors should not be allowed to obscure their importance as instances of a high order of behavior control. It is no trivial achievement to be able to produce, rapidly and at will, such novel and complex forms of behavior in animals. The fact that behavior can be so *shaped*, as Skinnerians would say, attests to the reliability and power of the seemingly simple principle of reinforcement. It might also be noted that the word *shape* is particularly appropriate in this context since it emphasizes that reinforcement results in a redistribution, or molding, of existing behaviors, rather than the acquisition of behaviors the subject was strictly incapable of emitting.

The principle of successive approximations can be seen in the development of natural life behaviors. Language acquisition is a good instance. The baby's first words, however imprecise, are usually greatly rewarded by the parents. If the first attempt to say "daddy" should come out as "dagum," this is nevertheless met by great praise and attention, and rightly so, since an important step in language acquisition is that the child simply attempt to name objects in the world. When labeling efforts are well established, the parent can then selectively praise more precise expressions. At a still later stage, of course, social attention is strictly withheld from the very "baby talk" that was at first so extravagantly rewarded.

Many forms of skill training involve an initial stage, more or less explicitly thought out, in which the learner is reinforced for approximate, but ultimately inappropriate, forms of the skill. The awkward and often ludicrous first steps of the would-be figure skater are indiscriminately encouraged because it is only when upright skating motions are well established that the skater is likely to emit the finer movements ultimately desired. Similarly, in teaching a child to write it is necessary to first reward any production of recognizable letters, however inappropriate in absolute size or elements of shape. These productions in turn become the basis for shaping the controlled movements that eventuate in graceful writing. In these and other instances of shaping, the basic principle is to secure new behaviors by strengthening existing responses that dispose the learner to emit the behavior finally desired. Note that the technique is based on the assumption that a few reinforcements of any behavioral "step" are sufficient to markedly increase the frequency of that behavior. In fact, an important concern in applying the method is that too much reinforcement of the first approximate forms of the behavior may retard acquisition of the ultimately desired behavior. Baby talk, however charming at age three, is a handicap at six.

Extinction and Spontaneous Recovery

In Skinner's system extinction is defined simply as decreased frequency of response accompanying withdrawal of reinforcement. The definition makes no reference to *why* behavior decreases in strength under withdrawal of reinforcement or to a mechanism underlying the decline. Although not interested in mechanism, Skinner paid great attention to the particular patterns of behavior accompanying extinction under differing experimental conditions.

Figure 6-5 shows a typical cumulative record generated during extinction of a reinforced bar press response. The animal had been conditioned as described above in connection with Figure 6-4 and had emitted about 100 bar presses, each of which was reinforced. Extinction began on the following day. The food magazine was disconnected so that movement of the lever was without effect except upon the recording device, and the subject was simply placed in the chamber for a one-hour period (the left half of Figure 6-5). It can be seen that at the outset of the extinction session, bar pressing was emitted at a high rate but thereafter declined through a series of cyclic fluctuations to the preconditioned rate (virtually zero) by the end of the hour. This pattern has proved typical of the extinction of behaviors that have been invariantly reinforced—the behavior declines in strength quite rapidly and is characterized by alternate bursts and pauses. These fluctuations, Skinner held, are due to emotional reactions (frequently readily observable) generated by the switch from 100 percent to zero reinforcement. Skinner was able to show that these cyclic variations in strength do not characterize extinction following

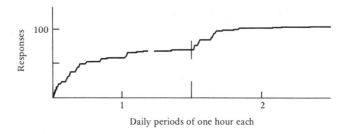

Figure 6-5. Cumulative record generated by a hungry rat during extinction of a regularly reinforced bar press response. The left- and right-hand sides show discrete one-hour extinction sessions separated by an interval of 48 hours.
(From B. F. Skinner, *The Behavior of Organisms: an experimental analysis*, p. 79. © 1938, renewed 1966. Reprinted by permission of Prentice-Hall, Inc., Englewood Cliffs, N.J.)

a *series* of *re*conditionings and *re*extinctions, since emotional reactions to nonreinforcement then become adapted out.

The right-hand portion of Figure 6-5 shows the behavior of the same subject during a second extinction session administered 48 hours after the first. It can be seen that the rate begins at a value many times that of the final value of the first extinction session and declines quickly to its preconditioned level. Such *spontaneous recovery* of an extinguished response upon reintroduction to the learning situation is, of course, a general characteristic of conditioned behaviors. Hull, it will be recalled, interpreted simple extinction as resulting from the buildup of an inhibitory state which dissipates with rest, thereby permitting renewed expression of the behavior upon reexposure to the situation. Skinner's explanation of spontaneous recovery stands in interesting contrast. He noted that the first conditioned bar press responses are proximate to stimuli unique to the outset of the session, e.g., those involved in introduction to the chamber and approach to the bar, and therefore these stimulus–bar press relations must be repeated a number of times without reinforcement before their rate declines to the preconditioned level, just as the bar pressing that characterizes later portions of the conditioning session must be repeated without reinforcement (Skinner, 1950). In this view, spontaneous recovery is not recovery of strength at all but rather a case of incomplete extinction. Note that Skinner's explanation of spontaneous recovery, like his explanation of extinction, is a matter of pointing to relations between particular experimental conditions and particular properties of behavior, whereas Hull's explanation involves appeal to inferred states of the organism.

Our brief discussion of extinction following regularly applied reinforcement makes clear that extinction must play a critical role in the method of successive approximations. The controller must not only strengthen successive new behavioral steps but must also weaken behavioral approximations rewarded earlier in the sequence but now deemed insufficient. Only by the manipulation of both reinforcement and extinction is behavior shaped to the newly desired form. The method of successive approximations thus highlights the two basic operations molding behavior—the giving and withholding of reinforcement.

Schedules of Reinforcement

The operant conditioning experiments reviewed to this point present the response-reinforcement relation as a powerful determinant of behavior. Behavior can be turned on and off, so to speak, and novel behaviors made to appear by the simple operations of giving and withholding reinforcement. In Skinner's philosophy of science, when you have got hold of an important relationship it is a good thing to analyze it exhaustively, and this is precisely what he proceeded to do with the behavior-consequence relation. If mere presentation or withholding of reinforcement can account for gross changes in behavior strength, how much more of the variability in behavior might be accounted for by variation in such conditions as the number, amount, immediacy, and scheduling of reinforcement. While Skinner undertook experimental analysis of each of these factors, his work on schedules constitutes the most distinctive contribution of his system to our knowledge of the response-reinforcement relation and hence will be singled out here. It should be noted that Skinner also placed great stress upon immediacy of reinforcement in the control of behavior. However, this emphasis was based less upon experimental study of the timing of reinforcement than upon the general observation that reinforcement most powerfully affects the behavior most contiguous to it, as seen in the phenomenon of superstitious behavior.

Before we review the research on schedules, Skinner's concept of reinforcement should be elaborated somewhat. To this point we have discussed reinforcement in terms of stimuli that strengthen responses when presented. But there are many stimuli that strengthen behavior when removed. For example, the removal of an electric shock will ordinarily strengthen any immediately preceding behavior. Skinner used the terms *positive reinforcement* to refer to the strengthening of behavior by stimulus presentation and *negative reinforcement* to refer to the strengthening of behavior by stimulus withdrawal. It is important to realize that the numerous stimulus events commonly thought of as "rewarding" and "punishing" in daily life will ordinarily function as positive and negative reinforcers, respectively, and that all reinforcers are assumed by Skinner to follow the same laws in their relations to behavior. It should be borne in mind, then, that the findings we review here on schedules and other parameters of reinforcement, though derived from study of but a few reinforcers in several lower species, are argued to hold for all reinforcing events in all organisms. For purposes of convenience, our treatment will focus on positive reinforcement, but analogous relationships hold for the case of negative reinforcement.

By schedule of reinforcement is simply meant the way that reinforcement is distributed in relation to behavior. Thus, reinforcement might follow every instance of a given behavior, or only every tenth response, or might occur after some variable number of responses, or might be available every other minute, or after passage of each hour, and so on through an infinite number of possibilities. Experimentally, schedules of reinforcement have been dealt with in terms of broad types of schedules defined by the amount of time or amount of behavior intervening between reinforcements. The research has focused on two questions which we shall consider here in relation to the major categories of schedules: (1) What pattern of behavior is observed while a given schedule of reinforcement is in effect? (2) What pattern of behavior is observed following termination of the schedule and withdrawal of reinforcement?

Continuous Reinforcement (CRF). In CRF, reinforcement is continuously available so that every instance of the behavior is reinforced. We saw the typical effects of this schedule illustrated in Figures 6-4 and 6-5, which show bar pressing behavior of the rat under CRF and following termination of CRF, respectively. The behavioral changes associated with application and withdrawal of CRF are quite abrupt. As the figures show, when CRF is applied behavior quickly increases to a fairly steady but quite moderate rate and then almost as quickly declines sharply in strength when CRF is terminated; however, the latter portion of the extinction period may be marked by occasional bursts of responses alternating with relatively long periods of nonresponse.

The effects of CRF are thus simple and straightforward—behavior is rapidly instated but has relatively little persistence when the schedule is terminated. But in the natural world it is rarely the case that reinforcement follows every instance of a given behavior. In fact a moment's reflection will confirm that nonreinforcement is all too frequently the fate of most complex everyday life behaviors. Of greater import to the understanding of human and animal activity, then, are experiments investigating the effects of *intermittent* reinforcement, i.e., schedules of reinforcement in which a behavior is not reinforced on every occasion. One basic form of intermittency is when reinforcement is available only after a certain interval of time has elapsed, an interval of either fixed or variable duration.

Fixed Interval (FI) Schedules. In an FI schedule each reinforcement is made available a fixed interval of time after the preceding reinforcement. For example, in a three-minute fixed interval schedule (FI 3), a response is reinforced and then reinforcement is withheld from any response emitted in the next three minutes. The first response after the three-minute interval has elapsed is reinforced and reinforcement is again withheld for three minutes, and so on through as many such cycles of reinforcement and nonreinforcement as the experimenter might wish. Note that what is fixed in this schedule is not the interval between actual reinforcements but the interval between points of *reinforcement availability*, since delivery of reinforcement must await the animal's response after the designated interval has elapsed. As we shall see, however, the reinforcement intervals generally come to approximate quite closely the interval of reinforcement availability.

Behavior under an FI schedule characteristically manifests three stages of development, as illustrated in Figure 6-6, which shows the effects of an FI 3 schedule of food reinforcement on bar pressing in the rat. At the outset of training (Figure 6-6A), the intervals between reinforcements are marked by small extinction curves that mimic the extended extinction curve seen in Figure 6-5 following a period of CRF training. Just as many reinforcements instate a behavior that weakens when reinforcement is withdrawn, so each single FI reinforcement produces an increased rate that dwindles during the interval until the next reinforcement. However, as training proceeds these small extinction curves disappear and interval behavior begins to appear at a relatively steady rate, resulting in a fairly straight-line record (Figure 6-6B). Finally, with prolonged FI training the response curves between reinforcements take on a scalloped appearance (Figure 6-6C) precisely the opposite of the form seen in the initial stage of training. What accounts for this reversal of response pattern? A fine-grain analysis of the experimental situation shows the changing pattern to be

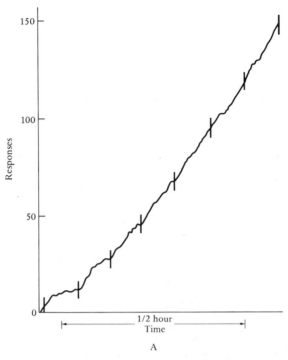

Figure 6-6. Graphs A, B, and C are taken from successive portions of FI 3 training of bar press behavior in a white rat. The vertical strokes in the curve indicate delivery of the food reinforcement. Section A shows behavior at the outset of FI

fully predictable from the basic principles of reinforcement and extinction.

As noted above, the beginning pattern of increasing rate following reinforcement and decreasing rate late in the interval (Figure 6-6A) would be expected under any initial reinforcement and extinction operations, respectively. However, the extinction and reinforcement in an FI schedule are not continuous but obtain only at certain points in the situation. Specifically, the organism is *never* reinforced immediately following receipt of a reinforcement but is always reinforced after some post-reinforcement interval has elapsed. Consequently, if there are differential stimuli unique to the beginning and end of the fixed interval, we could expect the weakening effects of extinction and the strengthening effects of reinforcement to become specific to those points, respectively. And there are such differential stimuli. Immediately after reinforcement there will be the relatively short-lived autonomic, taste, and tactual cues arising from responses

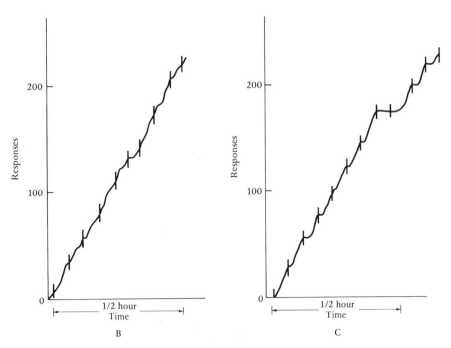

training. The portion of the record shown in section B was taken after about 60 reinforcements, and section C was obtained from the same rat after 17 days of training at 20 reinforcements per day.
(After Keller and Schoenfeld, 1950, p. 84. By permission.)

associated with receipt and ingestion of reinforcement. Bar presses in the presence of these cues are, of course, consistently extinguished in an FI schedule. In contrast, the latter portion of the interval will be characterized by the absence of the reinforcement-induced cues and the buildup of fatigue and kinesthetic cues induced by various "interval-filling" behaviors, e.g., movement around the experimental chamber, premature bar presses. Bar presses in the presence of the latter complex of cues *are* likely to be reinforced. Hence the scalloped pattern of behavior under prolonged FI training is readily accounted for by the empirical principles of reinforcement and extinction. The organism forms a temporal discrimination in the sense that its behavior comes to approximate the actual time of reinforcement. The anticipation is gradual and approximate in character rather than rapid and precise because the cues upon which it is based are of low salience and probabilistic rather than highly discriminable and certain.

Figure 6-7 shows the remarkably similar development of FI behav-

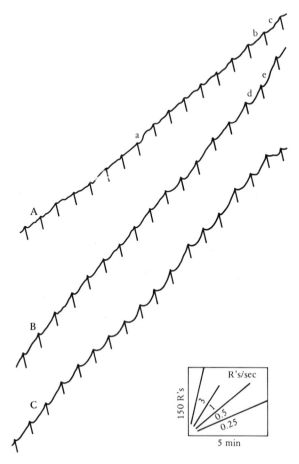

Figure 6-7. Three phases of FI 1 behavior in the pigeon. Reinforcement delivery is indicated by the nearly vertical slash marks on the cumulative records. Records A and B are taken from the beginning and end, respectively, of the second daily training session, and record C is from the fourth session. Note the similarity to performance of the rat as seen in Figure 6-6.
(After C. B. Ferster and B. F. Skinner, *Schedules of Reinforcement*, p. 143. © 1957. Reprinted by permission of Prentice-Hall, Inc., Englewood Cliffs, N.J.)

ior in a pigeon reinforced by access to grain reinforcement at one-minute intervals (FI 1). Occurrence of reinforcement is indicated by the nearly vertical slash marks in the cumulative records. Record A, the second training session, begins with a rate of about 0.4 pecking responses per second, and early in the session one can discern several small extinction curves between reinforcements. A slight tendency to pause after reinforcement appears at points b and c and becomes progressively greater in record B, the end of the second training session. An increase in terminal rate, the rate immediately preceding reinforcement availability, accompanies the development of pause after reinforcement in record B. The terminal rates at points d and e are about 1.25 and 1.5 responses per second, respectively, in contrast to the initial rate of 0.4 per second. In record C, the fourth session on FI 1, the intervals between reinforcement are consistently scalloped with terminal rates reaching 1.7 responses per second.

A somewhat more detailed view of final FI behavior can be seen in Figure 6-8, which shows the performance of a pigeon after 66 hours of training on FI 4. The six curves are from a single training session but for economy of space are taken from the original record and displayed side by side. At point 1 in the figure the response-recording pen moved to the bottom of the recording sheet to continue the cumulative record at point 2. (The beginning of the downward excursion of the pen is visible at point 1.) Similarly, point 4 continues the record interrupted by the resetting of the pen at point 3, point 6 continues the record from point 5, and so on. In general, the bird

Figure 6-8. Record of FI 4 performance of a pigeon after 66 hours of training. Note that in the period just prior to reinforcement availability, individual responses become imperceptible and merge into an essentially straight-line record as a result of the sustained high response rate in combination with the small response step of the recorder.
(After C. B. Ferster and B. F. Skinner, *Schedules of Reinforcement*, p. 159. © 1957. Reprinted by permission of Prentice-Hall, Inc., Englewood Cliffs, N.J.)

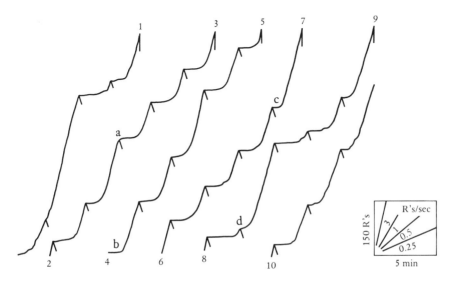

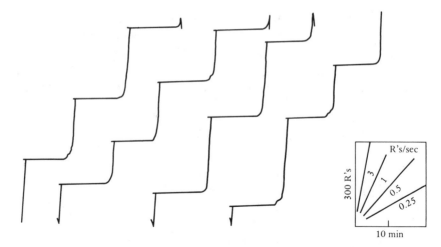

Figure 6-9. Performance of a pigeon after 25 hours of FI 10 training with a light "clock." Reinforcement delivery is indicated by a short horizontal *leftward* movement of the pen.
(After C. B. Ferster and B. F. Skinner, *Schedules of Reinforcement*, p. 269. © 1957. Reprinted by permission of Prentice-Hall, Inc., Englewood Cliffs, N.J.)

pauses about two minutes after each reinforcement and then gradually accelerates its pecking response to a terminal rate that may attain three responses per second. There are exceptions to this pattern—the bird occasionally shifts abruptly to the terminal rate, as at points b and c, or occasionally responds shortly after reinforcement, as at point a—but the scalloped pattern is certainly characteristic.

We have stated that response rate during the inter-reinforcement intervals of an FI schedule is controlled by cues arising from the organism's own behavior, and that the characteristic FI scallop reflects the gradual change from a cue complex associated with extinction at the beginning of the interval to a complex associated with reinforcement at the end of the interval. If this analysis is correct, it should be possible to sharpen the scallop pattern simply by giving the organism a more discriminable cue to how much of the interval has elapsed. Ferster and Skinner (1957) have done just this by training pigeons on FI with an added "clock" in the form of a continuous stimulus change that is perfectly correlated with the passage of time in the inter-reinforcement interval. A slit of light is projected on the response key and is made to lenghten at a constant rate as the interval progresses, reaching maximum length at the time of reinforcement availability. Coincident with the receipt of reinforcement, the slit rapidly contracts to its minimal value and then lengthens again throughout the next interval, and so on through repeated cycles of the FI schedule. Figure 6-9 shows the performance of a pigeon after 25 hours of FI 10 training with such a clock. It is apparent that the clock has eliminated responding during the first 7.5 to 8 minutes of the 10-minute intervals. Responding is essentially confined to the last minute of the intervals, with terminal rates accelerating to well above ten responses per second. We would not expect perfect temporal discrimination by the bird (i.e., zero responses during the interval followed by a single response at the point of reinforcement availability) because the final stages of the continuous light change would be difficult to distinguish from the terminal value and hence there would be anticipatory responding.

That response rate *is* controlled by the "clock" is well demonstrated simply by reversal of the direction of movement of the clock. Record A in Figure 6-10 shows final performance of a pigeon on an FI 3 + clock; it is apparent that the bird exhibits the same sharp temporal discrimination seen in the previous figure. Record B was taken immediately following record A, but the direction of change of the light slit length was reversed at the arrow. That is, from this point on in the training the light slit was projected at maximum length immediately after reinforcement and shrank as the interval progressed, reaching the smallest size at the time of reinforcement availability. The large slit heretofore correlated with the end of the inter-reinforcement interval now occurred immediately after reinforcement, and the small slit heretofore correlated with nonreinforcement now appeared just before reinforcement. It can be seen that the reversal of the clock produced a complete reversal of behavior—response rate over the next several FI segments is now maximal just after reinforcement and minimal just before reinforcement. After several intervals the bird's performance begins to adjust to the new correlation, as seen in the latter portion of record B. Further training with the reversed clock would be expected to produce again the pattern in record A, since the direction of the clock is of course arbitrary.

These clock experiments support Skinner's view that the pattern of behavior generated under ordinary FI, and as we shall see under other schedules of reinforcement as well, can be accounted for strictly in terms of specifiable stimuli in the situation and empirical principles of reinforcement.

Fixed interval reinforcement obtains in everyday human activity in such forms as payment at regular intervals and periodic assessments of performance, e.g., grades in courses. These instances may not display discernible scalloping of performance because ordinarily human

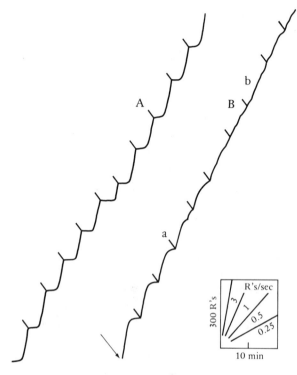

Figure 6-10. The effect of clock reversal on FI 3 performance in the pigeon. Reinforcement is indicated by the small vertical slash marks on the response record.
Record A shows well-developed FI 3 performance under original training with the
clock. Record B was generated following sudden reversal of clock direction, beginning at the point marked by the arrow.
(After C. B. Ferster and B. F. Skinner, *Schedules of Reinforcement*, p. 277. © 1957. Reprinted by permission of Prentice-Hall, Inc., Englewood Cliffs, N.J.)

activity affords opportunity for secondary reinforcers to sustain behavior in the interval and because humans have a variety of devices (cal-
endars, clocks, counting) that enable fine temporal discriminations.
Nevertheless, any college instructor could readily predict the curve
that would result if the daily frequency of student questions was plotted in a course having biweekly quizzes. And when experimental situations are arranged for human subjects that approximate the conditions of animal FI training, the results are similar to those we have
examined. The classic reference experiment in this vein is by Holland
(1957), who studied *visual observing responses*. His human subject
was seated in a small room before a dial. The pointer on the dial
occasionally deviated from a given setting and the subject's task was to

detect that deviation and restore the pointer to its initial setting by pressing a button. However, the room was dark and the subject could see the dial *only* by pressing another button, which flashed a light for a fraction of a second. Pressing this second button is the observing response since only by this means could the subject see the state of the pointer. Detecting and correcting a pointer deflection is the reinforcement in this situation, the reinforcing power of this event probably stemming from the fact that most human subjects are eager to show that they can attain the maximum score (i.e., detect all deflections) in such a situation. Holland had then only to schedule deviations of the pointer at fixed intervals while recording the rate of button-press observing responses to achieve an experiment comparable to animal FI training.

The lower curves in Figure 6-11 show the performance of one of Holland's subjects after several hours on FI 3. For comparison purposes, curves taken from pigeon FI training are presented in the upper portion of Figure 6-11. The similarity of the records is apparent. Like the pigeon, the human subject responded (flashed the light) infrequently just after a reinforcement and was likely to accelerate responses as the interval passed. Incidentally, Holland's experimental setup also has considerable practical significance since it provides a means of investigating conditions affecting the maintenance of vigilence in situations where one must monitor a scene for infrequent events, as in monitoring a radar screen.

What are the effects of an FI schedule of reinforcement upon extinction? Figure 6-12 shows some early data from Skinner's laboratory that illustrate the primary characteristics of FI extinction. Curves A and B in Figure 6-12 are extinction curves for individual rats following a series of daily FI training sessions involving bar pressing for food reinforcement. Curve C shows the extinction for one of these animals following a period of continuous reinforcement (CRF) for bar pressing given prior to the FI training. (The curve is that shown in larger size in Figure 6-5 and discussed in connection with CRF schedules.) Two points emerge clearly from comparison of curve C with curves A and B. First, the FI reinforcement greatly retarded extinction. While extinction after CRF (curve C) is essentially complete following the first session and after a total of about 100 responses, extinction after FI finds both subjects still responding at a fairly high rate after five sessions involving several hundred unreinforced responses. Second, the extinction curves following FI are considerably smoother than that following CRF. We noted earlier that the abrupt fluctuations in

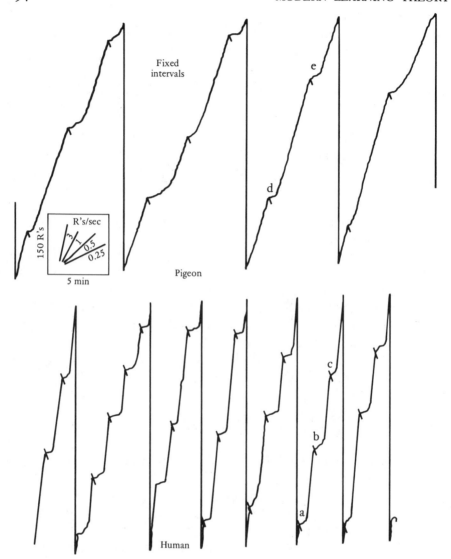

Figure 6-11. Fixed interval performance by a human subject compared with that of a pigeon.
(From Skinner, 1957a, p. 365. By permission.)

response rate during extinction after CRF were attributed by Skinner to emotional reactions stemming from the unaccustomed absence of reinforcement. Skinner argued that since FI training involves repeated instances of nonreinforcement in the context of occasional reinforcements that keep the animal going, the relatively smooth FI curve reflects a learned stability in the face of nonreward.

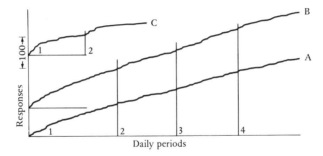

Figure 6-12. Extinction after FI training (curves A and B) compared with extinction after CRF training (curve C).
(After B. F. Skinner, *The Behavior of Organisms: an experimental analysis*, p. 135. © 1938, renewed 1966. Reprinted by permission of Prentice-Hall, Inc., Englewood Cliffs, N.J.)

It would be difficult to overemphasize the significance of Skinner's discovery of the persistence of behavior following intermittent reinforcement, as demonstrated by these data on FI training. The increased resistance to extinction generated by intermittent reinforcement is a very important aspect of normal behavior. Frustration tolerance and persistence in the face of failure are essential properties for effective coping in a world in which success is never guaranteed. In Skinner's words,

> The stability of reflex strength under [intermittent reinforcement] and the prolongation of the extinction curve following it are . . . responsible for a measure of equanimity in a world in which the contingency of reinforcing stimuli is necessarily uncertain. Behavior would be clumsy and inefficient if the strength of an operant were to oscillate from one extreme to another with the presence or absence of its reinforcement. (1938, p. 138)

Of course, strictly speaking, our comparison of curves A, B, and C is not pure since prior training and number of reinforcements differed for the CRF and FI subjects. However, the primary effects discussed here, which, as we shall see, are characteristic of extinction under other schedules of intermittent reinforcement as well, have been amply confirmed (Keller and Schoenfeld, 1950).

Variable Interval (VI) Schedules. In a VI schedule the time between successive points of reinforcement availability is varied in a fashion unpredictable to the organism but around some consistent average value. For example, a five-minute variable interval schedule

(VI 5) might involve inter-reinforcement intervals of from zero to ten minutes at one-minute steps of difference programmed in a scrambled order, e.g., 3, 7, 4, 1, 8, 5, 10, 0, 6, 2, 9. (In a zero-minute interval, reinforcement is made available immediately upon receipt of the last reinforcement.)

Several considerations are involved in setting up a VI schedule. First, most VI schedules will include zero or near-zero intervals in order to maintain moment-to-moment unpredictability of reinforcement. That is, when zero intervals are included in the schedule, the receipt of a reinforcement cannot become a signal that another reinforcement is not immediately available. Beyond the shortest interval, the inter-reinforcement steps can be determined in a wide variety of ways, e.g., the intervals might increase in equal steps as in the above example, or in geometric progression, or by the adding of the preceding two intervals. But the most important specification in a VI schedule is the *average interval between reinforcements*; the number, range, and magnitude of the intervals are usually practical matters limited by the programming equipment and the length of the training session. Finally, the specified intervals must be programmed to occur in an essentially random series that cannot be learned by the subject.

Sustained VI reinforcement characteristically leads to a *steady* rate of response that shows itself as a straight line in the usual cumulative curve. Typical VI behavior is illustrated in Figure 6-13, which shows the key pecking behavior of a pigeon during a nine-hour session on a VI 3 with food reinforcement. The successive curves displayed here were taken from the original record and placed close together for economy of space. It can be seen that the key pecking response was emitted at a remarkably steady high rate throughout the session and regardless of proximity to receipt of reinforcement (indicated by the small slash marks on the response record). This bird emitted approximately 35,000 responses during the session in a virtually continuous stream. There are no breaks or scallops, as we saw in FI behavior, since reinforcement availability is not correlated with temporal or behavioral features of the training situation.

The *absolute* rate exhibited in a food-reinforced VI schedule may be high or low depending chiefly upon the size of the average interval, the range of the intervals employed, and the level of deprivation. But it is the *stability* of the particular rate generated under a given set of conditions that is the distinguishing feature of VI schedules. With reference to natural life situations, one can see a similar behavioral effect whenever a reinforcing contingency operates in an unpredictable temporal pattern. To pursue the analogy we used earlier,

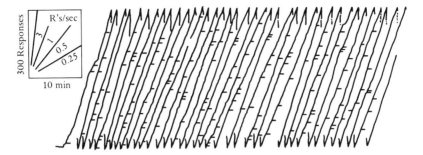

Figure 6-13. Sustained performance on VI during a nine-hour session.
(After C. B. Fester and B. F. Skinner, *Schedules of Reinforcement*, p. 339. © 1957. Reprinted by permission of Prentice-Hall, Inc., Englewood Cliffs, N.J.)

would the reader really question the effects on rate of studying of administering course quizzes on an unannounced, aperiodic basis as opposed to at fixed intervals?

The stable rate generated by VI schedules has proved particularly useful in studying the interaction of respondent and operant behavior. Some observations in this vein will be briefly considered since they indicate how schedules of reinforcement (and Skinnerian methodology in general) can further the analysis of complex processes. The procedure, devised by Estes and Skinner (1941), involves a coupling of food-reinforced VI training and explicit classical (respondent) conditioning of a fear reaction. Ferster and Skinner (1957) implemented the procedure in the following way. A hungry pigeon was reinforced for key pecking on a VI schedule until a stable response rate emerged. Then, while food-reinforced VI training continued, the general illumination in the apparatus was periodically dimmed for 30 seconds with each illumination change followed by a brief, unavoidable shock to the subject through the chamber floor. By the principles of classical conditioning the fear reactions elicited by the shock should become conditioned to the preceding light change. The investigators could then chart the development and strength of the conditioned anxiety, and its interaction with ongoing, organized behavior, in terms of change in rate of key pecking during the light signal itself.

Figure 6-14 shows the development of a lower rate of responding during the light change (CS). The small vertical lines below the response record refer to the onset and offset of the light CS, the offset being coincident with shock delivery. Response rate shows little change during the first CS presentation at the outset of the session, since at this point the light change simply has the status of a novel stimulus.

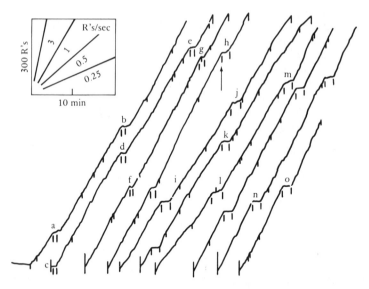

Figure 6-14. Development of response suppression to a preaversive stimulus during VI reinforcement.

(After C. B. Ferster and B. F. Skinner, *Schedules of Reinforcement*, p. 377. © 1957. Reprinted by permission of Prentice-Hall, Inc., Englewood Cliffs, N.J.)

But significant suppression of performance during the CS is evident at point *a* after only one association with the shock. On the fourth and fifth presentations of the CS (points c and d on the record) suppression is complete throughout the 30-second illumination change. At the arrow the duration of the CS was increased to 60 seconds, and after several presentations at this new duration, the VI performance is again completely suppressed throughout CS presentation, as at points n and o in the response record. Note, however, that the normal VI rate is usually resumed immediately after shock delivery and maintained to the next CS onset.

Figure 6-15 shows the effect of this conditioning procedure on lever pressing behavior in the rat. In this case the CS was a buzzer sound of five-minute duration, which occurred in the middle of daily sessions of continuous VI training with food reinforcement. The aversive stimulus was again a shock. The figure shows 14 successive conditioning sessions each with a single CS-shock pairing. It can be seen that lever pressing is almost completely suppressed throughout the latter part of each CS presentation, although nearly all segments show responding for the first few moments of the CS. The latter performance represents a temporal discrimination which forms under such

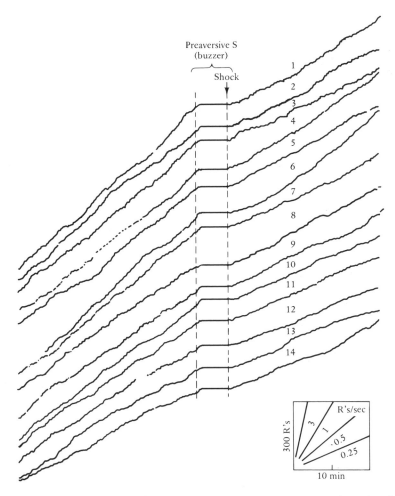

Figure 6-15. Development of response suppression to a preaversive stimulus imposed on VI training in the rat.
(After C. B. Ferster and B. F. Skinner, *Schedules of Reinforcement*, p. 384. © 1957. Reprinted by permission of Prentice-Hall, Inc., Englewood Cliffs, N.J.)

long-duration CS's, since CS onset is never *immediately* followed by shock. Note that the rat performs in essentially the same manner as the pigeon with respect to the relatively stable rates generated under VI training, the suppression of responding during the CS, and the restoration of the behavior following CS cessation. As seen in these studies, VI performance provides a useful tool for analyzing the interaction of emotional states and instrumental behavior. Such interaction is one of the most important clinical manifestations of fear and anxiety, namely, the disruption of ongoing, normal instrumental activities.

With respect to extinction after VI training, Figure 6-16 provides representative data. The figure shows extinction of a food-reinforced key peck response following extensive reinforcement on a VI 7 schedule. The curve is broken into two parts. Between record A and record B there intervened an eight-hour behavior segment (not shown) at a rate of about 250 responses per hour. The extinction session begins with a rate of about 1.5 responses per second, a rate that is indistinguishable from the final reinforced segment. This rate is maintained with but slight decline to point a, resulting in a total of about 14,000 responses to that point. Thereafter, the rate declines to about 0.2 responses per second at point b. Record B shows a brief return of a relatively high rate producing an additional 400 responses before falling off again at the end of the session. The entire behavior covered by Figure 6–16 involved a total of about 18,000 unreinforced responses, and further extinction sessions would doubtless have revealed a number of modest recoveries totaling several thousand additional responses.

It is apparent from examination of Figure 6-16 in relation to Figure 6-12 (showing extinction after FI training) that behavior following VI training is qualitatively more persistent. The reason is straightforward: extinction after VI more clearly resembles the conditioning phase itself. We noted earlier that behavior is more persistent after FI training than after CRF because FI trains the subject to continue responding in the face of periods of nonreinforcement. But we noted also that FI training produces cues to the timing of reinforcement, cues that are fairly quickly disconfirmed as extinction progresses. In contrast, VI training effectively minimizes temporal cues to the imminence of reinforcement; long VI intervals are identical to extinction. Therefore it is only after prolonged responding without reinforcement that the novelty of the extinction condition becomes discriminable to the organism.

FI and VI schedules are *time-based* categories of intermittent reinforcement in that availability of reinforcement is determined solely by the passage of time and nothing the organism does can hasten or deter reinforcement availability. We turn now to consider what may be termed *behavior-based* categories of intermittent reinforcement. In behavior-based schedules, reinforcement becomes available only after emission of a certain number of responses, which may be a fixed or

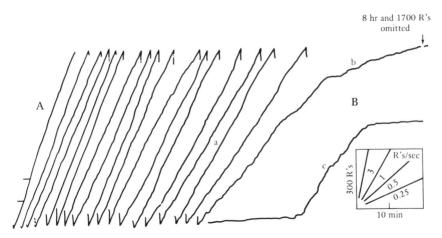

Figure 6-16. Extinction after 255 hours of VI 7.
(After C. B. Ferster and B. F. Skinner, *Schedules of Reinforcement*, p. 348. © 1957. Reprinted by permission of Prentice-Hall, Inc., Englewood Cliffs, N.J.)

variable number, and the passage of time is irrelevant to reinforcement availability.

Fixed Ratio (FR) Schedules. In FR schedules a response is reinforced each time some fixed number of responses have been emitted. The *ratio* refers to the ratio of unreinforced to reinforced responses. A ratio of 30:1 (FR 30) means that the organism is reinforced on every 31st response.

It is generally possible to introduce a fixed ratio of moderate size at the very beginning of training, since the organism will usually repeat a reinforced response at least a few times. For example, as soon as a rat has emitted a few reinforced bar presses, one might begin reinforcing on an FR 10 with confident expectation that stable bar pressing will develop. An important fact is that under such a schedule the rate of responding will soon pick up, as compared with responding on CRF. This is because under an FR schedule, the higher the rate at which responses are emitted, the sooner and the more frequent the reinforcement. As this contingency takes effect and the rate increases, one can again raise the ratio, producing more extended runs of responses and preparing the way for still larger ratios. In this way one can rapidly shape out quite high fixed ratios. Figure 6-17 shows data from an early study of this nature by Skinner. Rats were trained in the usual one-hour-per-day manner, progressing through ratios of 16, 24, 32, 48, 64, and 192 responses to one (the highest ratio permitted

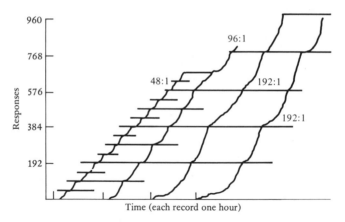

Figure 6-17. Reinforcement at several fixed ratios (rat).
(After B. F. Skinner, *The Behavior of Organisms: an experimental analysis*, p. 288. © 1938, renewed 1966. Reprinted by permission of Prentice-Hall, Inc., Englewood Cliffs, N.J.)

by the apparatus employed). The higher ratios were achieved by some animals after about only two weeks of training.

Figure 6-17 presents cumulative curves for one rat at the three highest ratios. The records generally show a pause after reinforcement followed by a sharp acceleration in rate until the next reinforcement. This pattern is accounted for by two factors. First, as in an FI schedule, responses in the presence of stimuli arising from receipt of reinforcement are never reinforced (at least at higher ratios) and hence drop out. Second, responses separated by short time intervals, i.e., responses within a run of responses, are more likely to be reinforced than those separated by long inter-response times; hence the animal will come to respond in runs, or bursts. The result is a scalloping effect like that seen under FI schedules but with a more abrupt transition from zero to high-rate responding because of the contingency that FR imposes between shorter inter-response times and reinforcement. In addition, the proprioceptive stimulation arising from high-rate responding will, because of its proximity to receipt of reinforcement, acquire secondary reinforcing properties. This stimulation will further promote an "all-at-once" character to FR responding, much in the manner detailed in Hull's analysis of response chaining. That is, as training progresses, responding rapidly creates those very conditions (proprioceptive cues) that have often preceded reinforcement and that therefore tend to perpetuate the responding. (Of course, in FI schedules some responding is necessary to procure reinforcement and thus the same secondary reinforcing mechanism will operate;

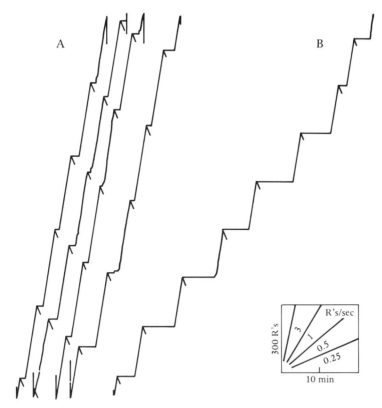

Figure 6-18. Final performance on FR 200 and FR 120.
(After C. B. Ferster and B. F. Skinner, *Schedules of Reinforcement*, p. 52. © 1957. Reprinted by permission of Prentice-Hall, Inc., Englewood Cliffs, N.J.)

however, the mechanism will operate in less direct fashion since high rate responding is clearly illustrated in Figure 6-18. Part A of the up receipt of reinforcement.)

Abrupt transition from pause after reinforcement to sustained high-rate responding is clearly illustrated in Figure 6–18. Part A of the figure shows performance of a pigeon on FR 200 after a history of almost 4,000 food reinforcements on FR's ranging from 50 to 180. The bird pauses after each reinforcement for periods ranging from a few seconds to about a minute and then shifts immediately to a rate of 3.5 to 4.0 responses per second until reinforcement is received. In part B another bird on FR 120 after a similar history shows somewhat longer pauses but the same instantaneous rate change from zero to the terminal rate.

A principle emerging from our review of schedules is that behavior during extinction is largely predictable from the contingencies operating under the preceding schedule of reinforcement. We would therefore expect many responses to be emitted during extinction after FR, far more than following a comparable number of reinforcements on CRF, because the organism has learned to repeatedly work for long stretches without reinforcement. We can also expect that extinction after reinforcement on high ratios will be characterized by long runs of high-rate responding alternating with sudden relatively prolonged breaks. The prolonged breaks follow from the consideration that the animal has almost never been reinforced for *starting off* a run of responses, and so there is little reason to begin again after an interruption. But when responding does begin again, there is reason to continue since reinforcement has often been given to a response that follows closely upon one or more other responses. And after training on high ratios, the longer the run in extinction, the more similar the proprioceptive and related cues become (within some limit) to cues that prevailed when reinforcement was received. Consequently the run continues at high rate.

Figures 6-19 and 6-20 show extinction after FR training in the rat and pigeon, respectively. The curves in Figure 6-19 were generated by individual animals in a one-hour extinction session that followed the ratio training illustrated in Figure 6-17. Consistent with the foregoing analysis, both rats made at least 1,400 responses in the first 20 minutes of extinction. In both curves there are runs of several hundred responses at the rate of 100 or more responses per minute, an extremely high rate for an effortful response like the bar press. Although neither animal is responding at the end of the work session, one should not assume that the effects of the FR conditioning have been eradicated. Further exposure to the situation would doubtless see a renewal of responding, as illustrated in Figure 6-20.

The bird whose extinction curve appears in Figure 6-20 had received extensive reinforcement on various FR schedules, the immediately preceding schedule being FR 170 for about 400 reinforcements. This extinction record reflects a well-developed contingency between high rate and reinforcement. At the outset of the extinction session approximately 1,000 responses occur at the rate of 200 per minute. A short break at point *a* is followed by another run of 800 or 900 responses. Then, despite a nearly three-hour pause as marked in the figure, there is another burst of some 600 responses before the session ends. Further extinction sessions would almost certainly reveal a number of additional such bursts before the behavior returned reliably to the preconditioned level.

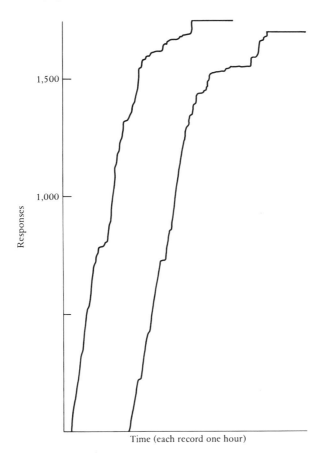

Figure 6-19. Extinction after reinforcement on fixed ratios. The two curves were generated by different animals in a one-hour extinction session that followed the ratio training illustrated in Figure 6-17.
(After B. F. Skinner, *The Behavior of Organisms: an experimental analysis*, p. 294. © 1938, renewed 1966, Reprinted by permission of Prentice-Hall, Inc., Englewood Cliffs, N.J.)

As illustrated in the foregoing figures, the major effect of FR schedules is the production of behavior at high rates and in all-at-once fashion. The analogy which comes most readily to mind in human affairs is the method of piece-rate pay whereby payment is made contingent solely upon the number of work pieces produced or work units completed. Piece rate is known to be an effective means of increasing productivity, at least over the short run. In fact, the very effectiveness of the system in inducing a high rate of work regardless of the toll on the worker has given the term a negative connotation in the working world. Skinner would argue, of course, that the mechanisms at work

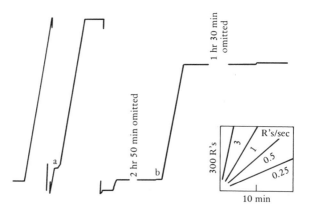

Figure 6-20. Extinction after FR training (pigeon).
(After C. B. Ferster and B. F. Skinner, *Schedules of Reinforcement*, p. 59. © 1957. Reprinted by permission of Prentice-Hall, Inc., Englewood Cliffs, N.J.)

in the high work rates generated under piece-rate schedules in industry are those at work in laboratory FR schedules of reinforcement.

Variable Ratio (VR) *Schedules.* In a VR schedule reinforcement occurs after a number of responses which varies unpredictably from reinforcement to reinforcement but around a fixed average. The schedule stands in the same relation to FR as VI does to FI. As in the design of VI schedules, various systems may be used to generate the numbers of responses in the VR series, but the smallest number is often zero, i.e., reinforcement is given to the very next response after the preceding reinforcement. This assures that stimuli arising from receipt of reinforcement cannot signal the immediate unavailability of further reinforcement.

Like FR, a VR schedule arranges for differential reinforcement of high rates, but unlike FR it also arranges for responding without significant pausing because of the inclusion of low or zero ratios in the series. The result is that very high, sustained response rates may be built up with a VR program. The fully established VR record will resemble the record obtained with a VI schedule in regard to the straight-line quality of the curve, but VR reinforcement usually leads to considerably higher response rates. This is because longer interresponse times are likely to be reinforced in a VI schedule, whereas the subject's delaying a response during a ratio schedule never increases the chances of a reward for the next response.

An instance of the high rates that can be generated with the VR schedule is given in Figure 6-21. This curve was produced by a pi-

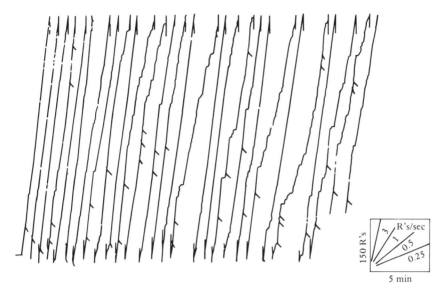

Figure 6-21. Late VR performance showing high overall rate.
(After C. B. Ferster and B. F. Skinner, *Schedules of Reinforcement*, p. 396. © 1957. Reprinted by permission of Prentice-Hall, Inc., Englewood Cliffs, N.J.)

geon reinforced on the average after 360 responses (a VR 360). The specific ratios in the series making up the VR 360 ranged from 1 to 720 responses. The bird had extensive training on the VR 360 as well as earlier training at lower VR schedules. The figure shows a segment from a session in which the overall rate is high. Sustained rates of responding, as in the first curves of the figure, are of the order of seven responses per second, with local rates reaching ten responses per second during 20 to 30 responses. During the segment shown here the bird emitted approximately 30,000 responses in the space of only 43 minutes! Such an extreme rate is not an invariable product of a high VR, and even the performance displayed here was not consistently maintained over all sessions. But that such peaks of prolonged effort can be elicited at all testifies to the very strong control exerted by VR reinforcement on response rate. If one considers, too, the relatively negligible return in actual reinforcement in the above performance (indicated by the slash marks), it may well be questioned whether the organism is not operating at a net loss in terms of the biological cost of the response to the return. In discussing analogies between schedules of reinforcement and human behavior in natural life situations, Skinner has pointed out that the makers of gambling devices and systems always build in a variable-ratio payoff. It is a

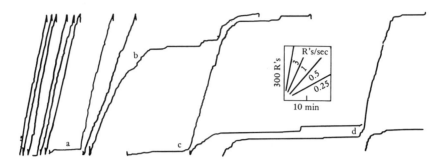

Figure 6-22. Extinction after VR 173.
(After C. B. Ferster and B. F. Skinner, *Schedules of Reinforcement*, p. 412. © 1957. Reprinted by permission of Prentice-Hall, Inc., Englewood Cliffs, N.J.)

compelling comparison, both in regard to the fact that gambling devices do in fact pay off on a variable-ratio basis and in regard to the extraordinary hold that such devices may exert on behavior despite sustained and predictable monetary losses.

In regard to extinction after VR reinforcement, there is not much that can be added to our previous discussions of the relations between schedules and extinction. Again, the principle holds that behavior during extinction preserves the character of the performance shaped under the schedule. This is further illustrated in Figure 6-22, which shows an extinction curve for a pigeon following a history of reinforcement on VR 173 and a number of earlier, smaller VR's. The VR 173 is in effect for the first several minutes of the record, but extinction prevails thereafter. Extinction begins with a run of about 5,000 responses at essentially the same high rate maintained under the VR reinforcement. As discussed in connection with FR extinction, such high-rate unreinforced responding is maintained because it generates the very cues that, during conditioning, were associated with imminent reinforcement. Pauses, as at point a, are followed repeatedly by segments at the beginning of which the bird responds at near the original variable-ratio rate. The session depicted in the figure covers about two hours; further sessions would probably show additional recoveries of the type seen in the latter portion of the figure. While extinction in this instance is relatively rapid, birds given longer training on more extreme VR's have been observed to make as many as 35,000 unreinforced responses before any prolonged break. This is the sort of behavior that, if the reinforcement history were not known, might well be termed pathological.

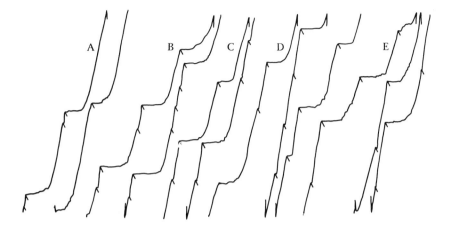

Figure 6-23.
(After C. B. Ferster and B. F. Skinner, *Schedules of Reinforcement*, p. 506. © 1957. Reprinted by permission of Prentice-Hall, Inc., Englewood Cliffs, N.J.)

Figure 6-24.
(After C. B. Ferster and B. F. Skinner, *Schedules of Reinforcement*, p. 510. © 1957. Reprinted by permission of Prentice-Hall, Inc., Englewood Cliffs, N.J.)

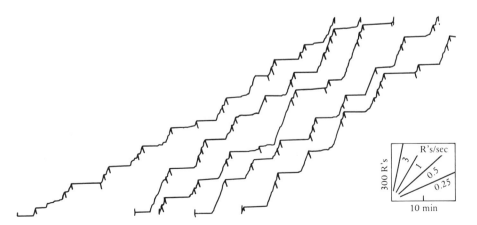

Other Schedules of Reinforcement. In addition to studies of FI, VI, FR, and VR, which are the basic scheduling operations, Skinner and his colleagues also experimented with various ways of combining these operations. A number of these complex contingencies of reinforcement will be briefly reviewed to further indicate the wide range of behavior patterns that can be accounted for by the manipulation of schedules.

Multiple schedules. A multiple schedule consists of two or more alternating schedules of reinforcement with a different stimulus present during each. For example, under a multiple FR FI a pigeon might be required to peck 50 times on a red key to procure reinforcement, at which point the key color changes to blue and the first response after the next ten minutes is reinforced, at which point the red key–fixed ratio schedule might again be in effect, and so on. Schedule changes may be made after each reinforcement, or the same schedule component may be repeated several times before alternation.

Figures 6-23 and 6-24 show multiple schedule performances of a pigeon and a rat, respectively, following extended training on the schedule with food reinforcement. The schedule components were correlated with different key colors for the pigeon and with different patterns of apparatus illumination for the rat. At this point the reader should pause and carefully examine the curves in the two figures. Drawing upon our previous discussions of the behavior patterns generated under various contingencies of reinforcement, the reader should have little difficulty in telling, simply by inspecting the figures, how many schedule components were involved for each subject, what the components were, and precisely when each was in effect. That is, the behavior displayed here tells us quite clearly what the environment is! That such diagnoses can be made with confidence attests to the validity of the principle that schedules of reinforcement account for a good deal of the patterning in behavior.

In confirmation (it is hoped) of the reader's perception, it should be noted that both subjects were involved in schedules that alternated between two components—FI and FR, with from one to three repetitions of each component between alternations. The curves of Figure 6-23 were taken from various points of training on multiple FI 5 FR; the ratio sizes ranged from FR 50 in record A to FR 300 in record E. Figure 6-24 depicts performance under multiple FI 5 FR 40 throughout. In both figures the scallop pattern peculiar to FI reinforcement and the all-or-none rate peculiar to FR reinforcement show

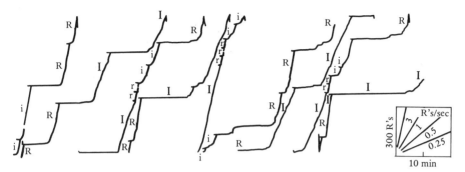

Figure 6-25. Performance of a pigeon after extended training on FR 50 FR 250 FI 2 FI 11.
(After C. B. Ferster and B. F. Skinner, *Schedules of Reinforcement*, p. 554. © 1957. Reprinted by permission of Prentice-Hall, Inc., Englewood Cliffs, N.J.)

exactly when each schedule component and its correlated stimulus were in operation.

Figure 6-25 shows the performance of a pigeon following extensive training on a four-component multiple schedule involving FR 50 FR 250 FI 2 FI 11. The longer intervals are marked I in the figure, shorter intervals i, larger ratios R, and smaller ratios r. The three larger-ratio segments near the start of the session show typical all-or-none high rates along with relatively long pauses after reinforcement (indicated here by the small horizontal leftward extensions from the curve). Next there is a block of longer-interval segments with zero response rates in the first half of each interval followed by acceleration of rate to the point of reinforcement. Two smaller-ratio segments follow at high rate and without pause, and then two shorter-interval segments with typical scalloped patterns, and so on throughout the remainder of the session. In general, the bird demonstrates four separate performances under the control of four key colors. Skinner (1957a) has reported obtaining as many as nine different performances under the control of nine different patterns on the key.

The dependence of multiple schedule performance on the correlated stimuli is nicely revealed in experiments by Ferster and Skinner (1957) involving reversal or sudden change in these stimuli. They found that if in the middle of fixed interval performance the key color is suddenly changed to the color consistently associated with ratio reinforcement the subject immediately begins to emit responses at a high rate. Similarly, interpolation of an interval key color during a ratio performance leads to immediate abandonment of high rate and assumption of the characteristic interval pattern. These observations show that just as a simple response may be switched on or off by the presentation of stimuli with reinforcement or extinction histories, so a *behavior pattern over time* may be switched on or off by the manipulation of stimuli correlated with different schedules of reinforcement.

Mixed schedules. Mixed schedules are the same as multiple schedules except that no external stimuli are correlated with the components. For example, under a mixed FR 100 FI 10 the subject is reinforced either on FR 100 or on FI 10 in random order under a constant external stimulus situation.

While mixed schedules provide no external cue to the reinforcement contingencies operating at any moment, they give rise, like any other schedule, to correlations between behavior-generated stimuli and the receipt of reinforcement, and these stimuli in turn may then become the basis for behavior patterns appropriate to the reinforcing contingency in operation. The directing role of response-produced stimuli in schedules of reinforcement is nicely evidenced in the mixed schedules phenomenon of priming. Priming is illustrated in Figure 6-26, which shows performance of a pigeon under a mixed FR 30 FR 190 at intermediate (part A) and advanced (part B) stages of training. Considering part A first, note that when the smaller-ratio schedule is in operation (as indicated by the short distances between successive reinforcement markings), responding occurs at a high, sustained rate between reinforcements. However, when the larger ratio is in effect, there are pauses in responding at points which correspond roughly to completion of the smaller ratio. Each long segment between reinforcements contains at least one of these pauses and sometimes as many as three, as at point a. It is obvious that the bird's own behavior is the stimulus for these pauses; the emission of approximately the number of responses in the smaller ratio "primes" the pause. With additional training (record B) the shorter runs under the larger ratio prime not only a pause but the subsequent sustained responding necessary to procure the reinforcement. To put the matter loosely, making approximately the number of responses in the smaller ratio "tells" the bird which schedule is in effect.

Figure 6-27 shows a similar behavior pattern in the rat following successive stages of training (records A through D) on a mixed FR 20 FR 160. Again, it is apparent that when the larger FR is in operation, emission of approximately the number of responses in the smaller ratio primes the rat into a pause and then into the sustained high rate

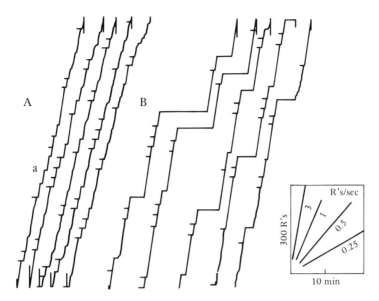

Figure 6-26. Performance of a pigeon at intermediate (part A) and advanced (part B) stages of training on mixed FR 30 FR 190. Priming is illustrated by the pauses in responding during the larger ratios.
(After C. B. Ferster and B. F. Skinner, *Schedules of Reinforcement*, p. 581. © 1957. Reprinted by permission of Prentice-Hall, Inc., Englewood Cliffs, N.J.)

Figure 6-27. Priming in the rat as manifested at successive stages of training on mixed FR 20 FR 160.
(After C. B. Ferster and B. F. Skinner, *Schedules of Reinforcement*, p. 591. © 1957. Reprinted by permission of Prentice-Hall, Inc., Englewood Cliffs, N.J.)

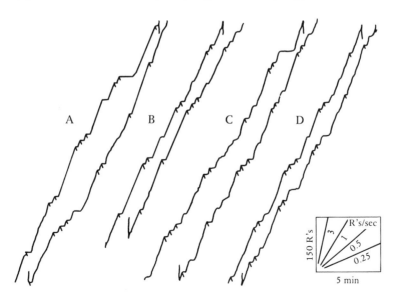

appropriate to the larger ratio. Similar priming obtains in mixed schedules with other components. For example, following a reinforcement on a mixed FI FR one might see a run of responses approximating the FR schedule which, if the FI is then in operation, primes the organism into a slower and perhaps scalloped response pattern suited to the FI component.

Tandem schedules. A tandem schedule is one in which reinforcement is programmed by two schedules acting in succession in fixed order and without correlated external stimuli. For example, in tandem FI 45 FR 10 the tenth response counted after 45 minutes has elapsed is reinforced, and this reinforcement requirement is repeated indefinitely. Since reinforcement is not given after completion of the first schedule component, there is no opportunity for response-produced stimuli in this component to develop a priming function, as in a mixed schedule. Instead, the behavior pattern that emerges under a tandem schedule generally takes the form of a composite of the patterns constrained by each component. For example, in tandem FR 350 FI 5 seconds, the FI component, while slight, will nevertheless considerably modify the differential reinforcement of high rate associated with the FR component. The result will be a relatively high sustained rate between reinforcements but a rate considerably lower than under FR 350 alone, because a high rate just prior to reinforcement is not selectively reinforced. Similarly, a tandem FI 10 FR 5 would likely generate the characteristic FI pattern because of the relatively large FI component, but the overall rate of responding would be higher than under FI 10 alone because of the differential reinforcement of high rate associated with the terminal FR component.

Figure 6-28 shows the pattern exhibited by a pigeon after extended training on tandem FI 45 FR 10 with a 15-minute "time out" following each reinforcement. (Time out refers to an experimenter-produced period of nonresponding and is usually brought about simply by turning off all light in the apparatus. The idea is that the stimuli arising from the subject's own behavior just before a time-out period will no longer be effective when the schedule resumes, and hence the contribution of such stimuli to schedule performance can be assessed by imposing the time out. The effect of the time out on the performance in Figure 6-28 was to extend somewhat the pause at the beginning of the intervals.) In general, Figure 6-28 shows a prolonged pause after reinforcement followed by a smooth overall acceleration in rate to a terminal rate which is then maintained with little varia-

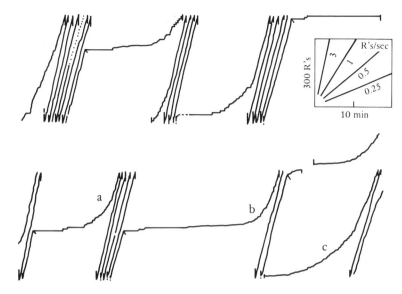

Figure 6-28. Performance on tandem FI 45 FR 10 with a 15-minute time out. (After C. B. Ferster and B. F. Skinner, *Schedules of Reinforcement*, p. 425. © 1957. Reprinted by permission of Prentice-Hall, Inc., Englewood Cliffs, N.J.)

bility until the next reinforcement. The effects of the FI and FR tandem components are apparent in the prolonged scallop and sustained terminal rate, respectively.

Chained schedules. In a chained schedule two or more components occur in a constant order, each correlated with a distinctive external cue but with only the last schedule component eventuating in primary reinforcement. For example, in chain VI 3 FR 10 the first response after three minutes produces a change in, say, key color and at the same time starts a ten-response countdown to food reinforcement. Since the cue that is present throughout the terminal component will be consistently associated with receipt of a primary reinforcer, it should acquire secondary reinforcing power and hence function to develop and maintain behavior during the preceding component schedule. Chained schedules, then, enable analysis of the development and operation of secondary reinforcement in extended patterns of behavior.

Figure 6-29 illustrates the typical outcome of experiments on chained schedules. The figure shows final performance of a pigeon on chain FI 7.5 VI 3. The second component was maintained at VI 3 throughout the experiment. The first component, which was

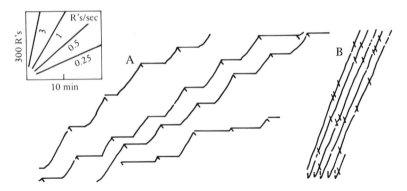

Figure 6-29. Final performance on chain FI 7.5 VI 3. Part A shows performance under the FI (first) component, and part B shows behavior under the VI component.

(After C. B. Ferster and B. F. Skinner, *Schedules of Reinforcement*, p. 665. © 1957. Reprinted by permission of Prentice-Hall, Inc., Englewood Cliffs, N.J.)

reinforced solely by the change to the stimulus appropriate to the second component, began at FI 2 but was increased as training progressed to FI 3.5, FI 5, and FI 7.5. Part A of the figure shows performance on the first (FI) component, and part B shows behavior under the VI component. The reinforcement lines through the curves of part A mark only change in key color and activation of the VI schedule, whereas the reinforcement lines in the curves of part B indicate delivery of food (as well as change in key color and activation of the FI schedule). As could be confidently expected, responding on the food-reinforced VI component displays the steady rate characteristic of VI reinforcement. But it is to be particularly noted that behavior under the secondarily reinforced FI clearly resembles the pattern we have repeatedly seen under food-reinforced FI, except for somewhat longer pauses after reinforcement and somewhat sharper acceleration of rate. These data indicate that the relationships between schedules of reinforcement and patterns of behavior, which we have been describing in relation to primary reinforcers, also hold for secondarily reinforcing stimuli.

Differential reinforcement of rate. Our analysis of schedules has noted that the response rate at the moment of reinforcement is an important variable. For example, a response that follows a pause is more likely to be reinforced in FI than in FR, and this relation contributes to the characteristically lower rate under FI. While schedules make likely certain rate-reinforcement relations, they do not guarantee a particular rate at the moment of each reinforcement. However, it is possible to arrange more direct control over rate within a schedule simply by making reinforcement contingent upon attainment of a rate criterion.

In *differential reinforcement of high rate* (or drh, as it is termed) rate is directly controlled by requiring that the response to be reinforced come within some specified time of the preceding response. The time requirement is ordinarily chosen on the basis of the prevailing average inter-response time (IRT). For example, suppose a rat is responding under a VI 1 schedule at the rate of 30 times per minute. The average IRT is then two seconds, with some pairs of responses separated by more than two seconds and some by less. The drh would be imposed by reinforcing the subject on the same VI schedule but now requiring in addition that the response to be reinforced come within two seconds of the response that preceded it. If this did not happen, reinforcement would be withheld until a pair of responses did occur within the specified period. The direct reinforcement of this IRT would soon reduce the *average* IRT and hence increase the number of responses per minute. The new average IRT could then be used to set up a new rate requirement at the moment of reinforcement, and so on. Under such a procedure, a considerably higher rate of responding would be built up than in the VI schedule alone. Under some conditions, the drh procedure will develop extremely high rates. Skinner (1957a) reports that by the differential reinforcement of high rates, pigeons have been made to respond as rapidly as 15 responses per second. At such high rates, however, considerations of response topography impose a limiting factor. That is, the subject's responses become so brief and minimal that it is not clear that they can be properly compared with the responses emitted at lower rates.

It is also possible to bring response rate to very low levels by similar procedures. In *differential reinforcement of low rate* (drl), responses are reinforced only when they are farther apart in time than a specified criterion, which is usually based on the prevailing average IRT. Since responses emitted at higher-than-criterion rate are consistently extinguished, the behavior gradually declines in rate until a relatively high proportion of responses meet the IRT criterion for reinforcement. The IRT requirement may then be lengthened to bring about a further drop in rate. As the IRT criterion is progressively lengthened, a point would inevitably be reached where the decreasing number of reinforcements per unit time would not be sufficient to maintain the behavior. Essentially, the animal would be operating at a loss, and the behavior could show signs of disappearing altogether. At this point the IRT requirement may be shortened somewhat until the rate recovers and the behavior stabilizes. By so adjusting the IRT requirement, it is possible to bring about very low rates while yet maintaining the behavior. Extremely persistent low-rate responding has been generated under such drl training. Skinner (1957a) reports that by appropriate adjustment of the reinforcement criterion, it was possible to keep a pigeon responding continually at IRT's in the range of several minutes for 1500 hours, i.e., 24 hours a day, seven days a week for approximately two months![4]

There remains a great deal that could be said about schedules of reinforcement. Many other scheduling operations are possible, and many questions invite study. What behavior patterns are generated under other or more extended combinations of schedule components? What are the behavioral effects of varying interval size or ratio size? What behavior is observed when a subject is allowed to choose between concurrently operating schedules of reinforcement? What happens if we arrange zero correlation between behavior and reinforcement?[5] Clearly, there is a wealth of data to be derived from the study of scheduling operations, and contemporary theoretical analyses indicate the complexity of this research area (Schwartz, 1978). For the present, the aim has been to describe the research on schedules in relation to Skinner's system, and the observations we have discussed are drawn in the main from Skinner's own pioneering work (1938) and his studies in collaboration with Ferster (1957). There can be no gainsaying the merits of that work as a distinctive and significant contribution of descriptive behaviorism. While other theorists had treated reinforcement and extinction as discrete operations, Skinner showed that the *patterning* of reinforcement and extinction significantly influences the form and persistence of behavior. In this vein, Skinner's own research delineated a number of new behavior-reinforcement relationships that reliably appear over a wide range of species and response systems, as we have illustrated here. These data suggest that schedules of reinforcement may account for a good deal of the variation in the patterning and persistence of everyday behaviors, variation that all too often is attributed to vaguely conceived motivational processes.

The Remaining Descriptive Principles

We have discussed Skinner's principle of reinforcement and related phenomena in some detail because this principle is so central and distinctive to his system. The remaining principles of descriptive behaviorism can be dealt with more briefly because we have met very similar principles in other theoretical contexts. Our primary purpose here will be to see how Skinner defined and interpreted these concepts within his system.

Secondary Reinforcement.

A stimulus which is not originally reinforcing can become reinforcing through association with one that is.

This principle is scarcely of less importance than the principle of reinforcement itself since it implies an extension of that principle throughout the S-R realm. The concept of secondary reinforcement is of particular importance in extending the reinforcement analysis to the human level since the bulk of human learning occurs in the absence of "obvious" primary reinforcers. Instead, the rewards for human learning are likely to take such forms as the approval of others, monetary returns, and sheer information acquisition, all of which, Skinner maintained, have derived their reinforcing power from earlier association with, or instrumentality to, some type of primary reinforcement.

A typical Skinnerian demonstration of secondary reinforcement is seen in Figure 6-30. The bar press data in this figure were generated by rats following the usual preliminary training to approach and eat out of the food dish upon sound of the food magazine operation. During the preliminary training the bar was absent and each animal received 60 combined presentations of sound and food (pellet delivery). The distinctive feature of the bar press conditioning is that, while the lever was inserted and connected so that each press operated the magazine, *no food was delivered at any time.* Figure 6-30 shows that pressing increased markedly even though presses were followed only by the sound of magazine operation. (Compare with Figure 6-4, which shows acquisition of the bar press response under food reinforcement.) It is this power to condition new behavior that marks the magazine sound as a secondary reinforcer. Since the primary reinforcer (food) is never forthcoming, the power of the magazine sound to strengthen and maintain the bar pressing becomes progressively weaker

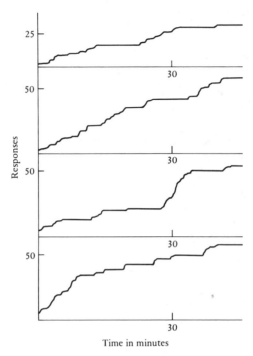

Figure 6-30. Bar press records generated by four animals under conditions of secondary reinforcement and upon their first exposure to the bar. Each bar press produced only the sound of food-magazine operation, but in a previous session the sound of magazine operation had been followed by the delivery of food. Here the sound is seen to exert a considerable reinforcing effect for each subject even though no food is given.
(After B. F. Skinner, *The Behavior of Organisms: an experimental analysis*, p. 83. © 1938, renewed 1966. Reprinted by permission of Prentice-Hall, Inc., Englewood Cliffs, N.J.)

as the session proceeds and is effectively dissipated after about a half hour. But it is possible, of course, to produce much stronger and longer-lasting secondary reinforcers within the operant methodology by manipulating such training conditions as the number and scheduling of associations between the primary and secondary reinforcing stimuli. We have already noted, for example, that secondary reinforcing stimuli conditioned under chained schedules of reinforcement will have behavioral effects similar to those produced under schedules of primary reinforcement. In general, Skinner assumed that all the laws of primary reinforcers would hold for secondary reinforcers as well.

The data of Figure 6-30 demonstrate secondary *positive* reinforcement, but secondary *negative* reinforcers may also be created by correlating neutral stimuli with noxious events. In the experimental study

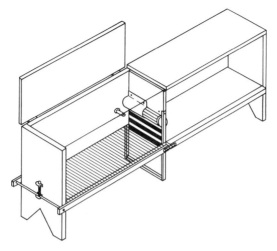

Figure 6-31. The apparatus used by Miller (1948) to study learning based upon conditioned aversive stimuli. The left compartment is painted white and has a grid floor for administering electric shocks. The right compartment is painted black and has a smooth solid floor; it is therefore the "safe" compartment. The door separating the compartments (painted with horizontal black and white stripes) drops out of the way when the rat turns the wheel or presses the lever.
(From Miller, 1948, p. 90. Copyright 1948 by the American Psychological Association. Reprinted by permission.)

of anxiety, discussed in connection with variable interval schedules, we saw that a cue which regularly preceded a brief, inescapable shock during VI conditioning acquired the power to suppress an ongoing operant much as the operant was suppressed by the shock itself. Would such a cue also be found to strengthen operant behavior by *removal*? Hefferline (1950) produced supporting evidence within the operant methodology by showing that bar pressing in rats could be maintained solely by removal of a cue that regularly preceded onset of an intense aversive light. But the classic study of conditioned negative reinforcement was conducted by Miller (1948) using the apparatus shown in Figure 6-31, and it is instructive to consider here because it affords quite contrasting interpretations of the phenomenon.

Miller first gave his subjects (rats) a series of escape conditioning trials by placing them in the "grid" side of the apparatus with the door to the "safe" compartment open and the grid electrified. The animals quickly learned to make rapid escape responses into the safe compartment. During this phase of training, the distinctive cues of the shock compartment (white walls and grid) should acquire secondary negative reinforcing power as a consequence of their association with shock. Each animal then received a second series of placements in the grid side, *but now with the door to the safe compartment closed and the shock turned off.* During these trials, the animal could simply remain in the grid compartment without ever receiving a shock; however, if the animal turned a wheel located above the closed door or pressed a bar mounted nearby, the door would open, allowing the subject access to the safe side. Miller found that under these conditions the majority of his subjects acquired, first, a stable wheel turning escape response, and when that response was rendered ineffective by the experimenter, a bar pressing escape response as well. Moreover, these responses showed characteristics of behavior acquired under primary reinforcement, specifically, decline in latency with repetition and extinction curves when no longer effective in opening the door. Miller's explanation of this new learning proceeded from the assumption that a state of fear was conditioned to the cues in the shock compartment during the initial escape training with shock. In the second phase of training, these cues elicited the fear state, which then functioned as a drive to motivate a variety of responses, including striking the wheel or lever, with consequent opening of the door. Movement into the safe side was then followed by reduction of the fear drive, which served as the crucial reinforcing event for acquisition of the wheel turning and bar pressing escape responses. As in Hull's theory, then, Miller's explanation assumes (1) a central drive state that motivates behaviors in the learning situation and (2) a drive reduction process that selects and strengthens particular responses. A Skinnerian interpretation, in contrast, appears either oversimplified or elegant, depending upon one's theoretical biases. It avoids, of course, any reference to intervening states or causal chains; the new escape behaviors are accounted for solely by the statement that they are responses followed by removal of secondary negative reinforcers. That is, the animal emits responses which result in its removal from the cues in the shock compartment, and by the principle of secondary reinforcement, removal of these cues strengthens whatever responses happen to precede removal.

Secondary reinforcement and behavior chaining. Like Hull, Skinner saw secondary reinforcement as making possible the establishment of extended behavior sequences whose final members only are reinforced in primary fashion. Such behavioral organization develops, Skinner maintained, because stimulus components of the series acquire secondary reinforcing power, which then operates both to extend the sequence and to integrate the component behaviors into a cohesive whole. The details of the analysis are essentially identical to Hull's account of response chaining and need not be repeated here (see pp. 84–87). Skinnerians, however, were more prone to test the analysis by attempting to set up novel response chains in lower animals. The Barnabus demonstration (Figure 6-32) is a good example of these efforts.

Barnabus, a white rat, was trained by two Barnard College psychologists to run off a complex series of manipulative and locomotor responses which included, among other behaviors, lowering a drawbridge, operating a "hand car," taking an elevator, and raising the college flag! The authors have described the major features of their demonstration:

> Barnabus' demonstration box looks a little like a doll's house, four feet high, with four floors, constructed from aluminum and transparent plastic, open at the front.
>
> The rat's [behavior chain] begins when he is dropped into the box on the first floor. A light goes on, the starting signal which initiates his series of responses. His first move is to mount a spiral staircase to a platform. . . .
>
> Now Barney runs to another platform, by pushing down a raised drawbridge and then crossing it. From here he climbs a ladder, summons a car by pulling an attached chain hand-over-hand, pedals the car through a tunnel, climbs a flight of stairs, runs through a tube, steps into a waiting elevator, and celebrates his progress by raising a Columbia flag over it. . . .
>
> The raising of the flag starts the elevator, on which Barney now triumphantly descends to the ground floor. There, at the sound of a buzzer, Barnabus rushes over and presses on a lever, for which action he receives a tiny pellet of food. When the buzzer stops, he whisks around facing the spiral staircase at the place where the chain of responses began. He is ready to go again.
>
> This outline of Barnabus' routine does not tell all that is involved. On the upper levels of the box, there is a platform on each side, and the rat must surmount some obstacle to get across from one to the other; he must perform some other act (climbing) to get to the next floor. (Pierrel and Sherman, 1963, p. 8–9)

In setting up this behavior chain, Barnabus was first given the familiar lever pressing training for food reinforcement in the presence of a continuous buzzer sound. The following stages of training are succinctly described by the experimenters.

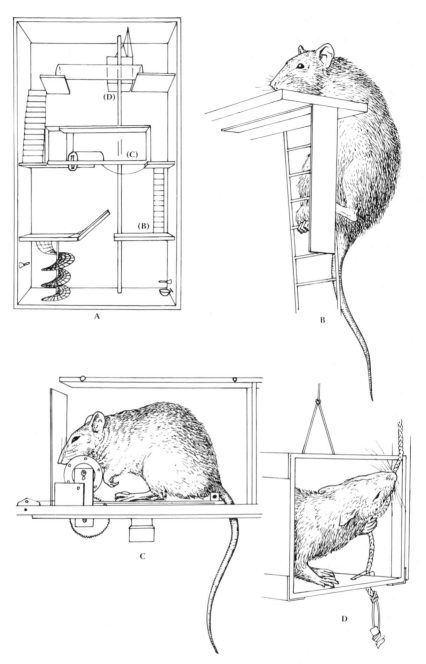

Figure 6-32. The Barnabus demonstration. Panel A shows a schematic of the demonstration apparatus. The behavior chain begins at the lower left and proceeds through the three flights, terminating with descent in the elevator to the lever at the lower right of the apparatus. The remaining panels show Barnabus executing various segments of the chain: (B) climbing a ladder between floors; (C) operating the "railroad car"; (D) raising the flag prior to descent in the elevator.

(From Pierrel and Sherman, 1963, p. 10. By permission. After photos from *The New York Times*, Feb. 19, 1958.)

Thus far Barnabus had learned to press the lever while the buzzer was on. It was then necessary to train him NOT to press the lever in the absence of the buzzer. The signal was terminated and remained off, until the rat stopped pressing and turned away from the lever. Then the buzzer was turned on again, his first lever-press was rewarded, and again the signal was terminated. This alternation was continued (rewards for lever-press when the buzzer was on, versus no reward for presses made during silence, when there was no signal). Finally Barney learned the discrimination and pressed only on signal. . . .

Next, Barnabus was placed in the elevator at the top of the box and rewarded with the buzzer for riding down to the first floor. Initially, he would jump out of the elevator as soon as he was placed in it. However, he was rewarded with the signal only when he rode to the bottom; the non-rewarded behavior of prematurely jumping from the elevator disappeared. This new response was trained by using the secondary reward of the buzzer rather than using food again. Now, when placed in the elevator, Barney would ride down the full three flights. His riding was rewarded with the buzzer. At the sound of that signal, he would run to the lever, depress it, and finally collect his pellet of food. . . .

Once Barnabus would ride down in the elevator fairly well, we introduced flag-raising. We wished to have him raise the flag by pulling on the chain inside the elevator cage. We "told" Barnabus what we wanted him to do by placing him in the elevator and waiting until he made some response to the chain. His first response to it was a sniff. This was immediately rewarded (secondary reward) by starting the elevator ride down, then by sounding the buzzer as he reached the first floor.

Since sniffing at the chain is far short of pulling on it hard enough to raise the flag, the next time that he was replaced in the elevator he was required to touch the chain. Barney did this first by sniffing and then by chewing it. On this trial he was rewarded for "chain-chewing" by the downward elevator ride. On successive trials we required successively more of the final response toward which we were training—a strong, sustained pull on the chain. On later trials we rewarded in succession: a chewing movement strong enough to pull on the chain slightly; chewing, plus holding the chain between his forepaws; a pulling at the chain with the teeth and forepaws. Gradually, his responses to the chain became strong enough to move the flag upward slightly; we continued this process of demanding a little more on each successive trial until the desired response was complete. (p. 10–11)

Continuing in this fashion, the experimenters added the remaining links of the behavior chain in backward order one at a time. Several of the additional links required a shaping process much like that accorded flag raising before the desired behavior was fully developed. The reinforcement for the conditioning of each new link of the chain was the production of the stimuli (secondary reinforcers) associated with the *following* (i.e., previously established) link in the chain, just as flag raising was rewarded by starting the elevator, and elevator riding was rewarded by the sounding of the buzzer. Finally, when all training was complete, exposure of the animal to the most remote stimulus (the starting light) was sufficient to immediately produce a rapid and smooth execution of the entire chain of behaviors.

Again, the point of establishing such demonstrations is a serious one. If the principle of secondary reinforcement can be used to create at will such novel behavior repertoires in the laboratory, can we not assume the principle also functions to produce behavior chains in natural life situations as well?

Generalization and Discrimination. Keller and Schoenfeld (1950) have provided definition of these phenomena from the operant viewpoint.

> *Generalization: An increase or decrease in the strength of one reflex, through reinforcement or extinction, is accompanied by a similar but smaller increase or decrease in the strength of other reflexes that have stimulus properties in common with the first.* (p. 116)

> *Discrimination: A reflex strengthened through generalization may be separately extinguished, so that an organism responds to one stimulus and fails to respond, or responds less strongly, to another.* (p. 117)

A study by Pierrel (1958) nicely illustrates the chief features of both generalization and discrimination. Following the usual preliminary training to bar press for food, rats received daily two-hour placements in the Skinner box in the presence of a continuous tone. The intensity (loudness) of the tone alternated every several minutes between two levels 40 decibels apart. Food reinforcement for bar pressing was available on a variable interval schedule throughout presentation of one intensity level (in Skinnerian terminology, the discriminative stimulus, or S^D), but reinforcement was withheld from all bar presses occurring during presentation of the other intensity (the S delta stimulus, or S^Δ). The reinforcement of an emitted behavior in the presence of the S^D coupled with extinction of that behavior in the presence of the S^Δ is, of course, simply discrimination training adapted to the procedures of the Skinner box, and after several weeks of such training the animals developed a much higher rate of bar pressing in the presence of the S^D than in the presence of the S^Δ. At this point a second phase of training was begun. The daily sessions were one hour in length and again involved alternating periods of the S^D and S^Δ (40-decibel change) stimuli and continued discrimination training. The S^D intensity was that used previously, and the VI schedule of reinforcement was continued in its presence as before. However, three *new* S^Δ stimuli were now introduced along with the old S^Δ intensity. The new S^Δ's were intensities 10, 20, and 30 decibels removed from the old S^Δ in the direction of the S^D. The overall sequence of intensity presentations was randomized and counterbalanced, that is, an S^Δ intensity was as often followed by another S^Δ intensity as it was by the S^D intensity. The subjects received 18 days of training under phase 2 procedures.

Figure 6-33 shows the number of bar presses per session as a function of tone intensity level during the beginning, middle, and final stages of phase 2 training. Let us first consider performance in the presence of the S^D (o-decibel) and initial S^Δ (40-decibel) stimuli. While it is clear a discrimination had been established by days 1–6 of the phase 2 training (largely as a result of phase 1 training), it can be seen that the further discrimination training of phase 2 considerably enhanced control of bar pressing by the S^D as compared to the S^Δ. On days 1–6 the ratio of S^D to S^Δ responses is about 7 to 1, whereas over days 13–18 the ratio is about 56 to 1. It should be noted that this improvement resulted from a lowering of response rate in S^Δ while the rate in S^D remained about the same. That is, these data indicate that the discrimination learning proceeded not by strengthening of response in S^D but by extinction of generalized responses in S^Δ. The idea that the major problem in establishing discriminative control is likely to be eliminating generalized reinforced behavior has been emphasized in Skinnerian analyses of the phenomenon (Keller and Schoenfeld, 1950).

The curves of Figure 6-33 provide a fairly complete picture of response strength to the S^D and S^Δ over the final stages of discrimina-

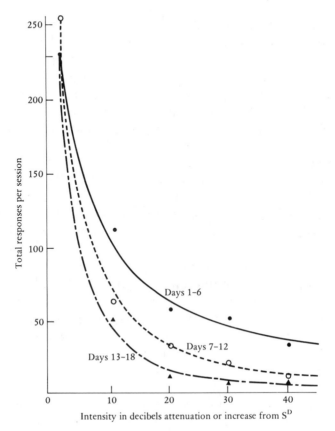

Figure 6-33. Total responses per session as a function of the auditory stimulus intensity level. Each point represents the mean of three animals taken over a block of six training days.
(From Pierrel, 1958, p. 307. Copyright by the Society for the Experimental Analysis of Behavior, Inc. Reprinted by permission.)

tion training, a picture that could easily be extended throughout the earlier stages of training to yield a larger family of curves of progressively decreasing slope. In addition, these curves capture progressively smaller changes in response strength to the S^Δ, which leads to an apparent asymptote of some small but detectable response strength in the presence of the S^Δ's. In these respects of continuity and sensitivity of measurement, the operant technique is superior, Skinner argued, to traditional approaches to the study of discrimination learning, such as the T-maze choice procedure, which provide only an all-or-none measure of response to the reinforced and nonreinforced stimuli.

Another feature of the operant approach is pertinent to consideration of the generalization test in Pierrel's study. Experimenters typically measure generalization of a conditioned response by presenting a series of test stimuli (stimuli other than the conditioned stimulus, or S^D) while withholding reinforcement in order to secure an uncontaminated measure of generalized response strength. Early studies of generalization frequently encountered the problem of maintaining a stable conditioned response throughout the test series, a problem exacerbated by the circumstance that the conditioned response was usually a continually reinforced response and therefore prone to rapid extinction. The operant approach affords an effective means of addressing this problem. We have noted that Pierrel reinforced her subjects in S^D according to a VI schedule, a schedule known to produce a stable response rate and high resistance to extinction. In fact, in the schedule used by Pierrel the inter-reinforcement interval occasionally exceeded the duration of an S^D presentation. By the end of phase 1, then, the animals not only had repeatedly experienced intervals in S^D without reinforcement but also had experienced a number of S^D presentations without reinforcement. This training, coupled with the continued discrimination training during phase 2, ensured that stable conditioned behavior would be maintained throughout the extended generalization test series.

With reference to Figure 6-33, responses to the S^Δ intensities (10-, 20-, 30-, and 40-decibel changes) reflect generalization of conditioning since these stimulus intensities have never entered into a reinforced relation with bar pressing. It can be seen that the power of the S^Δ intensities to evoke the conditioned response is directly proportional to their distance from the S^D. Of course, the generalization test stimuli are undergoing continual extinction, and as noted in the opening statement of the principle of generalization, the effects of extinction can also be expected to generalize to similar stimuli. Responsiveness along the entire intensity continuum in Figure 6-33, including the S^D, is thus *net response strength* after the interaction of generalization of conditioning and generalization of extinction. The effect of the continued discrimination training is evident in a progressive lowering of the generalization gradient over the three blocks of training. Again, we see evidence for the proposition that the establishment and refinement of stimulus control is a matter of extinguishing the generalization of conditioning to stimuli similar to the S^D. With still further discrimination training one would expect a further steepening of the generalization gradient until eventually responses would rarely be made even to the 10-decibel S^Δ.

Generalization and discrimination are inescapable parts of all learning. The facts of conditioning indicate that it is impossible to

strengthen or weaken behavior in one situation without simultaneously increasing or decreasing the power of similar situations to evoke
that behavior. And given that learned behavior is usually appropriate
only to a restricted range of stimuli, some discriminative control of
any learning is probably always necessary. It is not surprising, then,
that Skinnerian psychologists relied heavily on the principles of generalization and discrimination in extending the experimental analysis
of behavior to natural life situations. The following passage from Keller
and Schoenfeld (1950) argues that much of everyday life learning
reflects the pervasive and continual interaction of generalization and
discrimination.

> The process of education is greatly concerned with generaliza-
> tion and discrimination. It is possible that, at the beginning of
> life, all stimuli may generalize to produce mass and profuse re-
> sponses in the infant. As he matures and learns to discriminate
> objects in his environment, those generalizations which perse-
> vere during his pre-school years will probably be adequate for
> most of his gross adjustments, but they must be broken down
> later in the interests of his educational progress. The new gen-
> eralizations and discriminations will in turn undergo change as
> his schooling continues. From his science teachers, for example,
> he will learn that falling stones and falling paper obey the same
> law, that fish breathe in a way very similar to ours, that an alley
> cat and a lion have much in common. In fact, from one point
> of view, the whole business of science will be seen as the ar-
> rangement of nature's facts into new categories, with a stress upon
> the important but not obvious similarities and a disregard for the
> obvious but unimportant dissimilarities. Even changes in the
> fundamental theories of science are of this nature. The history
> of great discoveries is one of the reorganization of facts into new
> classifications on the basis of related properties. Planetary move-
> ments are tied into one general law with the motion of falling
> bodies. Electric currents are tied to the behavior of ions in
> chemical solutions. Nerve impulses are shown to be electrical
> phenomena. The behavior of *all* organisms follows the same basic
> laws. On simple and complex levels of animal and human be-
> havior, the operation of generalization and discrimination are
> among the most important phenomena with which we deal. The
> Army dog that growls at men in strange uniforms but greets its
> own soldiers; the child who learns to tell "Daddy" from
> "Mommy" (and then may call all men "Daddy"); the student

who learns that bats and whales are both mammals, not bird and fish; and a psychologist who cites three apparently different instances of behavior as examples of the same basic laws—all of these are doing very much the same thing. (pp. 161–162)

Punishment. Skinner also advanced some important ideas concerning the effects of punishment. Operationally, punishment may be defined as presenting a noxious stimulus contingent upon a particular response. Punishment is closely related to both escape conditioning, in which the organism learns a response that terminates a noxious stimulus, and avoidance learning, in which the organism learns a response that enables it to avoid receipt of a noxious stimulus. However, the focus in punishment is on the elimination of a previously acquired behavior, usually a behavior with a history of positive reinforcement.

In Skinner's bar pressing experiments with rats, punishment consisted of a sharp slap to the foot used in pressing the lever, delivered by the lever itself in the course of being depressed. The apparatus consisted of an electrically operated hammer striking forcefully against the lever in a direction opposite the downward motion of the lever. A classic study emerging from this technique examined the effects of a brief period of such punishment during extinction. Two groups of four rats each first received several days of bar press conditioning under an intermittent schedule of reinforcement. The conditioned reflex was then extinguished in both groups for two hours on each of two successive days. In one group all responses were slapped during the first ten minutes of the first extinction day, whereas the second group simply underwent normal extinction.

The result is shown in Figure 6-34. It can be seen that punishment sharply suppressed bar pressing during its application and for a similar period following its withdrawal. However, before the first extinction session had ended bar pressing had recovered to a stable rate and during the second extinction session even showed an increase in rate. The end result was that the total number of bar presses emitted by the punished group over the entire extinction process was the same as that emitted by the nonpunished group.

After several studies with findings similar to that of Figure 6-34 Skinner concluded that punishment, at least of the nature administered in his experiments, does not subtract from response strength in any long-term sense. Rather, punishment is likely to simply temporarily suppress behavior. The suppression, he argued, is really displacement of the punished behavior by the emotional reaction pro-

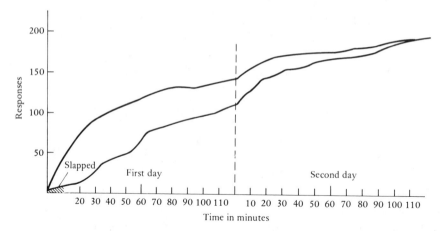

Figure 6-34. The effect of punishment on extinction of a food-reinforced bar press response. The lower curve was generated by a group of rats who received a sharp slap from the lever for each bar press emitted during the first ten minutes of the first extinction session and who thereafter underwent normal extinction. The upper curve was generated by a group of four animals who underwent normal extinction without any punishment. Punishment depressed rate of bar pressing for some time, but the behavior eventually recovered completely.
(After B. F. Skinner, *The Behavior of Organisms: an experimental analysis*, p. 154. © 1938, renewed 1966. Reprinted by permission of Prentice-Hall, Inc., Englewood Cliffs, N.J.)

duced by the punishing stimulus. This emotional reaction becomes conditioned to situational cues according to principles of Pavlovian conditioning and thus continues to suppress behavior for a time following cessation of punishment. With repeated exposure of the subject to the situation without punishment, i.e., without the UCS, the conditioned emotional reaction extinguishes, and the punished response is then free to reappear in all its conditioned strength. Consequently, punishment will be effective in "eliminating" a behavior only so long as the punishment and the emotional reaction stemming from it persist.

Estes (1944) followed up Skinner's studies using the same basic experimental procedure. Again, rats were punished at the outset of extinction of a food-reinforced bar press response and the effects of the punishment were measured against the course of extinction. However, in Estes's studies, punishment took the form of a brief electric shock administered through the floor of the Skinner box and extinction sessions were repeated until bar pressing had returned to a low stable rate such as might characterize the preconditioned level. Using a mild shock for bar pressing, Estes first replicated Skinner's observation that bar pressing was suppressed during the period of punishment but reappeared following withdrawal of the punishment, with no reduction in the total number of responses required to reach the extinction criterion. Estes then proceeded to explore the limits of recovery from punishment by varying conditions that, on general grounds, could be expected to be influential. In a series of experiments, he found that (1) either a longer period of mild shock or a period of severe shock at the beginning of extinction produced some reduction in the number of responses to extinction but *not* in the total time required to attain extinction; (2) the longer the period of positive reinforcement preceding the punishment treatment, the smaller the effect of any given level of punishment on the number of responses to extinction; and (3) intermittent shock for bar pressing during the punishment, as opposed to shocking every bar press, had a longer suppressive effect on bar pressing, but recovery from the punishment was nevertheless ultimately observed. In general, then, Estes's work demonstrated that recovery of a punished response is a very robust phenomenon. While under some conditions punishment reduced the number of responses to extinction, recovery was always observed, and in the final analysis extinction was no more rapidly achieved with punishment than without it. These observations are consistent with Skinner's conclusion that punishment does not weaken behavior in any permanent fashion.

Other observations by Estes bore upon Skinner's view that the suppressive effect of punishment results from a conditioned emotional reaction. In one experiment Estes employed the usual punishment-extinction sequence except that the animals were shocked *only when they were not pressing the bar*. This procedure was found to produce the same suppression of bar pressing and subsequent course of recovery as when shocks were given for bar pressing. That is, merely being shocked in the experimental situation had about the same effect upon bar pressing as being punished for making that response itself. This outcome, however, is exactly what would be expected under the hypothesis that the emotional reaction to punishment becomes conditioned to situational cues which then continue to elicit that reaction for a time, thereby inhibiting other behavior in the situation. This interpretation further suggests that the effects of *punishment for bar pressing* should be largely dissipated simply by exposing the animal to the punished situation for a period without further punishment and *with no bar present*, since such exposure would extinguish the power of situational cues to elicit the emotional reaction. Estes carried out precisely this operation and found almost full immediate recovery of bar pressing when the bar was subsequently reintroduced.

While both Skinner and Estes were careful to note the restricted nature of their punishment treatments, this set of experiments nevertheless gave the impetus to the view that punishment by itself is likely to be an ineffective means of changing behavior.[6] In addition, the research suggested that punishment has the undesirable side effect of conditioning an aversive emotional reaction to the punishing situation or agent (e.g., parent). The import of these propositions is summarized in the Skinnerian dictum, "Don't punish; positively reinforce."

But if punishment is an ineffective and costly means of behavior control, one may wonder why it nevertheless has always been so widely used in natural life situations. Keller and Schoenfeld (1950) provide two answers:

> First, if one disregards the restrictive and biologically uneconomical effects of punishment he may use it in depressing "wrong" behavior and thereby pave the way for the strengthening of "right" behavior [through positive reinforcement]. This is a technique still commonly met with in educational practice and often supported by experimental studies in human and animal learning (Bunch, 1928, 1935; Dodson, 1932; Warden, 1927; Gilbert, 1936, 1937; and others). Various researchers have shown that a combination of reward and mild punishment will reduce the time or the errors involved in the solution of problems.
>
> The second reason is not often mentioned: the use of punishment is *positively* reinforcing to the *user*. There is no gainsaying that punishment has its advantages in the control of others. Given the requisite physical strength, or the symbol thereof, we can always force others into a state of submission—at least temporarily. Whatever annoyance they have provided for us is thereby eliminated, and *we* are positively reinforced. On the basis of this fact alone, it is easy to see why we live in a world where punishment or threat of punishment is the rule; where power or the signs of power are considered all-important in achieving social control. (p. 111).

Motivation and Emotion. From time to time in dealing with Skinner's system we have made reference to the terms *motivation* (or *drive*) and *emotion*, but we have yet to state how Skinner would study these phenomena in their own right.

The term *motivation* has a long and confused history in the psychological literature, connoting variously the basic energy for organismic activity, an inferred state with discriminative properties, and an attribute of purpose or goal orientation in behavior, among still other meanings. The reader is by now prepared to accept that such conceptions would have no place in Skinner's system.

Skinner's handling of motivation is readily understood if it is remembered that in his system the aim is *always* to predict response probability, that response probability is *always* assumed to be a strict function of environmental conditions, and that controlling environmental conditions are *always* to be determined by test of specific stimuli in relation to reflex strength. To this point in the program, we have focused chiefly on response probability as determined by the stimulus *consequences* of behavior (reinforcement). But a rat that has received food for bar pressing will not press the bar under any circumstances; notably, it must also be hungry. Nor will the behaviors of seizing and eating food always occur when food is placed in front of an animal. Response probability is thus a function of conditions other than the stimulus consequences of behavior, and important among those other conditions are hunger, thirst, sex, pain, and other drives. Viewed from this perspective, then, to study motivation is simply to extend the experimental analysis of behavior to another class of conditions. Or to put the manner another way, motivation is to be understood as nothing more than a set of operations affecting reflex strength.

Figure 6-35 and 6-36 illustrate the analysis of motivation with reference to the hunger drive. The basic procedure in generating the data of both figures was to measure the rate of a food-reinforced bar press response in relation to variation in conditions defining degree of hunger. In the typical Skinner box experiment, such as we have discussed to this point, subjects are conditioned under a 23-hour hunger drive. This drive level is defined by giving the subject one feeding period per day, during which the animal is allowed to eat freely to satiation (generally about an hour of feeding). Food is then withheld for the next 23 hours, at which point the animal enters the Skinner box to begin the daily training session, which usually is an hour in length. The free-feeding period is given immediately after the experimental session, and 23 hours later the next training session is held, and so on throughout the experiment. A logical step in analyzing the role of drive would be to vary the hunger condition against this "normal" base level of hunger in the Skinner box. Figure 6–35 shows the result of just such an experiment with the the focus on the relation between drive and performance in extinction.

The curve labeled 0 in the figure shows the performance of a group

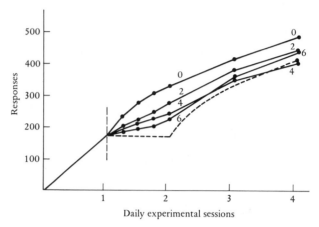

Figure 6-35. The relation between drive and extinction. On the first day of extinction (Session 2) the drive levels were reduced by prefeeding the amounts indicated
(2, 4, or 6 grams) for each curve. The following two days of extinction (Sessions 3
and 4) were conducted at normal drive level for all groups. The broken-line curve
is the assumed case in which drive is so low on the first day of extinction that no
responding occurs.
(Adapted from B. F. Skinner, *The Behavior of Organisms: an experimental analysis,* p. 384.
© 1938, renewed 1966. Reprinted by permission of Prentice-Hall, Inc., Englewood Cliffs,
N.J.)

of rats tested throughout the experiment under the typical 23-hour
deprivation schedule described above. Following daily training to bar
press under intermittent reinforcement, these animals underwent three
daily one-hour extinction sessions. The curves labeled 2, 4, and 6
were generated by groups of animals given the same daily training
and extinction sessions as Group o and maintained on the same daily
deprivation schedule *except* that they received either 2, 4, or 6 grams
of food by free feeding just prior to beginning the first extinction
session (about 20 grams would be a normal free-feeding intake under
23 hours of deprivation). These groups were returned to the normal
deprivation schedule for the second and third extinction sessions.

The various groups differed somewhat, of course, in their response
rates under reinforcement, and to correct for the contribution of these
differences to subsequent performance in extinction, the data for each
group were multiplied by a factor so as to bring the final reinforced
response rate (over the last two days of conditioning) to the same
value across groups. That value is plotted for session 1 of the graph.
Session 2 shows the following first day of extinction under the different degrees of drive with responding measured at four points during
the session in order to follow the change during the hour. Sessions 3

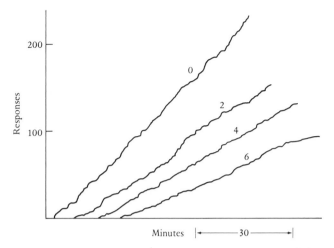

Figure 6-36. Daily conditioning records for an individual animal following prefeeding in the amounts indicated (o, 2, 4, or 6 grams).
(After B. F. Skinner, *The Behavior of Organisms: an experimental analysis*, p. 393. © 1938, renewed 1966. Reprinted by permission of Prentice-Hall, Inc., Englewood Cliffs, N.J.)

and 4 on the graph show the next two days of extinction with drive at the normal level for all groups, with only the end points of each session being measured.

With reference to the first day of extinction, it can be seen that the effect of the progressive reduction in drive by prefeeding is a progressive reduction in rate of response. But on the following day, with all groups at normal drive, Groups 2, 4, and 6 all show an *increase* in rate while in contrast Group o shows a decrease in rate. By the end of the second extinction session at normal drive, the initial cumulative response gap between Group o and the remaining groups is considerably reduced. In effect, the changes in drive postponed some responding from the first day of extinction and thus shifted the curves to the right. This relationship is shown in pure form in the broken-line curve, which represents Skinner's hypothesis as to what might be expected if the drive on the first day of extinction were assumed to be so low that no responding at all occurs. Since no response occurs, no extinction occurs; the extinction process is simply displaced one day and then proceeds in the normal fashion. With additional extinction sessions the broken-line curve would be expected ultimately to reach the height of the Group o curve. In sum, the experiment indicates that drive controls response rate during extinction but not the number of responses to extinction.

Figure 6-36 shows the outcome of a similar experiment directed to the relation between drive and conditioning. Again, animals were maintained on the 23-hour food deprivation schedule, and hunger drive was further manipulated by feeding either 0, 2, 4, or 6 grams immediately before the daily experimental sessions, which comprised one hour of periodic reinforcement for bar pressing. Each animal was tested a number of times at each prefeeding level, these levels being varied in random order at two-day intervals. The figure shows sample daily conditioning records for one rat under this procedure. Each curve is a cumulative response record for a single session, but the curves of four independent sessions at the prefeeding levels indicated have been placed within the same coordinates to facilitate compari-son. It is apparent that rate of response under reinforcement was di-rectly proportional to the degree of drive.

The basic analysis illustrated in Figures 6-35 and 6-36 could be extended to a wide range of questions concerning the role of motiva-tion in behavior. For example, if one were interested in whether learning is affected by drive level, one might measure the resistance to extinction of a bar press response following conditioning under different degrees of hunger; if one were interested in whether drives have an additive impact on behavior, one might measure rate of re-sponse during extinction under simultaneous food and water depri-vation as compared with food or water deprivation alone; if one were interested in whether there is such a thing as a curiosity drive, one might measure rate of bar pressing for stimulus novelty (i.e., rein-forcement consists of presentation of novel stimuli) as a function of prior deprivation (restriction) of stimulus variability; if one were inter-ested in the Freudian hypothesis that most behavior is sexual in na-ture, one might test the range of behaviors that vary in strength as a function of operations defining that drive.

Two features of this approach stand out. First, questions concern-ing the role of drive are seen to reduce to the measurement of specific behaviors in relation to particular operations performed on the orga-nism. Second, the analysis proceeds directly from the operation per-formed to the behavior itself. At no point is it assumed that the op-eration produces intermediate stimuli, states, or processes in the organism that serve an explanatory role in the ensuing behavior. In-stead, the topic of motivation is addressed, like all topics in Skinner-ian psychology, strictly at the level of demonstrable inputs and out-puts. In this light, the concept of drive or motivation "is not actually required in a descriptive system . . . [but] is useful, however, as a device for expressing the complex relation that obtains between various similarly effective operations and a group of covarying forms of behavior" (Skinner, 1938, p. 368).

The same remarks could be directed to the concept of emotion. In Skinner's system, we again find ourselves concerned with a class of operations (e.g., presentation of a painful stimulus, sudden withdrawal of reward) that affect certain reflexes, and the understanding of emotion proceeds by studying those operation-reflex relationships. In the study of emotion, as in motivation, Skinner preferred to measure conditioned reflexes affected by the operation, rather than the unconditioned reflexes involved. For example, we have already seen how the emotion of anxiety might be studied in terms of changes in conditioned operants. Figures 6-14 and 6-15 show that a stimulus (S^D) which regularly precedes an aversive stimulus acquires the power to suppress ongoing operant behavior, much as it is disrupted by the primary aversive event itself. The suppression of operant activity in the presence of the S^D can be taken to define anxiety, and this emotion can then be understood in terms of parametric investigation of the acquisition and extinction of S^D power. For example, is anxiety (the suppressive power of the S^D) greater when the S^D − aversive stimulus pairing is predictable or when it is unpredictable? Is anxiety so defined diminished by the administration of certain drugs? Is anxiety so defined best eliminated by sustained exposure to the S^D? Or by positive reinforcement in the presence of the S^D? And so on. Such analyses require no reference whatsoever to the internal state of the organism in accounting for the observed outcomes.

In short, emotion is to be studied in exactly the same manner as motivation.

Our discussion of motivation and emotion should assist understanding of Skinner's position on latent learning and the learning-performance distinction. As we have seen, theorists such as Tolman and Hull postulated a learning state of the organism which might or might not be faithfully reflected in performance, depending upon the motivational (or emotional) state of the organism. But a concern that performance may not reflect a learning state simply cannot arise in a formulation that deals *only* with performance. The quotation above from Skinner to the effect that the concept of motivation is not needed in a descriptive system could also be applied to the concept of learning. Instead, with reference to operant activity, there is only an observed probability of response which is a common function of a number of operations, notably reinforcement, motivation, and emotion.

The Descriptive Principles: A Behavior System

The principles reviewed above make up the core of Skinner's system. Strictly speaking, the *empirical* principles of Pavlovian conditioning should be added to those listed since, as noted earlier, Skinner accepted these as governing the learning of involuntary, or autonomic, responses (respondents), whereas the principles detailed here were held to govern the learning of all voluntary, or skeletal muscle, activity (operants). But while Skinner accepted Pavlovian principles in relation to respondent behavior, he saw the principles of operant conditioning, as developed within the Skinner box methodology, as the key principles in the understanding of human and animal activity.

To be sure, Skinner's principles deal with the same basic phenomena that occupied other theorists, but they nevertheless have a fundamentally different character. In part that different character arises from the fact that the principles were defined within a single experimental setting and measurement procedure. As we have seen, Skinner's own research showed that the basic phenomena of learning could be demonstrated and productively studied within the confines of the Skinner box and the cumulative curve, and his followers continued to find ingenious Skinner box analogs of complex behavior processes. In consequence, the findings and principles that emerged had an operational uniformity that set them apart as a distinctive system of treating behavior.

More important, Skinner's explanatory principles do not take the form of theoretical hypotheses as in other S-R theories but rather are summaries of experimentally observed stimulus-response relations. Of course, truly descriptive principles of behavior were to be expected from the author of the "Concept of the Reflex" paper, but nevertheless one cannot fail to be impressed by the consistency with which Skinner carried out his early ideas about the nature of explanation. And in Skinner's eyes at least, the principles that emerged were testimony to the merits of his inductive approach: "[W]hen lawful changes in behavior take place before our very eyes—or, at most, only one step removed in a cumulative curve—we lose the taste, as we lose the need, for imagined changes in some fanciful world of neurones, ideas, or intervening variables" (1958a, p. 99).[7]

The operant methodology and principles, then, can be seen as a remarkably self-contained universe where one may deal with the major problems of behavior in a coherent way. The scope, internal consistency, simplicity, and dogmatic positivism of Skinner's approach are appealing to many psychologists, particularly to those impatient with hypothetico-deductive theory. To such psychologists, the method is the message. And the crux of the message is to stay at the level of observables.

Since Skinner's principles did not carry hypothetical-deductive meaning, they looked not inward, so to speak, to the arena of conflicting theories, but rather outward to the real world. That is, they lent themselves to application. It is interesting to note that Skinner, upon completion of the book that presented his descriptive system, expressed a quite conservative view of application, noting that "the book represents nothing more than an experimental analysis of a representative sample of behavior. Let him extrapolate who will" (1938, p. 442). As events proved, if anyone willed, it was Skinner! Relying upon the principles of operant conditioning, in particular the principle of reinforcement, Skinner undertook extensive analyses of complex human activity and eventually became the preeminent spokesman for the relevance of S-R principles to everyday human affairs, so much so that in the popular mind he is now virtually equated with behaviorism.

We have suggested applications to human behavior in a limited way in the course of presenting the various principles. We shall now consider some systematic applications of the behavior system.

APPLICATION

Skinner's work on the application of S-R principles to human affairs is so extensive as to preclude full review here in any detail. His applied writings comprise some half-dozen books and scores of articles and cover virtually every area of human activity. Rather than provide a general summary of these writings, we shall illustrate Skinner's work in this vein by reviewing a few applications in sufficient detail that the reader can see the full continuity from the philosophy of science through the laboratory analysis of behavior to the explanation and control of human activity in natural life settings. Our review will focus on two broad and fundamental areas—language and education.

Verbal Behavior

In 1957 Skinner published *Verbal Behavior*, the results of a 20-year study of this most complex and important of all human activities. The text is a straightforward extension of the descriptive principles of operant behavior in a comprehensive analysis of everyday language and the world of literature.

In contrast to traditional approaches, which typically view language as the servant of meaning, Skinner maintains that language and meaning alike are to be understood by determining the variables of which verbal activity is a function. Like any voluntary activity, language is held to be composed of operants—emitted behavior segments which show a lawful relation to environmental variables. By this criterion, Skinner claimed, verbal operants may range in length from a single speech sound to lengthy phrases or sentences that appear in unitary fashion, e.g., *the truth the whole truth and nothing but the truth*. The task is to predict the emission of verbal operants in both their single and their combinatorial state, and as in the case of any operant the primary controlling factor will be reinforcement. But the distinctive and defining feature of verbal behavior is that the reinforcement is mediated by the action of others. That is, the function of language is to affect the environment in reinforcing ways by affecting the actions of others.

Within this framework Skinner proceeded to a broad interpretative analysis of language use. A brief outline of the analysis will be provided here, and Skinner's illustrations of the various principles involved will be used. He began with a new taxonomy of verbal behavior.

The Mand. Mands (as in com*mand*, de*mand*) are verbal operants that specify the nature of the reinforcement sought. *Wait! Sh-h! Get out! Stop! Water!* are instances of mands, and in each case a particular characteristic reinforcing consequence is clearly indicated. Since mands are emitted under conditions of deprivation or aversive stimulation for which relief is sought, they become specific to these states and can be made more likely by increasing the appropriate motivational state on the one hand and by supplying the appropriate reinforcement on the other. The existence, strength, and probability of emission of mands, then, are accounted for by their status in a three-term contingency involving *motivation→mand→reinforcement*.

Mands may take forms other than the imperative, as in requests (*Bread, please!*), entreaties (*Come, I beg of you!*), questions (*What is your name?*), advice (*Go west!*), warnings (*Look out!*), permissions (*Go ahead!*), and offers (*Take one!*), but in each case the defining feature of specifying a particular reinforcing consequence is evident. In addition, mands, like any operant, may be extended to nonreinforced situations through the mechanism of generalization, particularly when the motivation is strong. Such extension is seen when people mand the actions of inanimate objects, such as vehicles (*Come on, you, start!*). Occasional adventitious reinforcement may contribute to the maintenance of such manding, as in the superstitious "Come seven!" by the dice player. Wishing (*Oh to be in England, now that April's there!*) and cursing (*Bad luck to you!*) are similar extensions of manding which Skinner termed *magical* in the sense that the consequences sought never actually occur as the result of similar verbal behavior. Skinner noted that literature is rich in the use of magical mands, magical because the request is strictly rhetorical. He found that of the first lines of English lyric poems in a number of anthologies about 40 percent are characteristic of mands. Fifteen percent of these mand the behavior of the reader (*Tell me, where is fancy bred? Come live with me and be my love*), another 15 percent address someone or something besides the reader (*Go, lovely rose*), and the remaining ten percent are statements of wishes (*A book of verses underneath the bough*). In Skinner's words,

> The richness of these examples from literature exemplifies a general principle . . . "Poetic license" is not an empty term. Literature is the product of a special verbal practice which brings out behavior which would otherwise remain latent in the repertoires of most speakers. . . . Among other things the tradition and practice of lyric poetry encourage the emission of behavior under the control of strong deprivations—in other words, responses in the form of mands. (1957b, p. 50)

While mands are controlled by motivational states, verbal operants are generally under more specific forms of stimulus control. Many verbal operants are controlled by other verbal activity and can be classified by the nature of that activity.

Echoic Behavior. Here the response generates a sound pattern similar to that of the stimulus, e.g., subjects asked to report the first word they think of upon presentation of a stimulus word are likely to respond with words that rhyme or are otherwise similar to the stimulus word. Parents and teachers make use of explicitly instated echoic behavior in the teaching of language. The child is manded "Say X" and is reinforced by approval if it yields a matching sound pattern.

Textual Behavior. Here verbal responses are under the control of nonauditory verbal stimuli, e.g., printed letters or characters. In the main, textual behavior refers to reading. Like echoic behavior, textual behavior is first strongly reinforced in the process of formal education but eventually is sustained by the reinforcements implicit in the process, such as the acquisition of information, the induction of exciting states, and so on.

Intraverbal Behavior. Intraverbal behavior is characterized by the lack of a point-to-point correspondence with the controlling verbal stimuli (as exists in the case of echoic and textual behavior). For example, *How are you?* is usually a stimulus for *Fine, thanks,* but there is no structural correspondence between the elements of this stimulus and response.

For adults virtually any verbal response is subject to some degree of influence from a number of other verbal responses, and practically all verbal responses exert some stimulus control over a number of other verbal responses. These intraverbal relations are the result of thousands of reinforcements given to thousands of verbal responses in thousands of verbal contexts. The result is a vast network of associations among the verbal operants in an individual's repertoire. Verbal chaining, as in learning and reciting a poem from memory, provides evidence of both near and remote intraverbal control. Near control is shown by the circumstance that each word acts as the immediate stimulus for emission of the following word; remote control is seen in the usually successful strategy of aiding a faulty memory by "backing up" to the beginning of the piece to reinstate additional controlling stimuli. Similarly, the word association experiment and the free association procedure, in which the subject emits the train of words brought to mind successively by a single verbal stimulus, clearly display the influence of the subject's learning history. Of course, intraverbal stimulus control is not just a matter of individual words influencing the emission of other words; the controlling stimulus and response units may be quite complex. In general, the more complex the verbal stimulus, the more specific the resulting verbal response. Thus an expression such as *That has nothing to do with the* _____ will very likely produce the response *case,* or a very similar response.

The Tact. Turning now to verbal behavior under the control of nonverbal stimuli, many verbal responses are concerned with naming or labeling the objects and events of the world around us. A verbal response of this nature was termed a *tact* by Skinner, the term being intended to suggest behavior which makes con*tact* with the physical world. Tacts make up a large portion of a typical vocabulary and lie at the heart of the study of language and theories of meaning. But in Skinner's view the acquisition of tacts is no more mysterious than the acquisition of a discriminative bar press response. Each case involves the same three-term contingency:

$$stimulus \rightarrow response \rightarrow reinforcement.$$

In the presence of particular features of the world, the child emits verbal responses (names or descriptors) which are either reinforced (usually, by social approval or confirmation) or extinguished, depending upon the appropriateness of the response. Through countless shaping and differential reinforcement experiences, the individual builds up the tacting response repertoire of the verbal community.

Tacts may be extended through generalization in several different ways.

Generic extension. In this case the property of the novel stimulus which evokes the extension is the same property that determines the reinforcement of the original tact, e.g., a familiar species name is applied to a hitherto unobserved life form which exhibits the properties that are the basis for reinforcement of the species name. In general, generic extension is the recognition of a new instance of a prevailing *stimulus→response→reinforcement* contingency. Such extensions characterize the language of science.

Metaphorical extension. In this type of extension the property shared by the novel stimulus and the tact is not the property that determines reinforcement of the original tact. An example would be calling someone a mouse; the extension is clearly not based on the physical characteristics which ordinarily determine reinforcement of the label "mouse," but rather on other shared properties, such as timidity or fearfulness. The basis of metaphorical extension may be quite subtle. *The child is bright as a dollar* involves a very vague but nevertheless intuitively meaningful similarity between a bright child and a bright dollar. Moreover, the similarity underlying metaphorical extension need not be limited to common properties of the stimuli but may be based on a common *effect* of the stimuli. Thus, in *Juliet is the sun* we must suppose that Juliet and the sun affect the speaker in a similar way. The subtleties involved in metaphorical extension, which characterizes the language of literature, are commonly thought to reflect a special form of analogical thinking. But in Skinner's analysis no special process is involved; metaphorical thinking is simply another instance of generalization to a novel stimulus on the basis of properties shared by that stimulus and a previously reinforced stimulus.

Two related processes are *metonymical* and *solecistic* extensions of the tact. In metonymical extension a stimulus acquires control over the response merely because it accompanies the stimulus upon which reinforcement is based, e.g., we say *The White House denied the rumor* instead of *The President denied the rumor*. In solecistic extension the property which gains control of the response is but weakly related to the reinforced properties even to the point of illogic. Mrs. Malaprop exclaims *You go first and I'll precede you*; while *precede* is contradictory to *follow*, which should have been employed here, they are nevertheless alike in both dealing with the order of events. In both metonymical and solecistic extensions the broad principle of generalization is again evident.

Abstraction. Abstraction proceeds through the limitation of tact extension by the process of differential reinforcement. For example, suppose a child who has never seen a pyramid-shaped object before is shown a small red pyramid and is given some reinforcement for applying the label "pyramid" to it. This training will increase the probability that the child will say "pyramid" when shown any similar object, such as a small red cube, cone, or triangle. But if the community refrains from reinforcing such extensions while continuing to reinforce the response "pyramid" in the presence of any pyramid regardless of its color, size, or substance, then the response "pyramid" will eventually be given only to pyramid-shaped objects. We would say that the child has abstracted pryamidal shape, meaning by this that the child applies this tact in the presence of the defining property only and regardless of the accompanying properties. In similar manner, according to Skinner, we learn all abstractions, even those of the highest order. Thus, after sufficient experience in identifying Picasso's paintings as distinct from those of other cubist artists, one may learn to correctly identify a previously unseen painting as "a Picasso," basing that response upon properties of style that are unique to Picasso's works (Tighe, 1968). Basically the same process, Skinner claims, is observed in lower organisms. A pigeon reinforced for pecking at a small red triangle may peck at any small or red figure, but the peck response can be brought under the exclusive control of redness by reinforcing pecks in the presence of red regardless of accompanying size and shape values and extinguishing pecks to any figure that is not red.

The audience. Audiences are another important class of nonverbal controlling stimuli. An audience may consist of a single listener, as well as groups, and may include the speaker as a listener to his or her own behavior. Different audiences will reinforce different verbal behavior and hence audiences become discriminative stimuli. Among the verbal behaviors controlled by audiences are the very language utilized, as when a bilingual child uses one language when addressing its parents and another language when addressing schoolmates; the use of jargon or special vocabularies appropriate to particular groups; the subject matter discussed; the style of speech, as in the degree of metaphorical extension encouraged; and the tone of address, as determined, for example, by the relative social status of the audience. In addition, places may have an audience-like effect. We become talkative when we enter a place where verbal behavior tends to be strongly reinforced, as in a club room, or we become quiet upon entering a place where talk is extinguished or negatively reinforced, as in a church or library.

Having provided a taxonomy of verbal activity and its stimulus control, Skinner turned to a consideration of how these categories operate in the actual production of verbal behavior. The fundamental principle stressed by Skinner in this regard is that *any segment of verbal behavior is likely to be a function of many variables operating at the same time.* The principle of *multiple causation* is evident in the familiar procedures of *prompting* and *probing*, in which a controlling variable is added to existing sources of strength to make a given verbal behavior more likely.

In the *prompt* the desired verbal behavior is known to the controller. The classic case is that of the stage prompter who supplies the actor on stage with the beginning of the forgotten passage. This is an *echoic prompt*, and its repetition by the actor is sufficient to evoke the remainder of the passage, thereby showing the presence of another source of strength, in this case, intraverbal control. Echoic prompts are commonly used in education to bring about responses which may then be further reinforced until memorized, i.e., brought under full intraverbal control. A related procedure is the *textual prompt*, in which some form of text, e.g., notes or teleprompter, is used to supplement weak intraverbal behavior by a speaker. Advertising makes frequent use of both echoic prompts (*Say "Luckies please"*) and textual prompts, as when the name of a "notion" product appears by the cash register.

Finally, in a *thematic prompt* the desired verbal behavior is brought about by the supplying of intraverbal stimuli that are linked to the response in question. Thus, we might induce a host to ask, "Would you like another drink?" by talking about various beverages, telling jokes about drinking, or even rattling the ice in our glass (which action qualifies as verbal behavior under Skinner's definition).

In the *probe* the controller does not specify in advance the precise verbal behavior to be evoked. An *echoic probe* occurs when a weak or unstructured echoic stimulus fails to evoke a matching verbal response but nevertheless combines with some other variable to produce verbal behavior. Repeated sound patterns commonly function as echoic probes, e.g., the rhythmic clicking of train wheels is likely to suggest a verbal refrain. Skinner devised an echoic probe for experimental use which he called the "verbal summator," the term *summator* being used to emphasize the fact that echoic probes work through the cumulative impact of repetition. The verbal summator presents a repeated pattern of low-intensity, indistinct speech sounds, and the subject is asked to report anything meaningful that comes to mind. Since the echoic stimulus itself is too unstructured to permit verbal matching, the subject's statements must be attributable to other controlling variables, such as the subject's emotional or motivational states, and these controlling variables may be inferred from examination of the subject's verbalizations. The device, then, is a kind of projective test.

Similarly, a *textual probe* is a weak textual stimulus, as when a fleeting glimpse of a sign results in nonmatching verbal behavior. Skinner relates the incident of a motorist who, after a narrow escape from a serious accident, glimpsed a sign reading ONE MILE TO DEATH, but the sign actually read ONE MILE TO BATH. Subliminal advertising, in which messages are briefly flashed on a movie or television screen without the viewer's awareness, is based on the assumption that such stimulation will combine with motivational variables to induce the viewer to mand certain products.

The *thematic probe* is exemplified by the well-known word association test in which the subject must respond with the first word that comes to mind upon presentation of a verbal stimulus. On the principle of multiple causation, the response is likely to be influenced by additional stimulus variables and therefore can be used to make inferences about the nature of those variables, as when a subject responds slowly or in bizarre fashion to an emotionally significant stimulus word. Related clinical procedures are the Thematic Apperception Test, in which the subject is asked to make up a story about an ambiguous visual scene, usually social in nature, and the Rorschach Test, in which the subject is asked to state what she or he sees in "ink blot" patterns. Again, the subject's responses must result from the probing of variables other than the immediate stimulus. Like the verbal summator, all these thematic probes may reveal segments of verbal behavior that would not be allowed overt expression under normal stimulus control.

Many other phenomena of language were accounted for by Skinner in terms of the principle of multiple causation, or multiple strengthening of verbal responses. His topics in this vein include (1) the rich thematic interconnections and multiple meanings that characterize great literature, (2) the rapport between speaker (writer) and listener (reader) as a matter of the strengthening of shared verbal response tendencies ("We are especially reinforced by speakers and writers who say what we are almost ready to say ourselves"), (3) Freudian slips, where parallel S-R connections interact to reveal portions of normally inhibited verbal behavior, (4) wit and humor, which Skinner felt were largely accounted for by the supplemental strengthening of contextually weak (unexpected) verbal responses, (5) the combination, or blending, of word fragments which give rise to verbal errors, or sometimes to useful new words (*fog* and *smoke* into *smog*) or humorous Spoonerisms (*Hoobert Herver, muttered buffins*), (6) verbal games and puzzles, which can be shown to rely heavily upon complex arrangements of probes and prompts, as in riddles and crossword puzzles, and (7) puns, which obviously play upon multiple sources of verbal control. The basic point underlying Skinner's analysis of these and many similar topics is that the phenomena of verbal behavior are to be understood simply as the mechanical result of the combined influence of familiar conditioning processes.

The Descriptive Autoclitic. Since verbal behavior depends upon the responsive action of the listener, it must provide clear guidelines for the listener's interpretation of the speaker's statements. Hence the existence of verbal responses which function primarily to provide information about other parts of verbal behavior. Skinner termed such responses *descriptive autoclitics*, the term being intended to suggest behavior that is based upon other verbal behavior.

One type of autoclitic informs the listener of the kind of verbal operant it accompanies. For example, *I read that such and such is going to happen* informs the listener that the statement is a textual response. Similarly, *I hear, I demand, I recall, I call it, I observe* are all informative, at least in a probabilistic way, of the nature of the accompanying operant. Other autoclitics say something about the state of strength of the accompanying verbal behavior (*I guess, I imagine, I believe, I suppose, I know, I promise,* and so on) or the emotional or motivational condition of the speaker (*I am happy to say, I regret to inform you, I hate to say,* etc.) or permit the listener to relate the verbal behavior to other aspects of the situation (*I agree, I admit, I predict, I expect,* and similar expressions). Negative autoclitics may be used to imply the existence of an alternative point of view on the response at issue (*I would not go so far as to say, I won't admit, I don't doubt, I cannot deny,* etc.) or to permit the speaker to make a response that, if unaccompanied by negation, might be aversive to others (*I need not tell you, I don't suppose you would care to,* and so on). Some autoclitics describe the basis of an accompanying response (*I say, To my way of thinking, In my opinion, They say, It may be said, You mean that, In other words, For example*). Other autoclitics have important quantifying properties. *All, some, none,* and the articles *a* and *the* function to define limits of the listener's reaction to the accompanying statements. Still other autoclitics function as relatively imperative mands, as *Behold!, Know then, Listen, Note that, Ditto, Vice versa, Assume, Let X equal.* . . .

Statements of negation (the response *no* and related forms) or of assertion (*yes, is*) may also have an autoclitic function in intensifying the listener's response to accompanying assertions. Both types are likely to be used when there is some collateral condition which is likely to weaken the listener's response. For example, to state *That metal is NOT gold* or *It IS so!* implies the existence of some counter indicating condition (e.g., a denial by someone else) and has the effect of strengthening the listener's reaction to the statement in the face of that condition.

All of these autoclitics, as well as still other forms and instances, exist because they indicate something of the circumstances in which a response is emitted or something of the condition of the speaker and so function to modify the listener's reactions in important respects. And it is the listener's actions that ultimately determine the reinforcement or nonreinforcement of verbal behavior. Autoclitics, like the verbal operants they serve, are traceable to the all-pervasive law of reinforcement.

Grammar and Syntax. Our discussion of words and phrases as autoclitics prepares for Skinner's further proposal that grammar and syntax may also be treated as conditioned autoclitics. Grammar and syntax refer to those features of language which specify correspondences and relations between words in the fashioning of sentences. Skinner's treatment focused on (1) inflections, that is, the changes of form that words undergo to mark distinctions of number, tense, gender, and other properties (e.g., go-going, Mary-Mary's, mouse-mice); (2) the order and grouping of words in a sentence; and (3) the use of prepositions and conjunctions. The latter he termed *manipulative autoclitics*, and (1) and (2) were grouped as *relational autoclitics*.

Several relational autoclitics of inflection are illustrated by the sentence *The boys play with the girl's ball while the man works.* The final *s* in the sentence serves, among other functions, the autoclitic function of indicating that the man and the work go together, while the *s* in *boys* serves as a similar relational autoclitic in its agreement with the form of the verb (play); the *'s* in *girl's* denotes possession of the ball by the girl. Word order as an autoclitic is illustrated by the sentence *The car carries the boat*, where the predication that the boat is carried by the car (and not the other way around) is given by the order of the terms.

Prepositions and conjunctions are manipulative autoclitics in the sense that they direct the listener to perform certain operations on the input. For example, the word *but* enjoins the listener to make an exception of something, the word *and* directs the listener to add to what has already been said, *in* specifies that something is to be contained in something else, and *if* specifies a proposition to be considered conditional in relation to another.

Autoclitic behavior is a critical aspect of effective composition, that is, of putting words together to make meaningful sentences. But in Skinner's view much of our spoken and written language is not, strictly speaking, composed but rather involves the transfer of previously learned verbal behavior. *Good morning, how are you today? Fine, thank you. And yourself?* is an exchange of essentially unitary responses evoked by familiar discriminative stimuli and involves no autoclitic effort on the speakers' parts. With increasing verbal experience, relatively large segments of verbal behavior are likely to come under the control of a single contingency and hence come to be emitted in a repetitive and unitary manner. This is particularly true where the verbal behavior deals with a stable sequence of variables, as in the repeated conduct of interviews organized around a common aim, the repeated preparation of reports on a given subject, or repeated exchanges with the same clientele. In such cases, a format tends to emerge whereby the verbal behavior assumes a characteristic content and autoclitic structure, and "composing the language" amounts to little more than choosing words appropriate to the particular instance at hand.

In similar fashion, Skinner argued, general experience with language composition results in the conditioning of what he termed *autoclitic frames*, that is, conditioned response tendencies toward appropriate autoclitic behaviors. Autoclitic frames combine with local situational cues to evoke effective verbal behavior. Thus, when a child has learned that, say, two or more dogs, cars, balls, and fingers are to be termed dogs, cars, balls, fingers, it is likely to respond with the appropriate plural form the first time it sees two lions. If a child has been reinforced for writing such phrases as *my father's hat, my father's car, my father's cup*, it is likely to correctly employ the autoclitic of possession when writing for the first time about father's moustache. Language composition of this nature is something less than a full-blown affair.

But when the situation is novel or sufficiently complex and the transfer of past verbal behavior cannot do justice to the variables operating, the speaker or writer must indeed compose verbal behavior, modifying, relating, and arranging responses in a manner to best guide the listener in receipt of the "message." In such composition, errors in sentence construction and weaknesses in effective communication are likely to arise from inadequate autoclitic behavior, particularly from inadequacies in the higher-order autoclitics of grammar and syntax.

Self-Editing. An important activity of the speaker is the "editing" of his or her verbal productions. The editing of written material is a familiar process. The writer, or another party, reviews and revises the material in the interest of enhancing the desired effect upon the prospective audience. For the same reason, the speaker may test his or her verbal production *subvocally* before offering it to the listener.

Self-editing is prompted largely by the threat of punishment from the verbal community. Here, punishment refers to any of the variety of ways the verbal community has of indicating its disapproval of a speaker's productions, including physical punishment, verbal criticism or abuse, withdrawal by the listener, inattention, and gestures or facial expressions of impatience, derision, or scorn. Deficiencies of speech which may bring about such punishment include unclear or illogical statements, verbal behavior inappropriate to the audience, repetition of the obvious or commonplace, indulgence in the esoteric or the picayune, egocentric productions, unpleasant vocal sounds, vulgar speech, incompleteness, digressions, interruptions, exaggerations, and still other problems.

Drawing on the principles of operant conditioning, Skinner noted that since punishment does not eliminate or weaken behavior, the need for self-editing of verbal productions will persist despite social punishment for deficient verbal behavior. But the same principles also explain how punishment may operate in subvocal behavior to effect rejection of undesirable responses. As we have seen in animal experiments, punishment controls the punished behavior by turning stimuli which accompany that behavior into conditioned aversive stimuli. In the case of verbal behavior, these stimuli will usually be those arising from the behavior itself. Hence by rejecting the potentially offending verbal behavior the speaker terminates the conditioned aversive stimuli. To put the matter in more familiar terms, subvocal production of punished verbal behavior reminds the speaker of that punishment and this reminder is sufficient to arrest the offending speech at the subvocal level.

The relatively mild social punishment that usually accompanies inappropriate verbal behavior would be expected to generate aversive stimuli in subvocal behavior that function essentially as signals to guide effective speech. But when strong punishment has been involved, the stimulation associated with the offending verbalization may be so aversive as to reduce even *subvocal* production of the material. This, Skinner felt, is the mechanism underlying Freudian repression.

The subvocal editing of verbal behavior not only censors verbal responses that might be punished but may also serve to encourage emission of appropriate but initially weak verbal behavior. "In reviewing behavior at the covert level, one may for the first time 'see what one has to say, and judge it worth saying' " (p. 381). Subvocal editing, then, has a strengthening function as well as a screening role.

Of course, not all verbal behavior is subject to covert editing. For example, under strong motivation or emotion one may "speak without thinking," often with unfortunate consequences. On such occasions one may sometimes observe a kind of postproduction editing, as when the speaker immediately corrects an ambiguous statement or, after emitting a response that might well bring punishment, says, "I take that back." Unedited verbal behavior also occurs when the speaker cannot properly react as a listener to her or his own verbal behavior, as, for example, when a rapid response is called for or when the speaker is fatigued, subject to distracting stimuli, or otherwise unable to attend to the necessary verbal productions. In addition, environments which provide little reinforcement for verbal behavior or little punishment for inappropriate responses can be expected to develop speakers with weak self-editing behavior.

The amount and type of editing also vary significantly with the nature of the audience. For example, the "confidant" and the psychotherapist are nonpunishing audiences that explicitly encourage low levels of editing. In a broader vein, the literary and scientific communities reinforce distinctly different criteria for editing. In contrast to scientific writing, "[l]iterary effects upon the reader do not in general depend upon the maintenance of a correspondence between the writer's behavior and a given state of affairs. The reader does not take practical action, is therefore not seriously misled, and makes no effort to hold the writer to a strict stimulus control" (p. 396). The result is that literary productions are rich in forms of verbal behavior that are not tolerated in scientific writing, behaviors such as metaphorical extensions, manipulation of symbols, and development of multiple meanings. The scientific community, on the other hand, is at pains to develop and rigorously reinforce rules of description that preserve precise and unambiguous correspondences between verbal behavior and its controlling environmental variables.

Covert Verbal Behavior and Thinking. In the Skinnerian analysis verbal behavior must begin in overt form, for it could not otherwise affect the environment and be reinforced. But several factors operate to reduce verbal responses to the covert level. Our discussion of self-editing makes clear that punishment is one of these factors. The verbal community generally encourages and differentially reinforces the young child's initial overt speech, but at a somewhat later stage the child's verbalizations are likely to be considered annoying and are punished (consider the adage "children should be seen and not heard"). Various forms of punishment obtain throughout the adult years for faulty, inappropriate, or imprudent (to self-interest) overt verbal behaviors; the privacy of covert behavior thus comes to have a distinct practical value. That avoidance of social punishment is a factor in covert speech is shown by the common observation that people who live alone are quite likely to begin talking to themselves aloud.

Another factor encouraging covert speech is a principle of least effort, whereby any operant tends to assume the minimum energy level necessary to maintain the reinforcing contingencies. For example, a bar press response will eventually come to be emitted with the least movement and force required to secure reinforcement. Similarly, when a speaker is her or his own and only listener, it is simply easier not to speak aloud. Again, a developmental progression is seen in this form of covert behavior, since young children are frequently observed to talk aloud to themselves while engaged in solitary play or problem-solving activities. Responses which become covert on the minimum effort principle are likely to become overt again when higher energy levels aid the maintenance of the controlling relations. For example, in working on a difficult problem, one is likely to begin talking aloud to oneself, since overt speech is likely to improve intraverbal chaining.

Finally, verbal behavior may be covert simply because it is weak. For example, we may be stimulated to tact a certain feature of the environment, but conditioning of the tact may not have been sufficient to elevate the response to the strength required for overt emission.

In examining these several variables that determine whether a verbal response will be overt or covert, Skinner noted that they do not affect any *other* properties of the response nor do they point to any significant difference between the two levels of behavior. Instead, they suggest that overt and covert verbal behavior are parts of a continuum. Therefore, Skinner concluded, nothing is gained by identifying covert speech with thinking, as was done by Watson. "There is no point at which it is profitable to draw a line distinguishing thinking from acting. . . . So far as we know, the events at the covert end have no special properties, observe no special laws, and can be credited with no special achievement" (p. 438). This, of course, is precisely the logic underlying the response-reinforcement theorist's use of the r_g.

But if thinking is not to be identified with covert speech, how are we to define it? In Skinner's words,

A better case can be made for identifying thinking with behavior which automatically affects the behaver and is reinforced because it does so. This can be either covert or overt. We can explain the tendency to identify thinking with covert behavior by pointing out that the reinforcing effects of covert behavior *must* arise from self-stimulation. But self-stimulation is possible, and indeed more effective, at the overt level.

When a man talks to himself, aloud or silently, he is an excellent listener. . . . He speaks the same language or languages and has had the same verbal and nonverbal experience as his listener. He is subject to the same deprivations and aversive stimulations, and these vary from day to day or from moment to moment in the same way. As listener he is ready for his own

behavior as speaker at just the right time and is optimally prepared to "understand" what he has said. Very little time is lost
in transmission and the behavior may acquire subtle dimensions.
It is not surprising, then, that verbal self-stimulation has been
regarded as possessing special properties and has even been identified with thinking. (pp. 438–439)

What are the automatic reinforcing effects of verbal behavior that
Skinner would classify as thinking? As one instance we have already
noted that self-editing provides significant reinforcement in guiding
more effective speech. Certainly, the maxim "Think before you speak"
has considerable general utility.

Soliloquies also incorporate automatic reinforcement. The simplest
form arises from the circumstance that any verbal behavior supplies
stimuli for other verbal behavior by the speaker, which in turn triggers
still further responses, the result being a "flight of ideas" which may
be rewarding to the speaker as listener. Similarly, speakers who construct for themselves verbal fantasies or verbal daydreams are saying
what, by definition, they like to hear. At a higher level of composition this mechanism also plays a role in the production of poetry and
fiction. The writer's compositions may ultimately prove to be rewarding to others, but they are first reinforcing to the writer, and that self-
reinforcement may sustain writing with little or no contribution from
others. Emily Dickinson, now acclaimed a great lyric poet, spent years
in relative seclusion writing hundreds of poems which no one else
saw until after her death.

Aside from artistic composition, verbal behavior may be automatically reinforced by practical consequences. Self mands, e.g., *Get up!
Keep going! Remember to look at the audience!*, may be effective by
induction from previous use as externally directed mands, e.g., *I succeeded because I told myself what my coach always tells me.* One may
also make use of self-addressed texts, such as diaries or memoranda,
as stimuli for one's own subsequent behavior. The self-tact, that is,
naming or classifying an object or event to oneself, serves to reduce
uncertainty and define appropriate behavior to the stimulus, particularly when the object or event is new or relatively unfamiliar.

Automatic reinforcement also plays an important role in problem
solving. Here, the speaker may generate intraverbal responses relevant
to the problem and verbally represent different solution possibilities
and probable outcomes. Such self-review eventually enables more effective manipulation of the environment, but there is also immediate

reinforcement for the speaker as listener in the differential strengthening of solution behaviors.

The Freudian defense mechanisms can also be seen as a form of automatic verbal reinforcement because they enable the speaker to escape or avoid aversive self-stimulation. Rationalization is an example. If one is forced to accept the least preferred of several alternatives, one may make the situation more palatable by defining for oneself all possible favorable features of the constrained alternative, thus reducing its aversiveness.

Finally, decision making of ethical significance may be assisted by self-review of relevant reinforcing contingencies. To say to oneself "I ought to do such and such" implies the approval of some agency for carrying out the behavior at issue, and "I ought not to do that" implies punishing consequences. The verbal representation to oneself of these contingencies then may determine subsequent behavior. As in problem solving, the power of such verbal representations to reinforce derives ultimately from their associated external conditions, but they have an immediate automatic reinforcing effect in the differential strengthening of alternative response tendencies. "The vicar of society within the individual, the Freudian superego, the Judaeo-Christian conscience, is essentially verbal. It is the 'still small *voice*' " (p. 444).

We can, then, discern many occasions on which verbal behavior provides automatic reinforcement. We have also noted that verbal behavior may be especially effective upon the speakers themselves. In addition, verbal behavior enables us to abstract and deal with one property of nature at a time, and this is a critical intellectual function. Should we therefore identify thinking with verbal behavior, especially with covert verbal behavior? No, concluded Skinner, because (1) as we have seen, we cannot distinguish special properties of covert as opposed to overt behavior and (2) not all problem-solving activity, and in particular not all covert problem-solving activity, *is* verbal. For example, consider the following problem. A subject is shown the letter R rotated 180° from its normal upright position, Я, and is given a brief time to decide whether it is a mirror image of the letter R or a true letter R. Performance on this task suggests that subjects arrive at correct solutions by mentally rotating the presented letter and matching to the normal standard; subjects' response times tend to be proportional to the amount of rotation of the presented letter (Cooper and Shepard, 1973). In addition, careful examination of verbal reports of covert problem-solving activity shows that what is described is often nonverbal in nature. After solving a problem which requires that a pile of boxes be used as a ladder to reach an object otherwise out of reach, the subject might say, "I thought of trying to use the boxes as a stepping stool." While this statement has the ring of verbal activity about it, the solver is actually *reporting* the appearance of *a nonverbal act.*

In summary, Skinner argued that we can find no basis to define thinking in terms of a particular class of activity, neither as covert activity, nor as verbal activity, nor even that traditional and intuitive favorite, covert verbal activity. Instead,

> The simplest and most satisfactory view is that thought is simply *behavior*—verbal or nonverbal, covert or overt. It is not some mysterious process responsible for behavior but the very behavior itself in all the complexity of its controlling relations, with respect to both man the behaver and the environment in which he lives. . . . When we study human thought, we study behavior. In the broadest possible sense, the thought of Julius Caesar was simply the sum total of his responses to the complex world in which he lived. . . . When we say that he "thought Brutus could be trusted," we do not necessarily mean that he ever said as much. He behaved, verbally and otherwise, as if Brutus could be trusted. The rest of his behavior, his plans and achievements, are also part of his thought in this sense. . . . So far as a science of behavior is concerned, Man Thinking is simply Man Behaving. (pp. 449–452)

Our review of *Verbal Behavior* has of necessity been selective. Skinner treated a number of additional topics as well as special conditions modifying the broad principles described here, and in so doing accounted for many other phenomena of language behavior. The rich and comprehensive nature of Skinner's analysis is suggested by the consideration that the illustrations used in this review, taken largely from Skinner, are but a small fraction of his applications to natural language usage.

It should be apparent that Skinner's interpretation of language is essentially a wholesale application of the principles of operant conditioning derived from his study of animal behavior. The concept of reinforcement is central to the analysis, and signs of social approval or disapproval are the significant reinforcers for language behavior.[8]

But while the nature of the reinforcement differs from that in animal behavior, the principles are identical. Verbal operants of all forms are shaped, acquired, maintained, combined, and brought under stimulus control through the processes of reinforcement, generalization, and discrimination.

Verbal Behavior stands as *the* S-R theory of language and as a significant step in the scientific analysis of what is generally regarded as the most complex and mysterious of human activities. In particular, Skinner's analysis was a welcome change of focus from vaguely conceived central determinants of language to experimentally analyzable external controls. In Skinner's words,

> Whenever we demonstrate that a variable exerts functional control over a response, we reduce the supposed contribution of any inner agent. For example, if we can show that the occurrence of a [verbal] response is due to the presence of a stimulus of specified properties, then it is not necessary to say that a speaker uses the response to describe the stimulus. If we can show that a response is stronger when we deprive the individual of food, then we do not need to say that a speaker uses the response to describe or disclose his need. If metaphorical extension can be shown to take place because a particular stimulus property has acquired control of a response, we do not need to say that a speaker has invented a figure of speech to express a perceived similarity between two stimuli. If an audience can be shown to strengthen a particular subdivision of a verbal repertoire, we do not need to say that a speaker chooses words appropriate to his audience. (p. 311)

To this we might add that if we can show that autoclitic behaviors are a function of environmental variables, we need not refer to the speaker's intentions or to an inner organizing and directing agent; and to the extent that we can show a correspondence between abstract verbal concepts and properties of the world, we demystify the realm of meaning.

Skinner's emphasis upon isolating variables that control language behavior is significant. Other theorists, notably Chomsky (1968), have emphasized regularities or rules that appear to characterize language use quite apart from a consideration of environmental variables. To theorists of Chomsky's persuasion, the presence of certain apparently universal rules of language behavior, together with the capacity of human beings for a seemingly infinite variety of verbal compositions, strains an S-R theory and bespeaks instead a major biological determination of language. A review of this controversy[9] is beyond the scope of this book, but it can be concluded that *Verbal Behavior* has been and remains a very influential book both among adherents of an S-R approach and, in catalytic fashion, among adherents of the Chomskyian view.

Teaching Machines and Programmed Learning

Suppose you are given the task of training a large group of rats, say 30 or so, to stable performance on a chain FI 5 FR 50. The development of such behavior involves, of course, a number of steps: adaptation to the test chamber, magazine training, bar press training, shaping of fixed ratio behavior, establishing of discriminative (S^D) control of the ratio behavior, utilization of the S^D as a conditioned reinforcer for developing FI 5 behavior, and melding of the successive response patterns into a stable behavior chain. Suppose further that you are given only one Skinner box and such limited time for the project that you have no alternative but to train all subjects at once (we may assume the Skinner box is comfortably large!). You are determined to make the best of these conditions and decide to divide the allotted training time equally among the seven steps. Under this procedure, your training would probably progress as follows. Adaptation to the test chamber might be satisfactorily achieved in group fashion since it poses no specific response requirements. However, during a brief period of magazine training some subjects would probably never gain access to the food dish, and among those who do a number would doubtless fail to acquire reliable approach behavior at the sound of magazine operation. Since time is limited you have no choice but to pass all subjects to the following stages of training. When bar press training begins, subjects with good magazine behavior have a reasonable chance of learning to bar press since their behavior is concentrated in the vicinity of the bar and appropriate responses are likely to be followed quickly by reinforcement. In contrast, subjects with weak magazine behavior are likely to be crowded out during bar press training and simply never emit the desired behavior. Of course, some of the latter subjects might "catch up" if approach behavior to the bar were reinforced, but the training schedule permits little opportunity for shaping. Similarly, when time pressures force start of the remaining training phases, only subjects with bar press behavior can develop ratio behavior, and only those with reliable ratio behavior would survive the extinction involved in developing discriminative control of ratio behavior. And, of course, if the S^D is not well established it cannot serve as a conditioned reinforcer to shape the FI behavior needed to complete the chain. The end result is that only a few of the 30 subjects would be likely to attain the full FI 5 FR 50 response chain, and the remainder would acquire at best a hodge-podge of weak and poorly directed behaviors.

The analogy to classroom teaching is admittedly crude, but there are a number of instructive common points. The teacher, too, must lead students through a hierarchy of steps that build competence in a subject matter. The teacher, too, is charged to develop that competence within a limited time. The teacher, too, must cope with group sizes that preclude individualized instruction. The teacher, too, frequently has no alternative but to accept substandard performance and must often endure a low success rate.[10]

The analogy points up two fundamental difficulties inherent in any group teaching situation, namely, the difficulty of securing the desired responses and the difficulty of arranging prompt reinforcement of those responses. Yet from the viewpoint of Skinner's system, controlling the reinforcement of emitted behavior is *the* fundamental principle of learning. It was consideration of the teacher's limited ability to implement this principle in the typical classroom situation that prompted Skinner's interest in educational technology.

What is required, Skinner maintained, is a device that will do what the teacher cannot do—provide immediate and frequent reinforcement for the thousands of responses that must be made in the course of developing competence in a basic discipline. An important step toward such a device was Pressey's (1926) apparatus for the automatic administration and scoring of tests of the multiple-choice variety. In Pressey's device the test is mounted on a mechanically rotated drum encased in a metal box in such a manner that one test question at a time is exposed through an aperture in the face of the box. The student answers by pressing one of several buttons corresponding to the multiple-choice alternatives. Pressing the correct button records the student's answer and activates the drum so that the next test item becomes exposed. By a simple adjustment the rotation mechanism can be preset to hold the drum stationary until the correct alternative had been chosen. Pressey noted that this machine also serves an important teaching function in that it gives the student immediate knowledge of right and wrong answers. But the primary function of the machine was automatic testing, and its teaching applications were conceived in relation to simple "drill and informational material." Skinner had in mind a much bolder conception of an automatic teaching device. He sought a device that would teach a full subject matter from beginning to end.

Building upon Pressey's device, a variety of machines have now been constructed in relation to Skinner's goal. All of these machines arrange steplike presentation of sets of educational material, usually visual and verbal material, stored in the machine on disks, cards, or tapes. A discrete portion, or "frame," of the material is presented to the student, who must compose a response to that material in some way, e.g., by moving printed figures or letters, by using a mechanical or electronic stylus, or simply by writing. The machine is then activated so that the student's response cannot be altered (e.g., it might be moved behind a clear plastic screen), and a correct, teacher-defined response is presented for comparison. If the match between responses is acceptable, this fact is signaled and the frame is automatically removed from the set so that it will not reappear upon a subsequent run through the material by the same student. When an unacceptable match of responses is signaled, the frame remains in the set. The student then progresses through each frame of the material in this manner until the set is completed. If errors have been made, the set is repeated, perhaps in several sessions, until the student has correctly responded to all frames. The student then moves to a new set of material, continuing in this fashion until all sets defining the subject matter have been completed without error. Both the sets and the frames within sets are arranged in hierarchical manner so that each set prepares the student for the next.

Two important differences from the Pressey procedure stand out. First, it is significant that the student is required to compose responses rather than merely to choose a response, since what is sought is not simply an ability to recognize a correct alternative but to produce from scratch and fully the behaviors that define competence in a subject matter. Second, the items in a Skinner-type teaching machine are not unordered samples of previous learning but new material presented in a carefully defined sequence of steps, each small enough that it can always be responded to by the unaided student but each advancing in some degree the student's ability to take the following steps.

A device of this sort can bring the writer of the material into contact with an indefinite number of students and at the same time have an effect upon each student much like that of a private tutor. As ennumerated by Skinner (1958b), the tutorlike properties of teaching machines are:

1. The machine ensures emission of responses to the material. In

Table 6-1. A set of frames designed to teach a third- or fourth-grade pupil to spell the word *manufacture* (from Skinner, 1958b)

1. Manufacture means to make or build. *Chair factories manufacture chairs.* Copy the word here:

□ □ □ □ □ □ □ □ □ □ □

2. Part of the word is like part of the word factory. Both parts come from an old word meaning *make* or *build*.

m a n u □ □ □ □ u r e

3. Part of the word is like part of the word manual. Both parts come from an old word for *hand*. Many things used to be made by hand.

□ □ □ □ f a c t u r e

4. The same letter goes in both spaces:

m □ n u f □ c t u r e

5. The same letter goes in both spaces:

m a n □ f a c t □ r e

6. Chair factories □ □ □ □ □ □ □ □ □ □ □ chairs

The word to be learned appears in bold face in frame 1, with an example and a simple definition. The pupil's first task is simply to copy it. When he does so correctly, frame 2 appears. He must now copy selectively: he must identify *fact* as the common part of *manufacture* and *factory*. This helps him to spell the word and also to acquire a separable "atomic" verbal operant. In frame 3 another root must be copied selectively from *manual*. In frame 4 the pupil must for the first time insert letters without copying. Since he is asked to insert the same letter in two places, a wrong response will be doubly conspicuous, and the chance of failure is thereby minimized. The same principle governs frame 5. In frame 6 the pupil spells the word to complete the sentence used as an example in frame 1. Even a poor student is likely to do this correctly because he has just composed or completed the word five times, has made two important root-responses, and has learned that two letters occur in the word twice. He has probably learned to spell the word without having made a mistake. (1958b, p. 971) contrast, traditional teaching procedures (lectures, textbooks, audio-visual aids) merely present material to the student, who may remain passive or even unattentive throughout.

2. The machine insists that each element of the material be understood before it allows the student to move on. Textbooks and lectures proceed without ensuring understanding.

3. The machine presents only that material which the student is likely to master readily, as indicated by prior empirical analyses of the learning program.

4. The machine helps the student to make the correct answer on every occasion. This is accomplished in part by careful programming of the material and in part by incorporation of techniques of verbal prompting and discriminative control (see below).

5. The machine unfailingly reinforces the student for each and every response, and does so immediately. Knowledge of incorrect responses serves as punishment that aids emission of alternative responses upon successive presentations. Knowledge of correct response strengthens the specific behavior at issue and helps maintain motivation and attention to the subject.

It is clear that the effectiveness of any teaching machine is critically dependent upon the material used in it, and a great deal of work has now been done on techniques of programming material for auto-mated presentation. The best way to get a sense of programming procedures, as well as a sense of what can be accomplished by a skillfully constructed program, is to work through several samples of programmed material. Table 6-1 shows a set of frames designed to teach third- or fourth-grade children to spell the word *manufacture*. Of course, the machine presents only one frame at a time to the student, who must respond to that frame before appearance of the next. Skinner's commentary on the set follows the table.

It is important to bear in mind that the student receives immediate reinforcement as soon as each copying response is completed; in this case, the student moves sliders to expose the correct letters in the open squares.

The foregoing set illustrates the technique of *vanishing*, whereby prompts, or cues, for correct response are progressively diminished until the response is emitted solely under internal control, or, as one might say, from memory. Thus, the student is first asked to spell *manufacture* with the full word in view, then with only some of the correct letters available, and finally in the absence of any visual support. Vanishing is an important programming technique since the object of much classroom teaching is to transfer new material from textual control to intraverbal control.

The frames of Table 6-1 teach the spelling of only one word. Skinner estimated that, at five or six frames per word, the teaching of four grades of spelling might require as many as 25,000 frames but that only about 15 minutes per day on a machine should suffice to mediate that many responses. A teacher could hardly provide that many reinforcements to a single student.

Table 6-2 shows a set of frames designed to teach a high school physics student about the emission of light from an incandescent source. Again, the machine presents only one item at a time and requires a response from the student before it will progress to the next item. The student writes the word judged to best complete the item and then moves a lever that uncovers the correct word, shown in the right-hand column of Table 6-2. The reader should work through the program frame by frame, covering the correct entries with a card and uncovering each entry only after completing each frame. Skinner's commentary follows the table.

Table 6-2. Part of a program in high school physics. The machine presents one item at a time. The student completes the item and then uncovers the corresponding word or phrase shown at the right (from Skinner, 1958b)

Sentence to be completed	Word to be supplied
1. The important parts of a flashlight are the battery and the bulb. When we "turn on" a flashlight, we close a switch which connects the battery with the ——.	bulb
2. When we turn on a flashlight, an electric current flows through the fine wire in the —— and causes it to grow hot.	bulb
3. When the hot wire glows brightly, we say that it gives off or sends out heat and ——.	light
4. The fine wire in the bulb is called a filament. The bulb "lights up" when the filament is heated by the passage of a(n) —— current.	electric
5. When a weak battery produces little current, the fine wire, or ——, does not get very hot.	filament
6. A filament which is *less* hot sends out or gives off —— light.	less
7. "Emit" means "send out." The amount of light sent out, or "emitted," by a filament depends on how —— the filament is.	hot
8. The higher the temperature of the filament the —— the light emitted by it.	brighter, stronger

Table 6-2. (*Continued*)

Sentence to be completed	Word to be supplied
9. If a flashlight battery is weak, the —— in the bulb may still glow, but with only a dull red color.	filament
10. The light from a very hot filament is colored yellow or white. The light from a filament which is not very hot is colored ——	red
11. A blacksmith or other metal worker sometimes makes sure that a bar of iron is heated to a "cherry red" before hammering it into shape. He uses the —— of the light emitted by the bar to tell how hot it is.	color
12. Both the color and the amount of light depend on the —— of the emitting filament or bar.	temperature
13. An object which emits light because it is hot is called "incandescent." A flashlight bulb is an incandescent source of ——.	light
14. A neon tube emits light but remains cool. It is, therefore, not an incandescent —— of light.	source
15. A candle flame is hot. It is a(n) —— source of light.	incandescent
16. The hot wick of a candle gives off small pieces or particles of carbon which burn in the flame. Before or while burning, the hot particles send out, or ——, light.	emit
17. A long candlewick produces a flame in which oxygen does not reach all the carbon particles. Without oxygen the particles cannot burn. Particles which do not burn rise above the flame as ——.	smoke
18. We can show that there are particles of carbon in a candle flame, even when it is not smoking, by holding a piece of metal in the flame. The metal cools some of the particles before they burn, and the unburned carbon —— collect on the metal as soot.	particles
19. The particles of carbon in soot or smoke no longer emit light because they are —— than when they were in the flame.	cooler, colder
20. The reddish part of a candle flame has the same color as the filament in a flashlight with a weak battery. We might guess that the yellow or white parts of a candle flame are —— than the reddish part.	hotter
21. "Putting out" an incandescent electric light means turning off the current so that the filament grows too —— to emit light.	cold, cool
22. Setting fire to the wick of an oil lamp is called —— the lamp.	lighting
23. The sun is our principal —— of light, as well as of heat.	source
24. The sun is not only very bright but very hot. It is a powerful —— source of light.	incandescent
25. Light is a form of energy. In "emitting light" an object changes, or "converts," one form of —— into another.	energy

26. The electrical energy supplied by the battery in a flashlight is converted to ―― and ――.

heat, light;
light, heat

27. If we leave a flashlight on, all the energy stored in the battery will finally be changed or ―― into heat and light.

converted

28. The light from a candle flame comes from the ―― released by chemical changes as the candle burns.

energy

29. A nearly "dead" battery may make a flashlight bulb warm to the touch, but the filament may still not be hot enough to emit light—in other words, the filament will not be ―― at that temperature.

incandescent

30. Objects, such as a filament, carbon particles, or iron bars, become incandescent when heated to about 800 degrees Celsius. At that temperature they begin to ―― ――.

emit light

31. When raised to any temperature above 800 degrees Celsius, an object such as an iron bar will emit light. Although the bar may melt or vaporize, its particles will be ―― no matter how hot they get.

incandescent

32. About 800 degrees Celsius is the lower limit of the temperature at which particles emit light. There is no upper limit of the ―― at which emission of light occurs.

temperature

33. Sunlight is ―― by very hot gases near the surface of the sun.

emitted

34. Complex changes similar to an atomic explosion generate the great heat which explains the ―― of light by the sun.

emission

35. Below about ―― degrees Celsius an object is not an incandescent source of light.

800

Several programming techniques are exemplified by [this] set of frames. . . . Technical terms are introduced slowly. For example, the familiar term *fine wire* in frame 2 is followed by a definition of the technical term *filament* in frame 4; *filament* is then asked for in the presence of the nonscientific synonym in frame 5 and without the synonym in frame 9. In the same way *glow*, *give off light*, and *send out light* in early frames are followed by a definition of *emit* with a synonym in frame 7. Various inflected forms of *emit* then follow, and *emit* itself is asked for with a synonym in frame 16. It is asked for without a synonym but in a helpful phrase in frame 30, and *emitted* and *emission* are asked for without help in frames 33 and 34. The relation between temperature and amount and color of light is developed in several frames before a formal statement using the word *temperature* is asked for in frame 12. *Incandescent* is defined and used in frame 13, is used again in frame 14, and is asked for in frame 15, the student receiving a thematic prompt from the recurring phrase "incandescent source of light." A formal prompt is supplied by *candle.* In frame 25 the new response *energy* is easily evoked by the words *form of* . . . because the expression "form of energy" is used earlier in the frame. *Energy* appears again in the next two frames and is finally asked for, without aid, in frame 28. Frames 30 through 35 discuss the limiting temperatures of incandescent objects, while reviewing several kinds of sources. The figure 800 is used in three frames. Two intervening frames then permit some time to pass before the response 800 is asked for.

Unwanted responses are eliminated with special techniques. If, for example, the second sentence in frame 24 were simply "It is a(n) _____ source of light," the two *very*'s would frequently lead the student to fill the blank with *strong* or a synonym thereof. This is prevented by inserting the word *powerful* to make a synonym redundant. Similarly, in frame 3 the words *heat and* preempt the response *heat*, which would otherwise correctly fill the blank.

The net effect of such material is more than the acquisition of facts and terms. Beginning with a largely unverbalized acquaintance with flashlights, candles, and so on, the student is induced to talk about familiar events, together with a few new facts, with a fairly technical vocabulary. He applies the same terms to facts which he may never before have seen to be similar. The emission of light from an incandescent source takes shape as a topic or field of inquiry. An understanding of the subject emerges which is often quite surprising in view of the fragmentation required in item building. (1958b, pp. 972–974)

Tables 6-1 and 6-2 are examples of *linear programming*, the format favored by Skinner. A linear program moves the student continuously along a single path of learning from beginning to end of a subject matter, requiring the student to supply the response to each item but using such small, carefully constructed steps that, ideally, the program may be completed without error. In contrast, a *branching program* (e.g., Crowder and Martin, 1961) provides the student with several response alternatives to choose among and, depending upon the response elected, directs the student on different paths through the material. A correct choice leads to a new step in the material (perhaps even skipping several steps) and poses new response alternatives. An incorrect choice requires the learner to make another choice, perhaps with a helpful comment on the incorrect response, or returns the learner to an earlier point in the program. The linear program assumes that errors are never helpful other than to indicate the teacher's failure to properly shape the step in question, whereas a branching program assumes that errors may be helpful in diagnosing the specific nature of the learner's deficiency and the remedial steps required.

But whether the program is linear or branching the aim is to allow each student to proceed on his or her own through the subject matter, and therefore at a rate best suited to individual motivation, interests, abilities, and prior learning. "Fast" and "slow" learners are then treated as individuals in the instructional process but without the stigma that often attaches to grouping classes under these labels. Since all learners, regardless of rate, are ultimately brought to mastery of the subject, as defined by errorless performance on the programmed material, the traditional concept of a course grade becomes meaningless. All students eventually earn a grade of A; lower grades merely indicate the amount of material covered by the student at a particular time.

The program samples of Tables 6-1 and 6-2 are brief but nevertheless make clear that writing a program is not an easy matter. In contrast to traditional teaching formats, program construction demands much greater specificity in dealing with the learning process. The programmer must (1) specify the actual behaviors (usually verbal) to be achieved in the course, (2) specify the responses the learner is likely to bring to the course at the outset, (3) specify the precise stimulus-response steps that will move the learner from the starting repertoire to the behaviors desired, and (4) design each step so as to maximize positive transfer throughout the program. Indeed, those who advocate programmed learning argue that one of its primary benefits is that it forces the teacher to deal with course goals and procedures in planned and concrete ways.

In writing a program the instructor can, of course, draw upon the experimental analysis of behavior and in particular upon the analysis of verbal behavior. But equally important is the consideration that programmed learning ensures that each student will leave behind a record of correct and incorrect responses, and these data are the ultimate testimony to the effectiveness of any program segment. In fact, writing a program is usually a protracted process involving continual interaction between principles and empirical results until a high state of program effectiveness is achieved.

Perhaps the most common objection to teaching machines and programmed learning is that they may make learning too easy for the student. The argument runs that we want students not merely to make correct responses but to learn to think about the subject matter, and thinking is unlikely unless the student is confronted by hard, error-inducing material. In Skinner's view the argument that the material must be made difficult in order to teach thinking is essentially a rationale for a confusing presentation or for the use of the threat of failure as a motivating device to keep the student at an otherwise uncompelling task. Students who learn to deal with difficult and confusing situations may be better students not because they have benefited from such hurdles, but because they were better students at the outset. Rather than deliberately make education difficult in order to teach thinking, Skinner argued we should analyze the behavior called thinking and incorporate that analysis into our instructional procedures. In this regard he claimed,

> The machine has already yielded important relevant by-products. Immediate feedback encourages a more careful reading of programmed material than is the case in studying a text, where the consequences of attention or inattention are so long deferred that they have little effect on reading skills. The behavior involved in observing or attending to detail—as in inspecting charts and models or listening closely to recorded speech—is efficiently shaped by the contingencies arranged by the machine. And when an immediate result is in the balance, a student will be more likely to learn how to marshall relevant material, to concentrate on specific features of a presentation, to reject irrelevant materials, to refuse the easy but wrong solution, and to tolerate indecision, all of which are involved in effective thinking." (1958b, p. 975)

As the analysis of verbal behavior revealed, Skinner viewed thinking as simply another class of behavior to the world. There should be nothing different in principle about facilitating the specific behaviors that make up what is called thinking than about facilitating the specific behaviors that define rote learning or, for that matter, any other psychological process.

The impetus which Skinner gave to the concept of automated teaching continues in force, although developments have not been entirely as he envisaged. In general, the past 20 years have seen a decreasing emphasis upon the machine aspects of programmed learning but an increasing acceptance of the principles that underlie programmed learning. As we have noted, the machine is clearly secondary to the material within it, and much the same benefits as provided by a teaching machine can be gained whenever the student can be trusted to follow the rules of a programmed text. Consequently, programmed textbooks and study guides have been developed for a wide variety of subjects and the field of programmed learning is today a large-scale educational and commercial enterprise. Skinner himself co-authored, with J. C. Holland (1961), a programmed text in introductory psychology (a purely descriptive introductory psychology, of course!). Another factor contributing to the demise of teaching *machines* of the type suggested by Skinner is the development of computer technology, which has far outstripped the capabilities of the earlier machines. For example, time-sharing computers can control the learning of many students simultaneously and have far greater power and flexibility in both presentation of materials and interaction with the operator. Most important, among those who today espouse programmed learning, technology of any sort is seen simply as an aid to implementing the principles stressed by Skinner, namely, individualizing the teaching process, reducing the traditional aversive control of the learner (e.g., the threat of failure), increasing the learner's responsiveness, and providing frequent and immediate reinforcement for appropriate responses by the learner. These ideas, which lie at the heart of teaching machines and programmed learning, are very influential in educational psychology today.

Other Applied Writings

The bulk of Skinner's other applied writings was intended for a broader audience. *Science and Human Behavior* (1953) is addressed to the general reader and provides a comprehensive explanation of human activity—individual, social, and institutional—from the perspective of the descriptive principles. His novel *Walden Two* (1948b) depicts a modern utopia fashioned upon reinforcement principles and presents in a highly provocative way both the promise of scientific psychology for the betterment of society and its challenge to traditional values. In *The Technology of Teaching* (1968) Skinner pursues the ideas behind programmed learning to a general treatment of the problems of education. *Beyond Freedom and Dignity* (1971) defines the implications of behavioral science, and in particular the assumption of environmental determinism, for the time-honored concepts of individual freedom, responsibility, and dignity. Skinner also contributed an important series of papers devoted to behavioral analyses of psychiatry and mental illness (e.g., 1954, 1959). Numerous other applied papers cover a potpourri of topics ranging from a "Skinner box" for the care of babies to the management of behavior in business and industry.

Throughout these writings Skinner pursued two fundamental propositions. The first is that if progress is to be made in the control of human activity we must abandon explanations based on appeals to inner processes and seek instead the causes of human behavior, and the means of its control, in environmental variables. All "inner" explanatory concepts—instinct, trait, will, psychic energy, need, memory, id, personality, excitatory tendency, and the like—are inherently metaphysical and divert attention from the true controllers of behavior, its environmental determinants.[11] The second proposition is that reinforcement is the paramount determinant of human behavior, and therefore the starting point of any effort to alter a given behavior should be the systematic examination and manipulation of its reinforcing consequences. The first proposition is the continuing expression of the philosophy of science and psychology that Skinner arrived at in his very first paper on the concept of the reflex. The second derives from the experimental analysis of behavior and incorporates the behaviorist's faith that the principles of conditioning hold for the natural life behavior of all organisms. One may debate these ideas, but it is certainly true that Skinner's systematic, vigorous, and ingenious application of these propositions has done a great deal to persuade the general public of the relevance of scientific psychology to everyday affairs.

While we shall not further review Skinner's popular writings here, a sense of their central concern, and of his eloquence as spokesman for behaviorism, is conveyed by these closing passages from *Beyond Freedom and Dignity*. Skinner is here concerned with those who resist a scientific analysis of human behavior because it threatens the traditional view of humans as self-directing beings who freely choose between alternative courses of action and thereby attain a dignity denied to lower beings.

> Science has probably never demanded a more sweeping change in a traditional way of thinking about a subject, nor has there ever been a more important subject. In the traditional picture a person perceives the world around him, selects features to be perceived, discriminates among them, judges them good or bad, changes them to make them better (or, if he is careless, worse), and may be held responsible for his action and justly rewarded or punished for its consequences. In the scientific picture a person is a member of a species shaped by evolutionary contingencies of survival, displaying behavioral processes which bring him

under the control of the environment in which he lives, and largely under the control of a social environment which he and millions of others like him have constructed and maintained during the evolution of a culture. The direction of the controlling relation is reversed: a person does not act upon the world, the world acts upon him.

It is difficult to accept such a change simply on intellectual grounds and nearly impossible to accept its implications. The reaction of the traditionalist is usually described in terms of feelings. One of these, to which the Freudians have appealed in explaining the resistance to psychoanalysis, is wounded vanity. Freud himself expounded, as Ernest Jones has said, "the three heavy blows which narcissism or self-love of mankind had suffered at the hands of science. The first was cosmological and was dealt by Copernicus; the second was biological and was dealt by Darwin; the third was psychological and was dealt by Freud." (The blow was suffered by the belief that something at the center of man knows all that goes on within him and that an instrument called will power exercises command and control over the rest of one's personality.) But what are the signs or symptoms of wounded vanity, and how shall we explain them? What people *do* about such a scientific picture of man is call it wrong, demeaning, and dangerous, argue against it, and attack those who propose or defend it. They do so not out of wounded vanity but because the scientific formulation has destroyed accustomed reinforcers. If a person can no longer take credit or be admired for what he does, then he seems to suffer a loss of dignity or worth, and behavior previously reinforced by credit or admiration will undergo extinction. Extinction often leads to aggressive attack.

Another effect of the scientific picture has been described as a loss of faith or "nerve," as a sense of doubt or powerlessness, or as discouragement, depression, or despondency. A person is said to feel that he can do nothing about his own destiny. But what he feels is a weakening of old responses which are no longer reinforced. People are indeed "powerless" when long-established verbal repertoires prove useless. For example, one historian has complained that if the deeds of men are "to be dismissed as simply the product of material and psychological conditioning," there is nothing to write about; "change must be at least partially the result of conscious mental activity."

Another effect is a kind of nostalgia. Old repertoires break through, as similarities between present and past are seized upon and exaggerated. Old days are called the good old days, when the inherent dignity of man and the importance of spiritual values were recognized. Such fragments of outmoded behavior tend to be "wistful"—that is, they have the character of increasingly unsuccessful behavior. . . .

The traditional conception of man is flattering; it confers reinforcing privileges. It is therefore easily defended and can be changed only with difficulty. It was designed to build up the individual as an instrument of counter-control, and it did so effectively but in such a way as to limit progress. We have seen how the literatures of freedom and dignity, with their concern for autonomous man, have perpetuated the use of punishment and condoned the use of only weak nonpunitive techniques, and it is not difficult to demonstrate a connection between the unlimited right of the individual to pursue happiness and the catastrophes threatened by unchecked breeding, the unrestrained affluence which exhausts resources and pollutes the environment, and the imminence of nuclear war.

Physical and biological technologies have alleviated pestilence and famine and many painful, dangerous, and exhausting features of daily life, and behavioral technology can begin to alleviate other kinds of ills. . . . There are wonderful possibilities—and all the more wonderful because traditional approaches have been so ineffective. It is hard to imagine a world in which people live together without quarreling, maintain themselves by producing the food, shelter, and clothing they need, enjoy themselves and contribute to the enjoyment of others in art, music, literature, and games, consume only a reasonable part of the resources of the world and add as little as possible to its pollution, bear no more children than can be raised decently, continue to explore the world around them and discover better ways of dealing with it, and come to know themselves accurately and, therefore, manage themselves effectively. Yet all this is possible, and even the slightest sign of progress should bring a kind of change which in traditional terms would be said to assuage wounded vanity, offset a sense of hopelessness or nostalgia, correct the impression that "we neither can nor need to do anything for ourselves," and promote a "sense of freedom and dignity" by building "a sense of confidence and worth." In other words, it should abundantly reinforce those who have been induced by their culture to work for its survival.

An experimental analysis shifts the determination of behavior from autonomous man to the environment—an environment responsible both for the evolution of the species and for the repertoire acquired by each member. Early versions of environmentalism were inadequate because they could not explain how the environment worked, and much seemed to be left for autonomous man to do. But environmental contingencies now take over functions once attributed to autonomous man, and certain questions arise. Is man then "abolished"? Certainly not as a species or as an individual achiever. It is the autonomous inner man who is abolished, and that is a step forward. But does man not then become merely a victim or passive observer of what is happening to him? He is indeed controlled by his environment, but we must remember that it is an environment largely of his own making. The evolution of a culture is a gigantic exercise in self-control. It is often said that a scientific view of man leads to wounded vanity, a sense of hopelessness, and nostalgia. But no theory changes what it is a theory about; man remains what he has always been. And a new theory may change what can be done with its subject matter. A scientific view of man offers exciting possibilities. We have not yet seen what man can make of man. (pp. 211–215)

SKINNER'S SYSTEM: SOME CONCLUDING COMMENTS

[T]he most rapid progress toward an understanding of learning may be made by research which is not designed to test theories. An adequate impetus is supplied by the inclination to obtain data showing orderly changes characteristic of the learning process. An acceptable scientific program is to collect data of this sort and to relate them to manipulable variables, selected for study through a common sense exploration of the field. (Skinner, 1950, p. 215)

Can it be said that the experimental analysis of behavior has demonstrated the truth of these statements? Is the behavior system superior to the other theoretical approaches we have considered? It is for the reader to decide these questions, but some relevant considerations will be underscored here.

Certainly the experimental analysis of behavior has resulted in extremely important methodological and empirical contributions. It is widely recognized that, as compared with the maze and related apparatus, the Skinner box and its associated technology provide superior experimental control and greater objectivity and efficiency of data collection. Operant test procedures have become standard research tools of the experimental learning psychologist, regardless of theoretical persuasion and, given the adaptability of these procedures, regardless of subject population. The advances in measurement and in standardization of research afforded by the Skinnerian methodology are most significant contributions.

Similarly, Skinner and other workers in the operant framework have added greatly to our knowledge of basic conditioning phenomena, especially in the area of reinforcement. The utility of the descriptive principles is widely acknowledged both among basic researchers and among those interested in practical problems of behavior. In fact, it might be noted that there may be a tendency, particularly among workers in applied fields, to credit Skinner's system too much as a source of principles of conditioning. As we have noted, the principles encountered in Skinner's system are largely antedated in the work of other investigators. The difference, and it is a critical one, is that they appear in Skinner's system divested of theoretical entanglements and with a simplicity and clarity imparted by their definition within a single research setting. In consequence, the empirical principles of conditioning appear in Skinnerian psychology in their most *communicable* form as the legacy of S-R research.

The fact that Skinner expressed his principles in empirical form means that the system cannot be assessed by the test applied to other S-R formulations, namely, does the system predict new knowledge that is consistent with its principles? In theories of the hypothetico-deductive type, the principles come first and the test data second, whereas in Skinner's system the data come first, the principles second. The principles thus must *always* be right, at least when applied within the arena in which their empirical base is generated. That Skinner's principles are essentially immune to the usual test of theory is frustrating to many, but nontestability is not a fair criticism to direct against a system that was explicitly designed to avoid theory-testing and that sees the issue of testability only in terms of the improved control of behavior.

However, a question that may properly be raised in relation to an empirical behavior system concerns the heavy burden it places on the observational skills of the systematist. Since the principles of the system go no further than its observations, it becomes a critical matter which behaviors are selected for analysis and which aspects of the environment are manipulated "through a common sense exploration of the field." Only if the observational base is sound can the principles be properly inclusive and properly expressed. *The Behavior of Organisms*, reporting as it does the development of the free operant and Skinner box technology, the pioneering work on the principle of reinforcement, schedules of reinforcement, response differentiation, generalization, discrimination, and still other phenomena, testifies to Skinner's rare observational skills and indeed reflects a most *uncommon* sense exploration of the field. But the descriptive principles are meant to constitute a behavior *system*, that is, an approach not just for an individual scientist but for the field as a whole, and it is by no means clear that the behavior system engenders in its adherents the observational skills it prizes.

One might fall back here on the test that Skinner would apply— does the system enable improved control of laboratory and natural life behaviors? Few would dispute that Skinner's system has taught us more than any other S-R formulation about the reliable and efficient production of laboratory behavior or has provided more useful suggestions for the control of behavior in natural life situations. But again, the program for the control of laboratory behavior and the extrapolation from descriptive principles to practice have been largely *outlined* by one person—Skinner—and so we face again the same question of the merits of the system as a system. In sum, to what degree are the accomplishments of the behavior system unique to Skinner versus to the system?

One may also question Skinner's reliance upon the control of behavior as the criterion of progress in understanding learning. The phrase *control of behavior* is meant to imply not only the ability to bring about desired behaviors but also an emphasis upon the efficient production of behavior. Is the isolation of principles and variables that are of import for the practical control of behavior a clear sign that a system is making "the *most rapid* progress toward an understanding of learning"? To answer this question let us consider one of the fundamental theoretical issues bypassed by Skinner—latent learning. As noted earlier, a system that deals only with performance has no room for "latent" learning, but one may nevertheless ask why Skinner has shown no interest in a phenomenon which, after all, does bear upon performance changes, however delayed the changes might be. The reasons are revealing of the priorities in Skinner's system. The focus on the control of behavior leads in turn to a strong concern with operations that have readily demonstrable effects on probability of response. Given this concern, it is difficult to imagine Skinner ever selecting for investigation an operation (nonreinforced exposure to the elements of the learning situation) so unlikely on a priori grounds to affect behavior in any clearly discernible fashion. Moreover, the exposure manipulation is an operation that has no effect upon behavior until it is coupled with a drive or reinforcement operation. Since exposure by itself has no pragmatic consequences, its separation from other operations can be seen as an inefficient means of achieving behavior change.

Such a view of latent learning could be said to be shortsighted in that there might be cases in which exposure followed by an appropriate operation yields effects that differ in important respects from those following conjoint manipulation of these operations. Consider the following possibility. Suppose that an experimentally naive animal fully satiated for food is placed in a Skinner box for a brief period with the bar in place and the food magazine operative. The animal's movements would likely result in several bar depressions and pellet deliveries which the subject, being satiated, would ignore. If the animal is now made hungry and returned to the Skinner box, the latent learning research suggests we would probably observe some facilitation of bar press behavior as a result of the prior nonreinforced experience. But it is at least conceivable that bar press acquisition would proceed even *more* rapidly, in terms of overall training time, than under the usual procedure, which typically involves successive periods of apparatus exposure, magazine training, and bar press training. The comparison has not been made experimentally,[12] but it poses the possibility of a potent effect from a seemingly inefficient behavior change operation.

Another and more generally significant possibility is that exposure may significantly affect the strength of a subsequently reinforced behavior. This could occur if exposure to a new learning situation allowed completion of exploratory or habituation processes that would otherwise be suppressed by, though remain competitive with, reinforced behavior in that situation. If so, the most patently efficient means of shaping behavior in a novel surround, namely, early and sutained reinforcement in the learning situation, would lead to less stable behavior than when a period of stimulus exploration is first allowed. This possibility has been suggested by Pereboom (1957) to account for variable outcomes in latent learning-type experiments, and Zeaman (1976) has summarized data indicating that reinforcement does indeed compete with stimulus novelty in the manner indicated.

However, such possibilities are not likely to have a high priority for investigation when the focus is on immediately demonstrable changes in behavior. The issue of latent learning thus throws into relief Skinner's emphasis on effecting the control of behavior at the sacrifice of possible broader understanding of behavior-environment relations. In this light, Skinner could be said to be more interested in the control of behavior than in the understanding of behavior. To this assertion, Skinner would doubtless reply that we know we understand behavior when we can control it. And to this it might be replied that we should seek to know the effects of all alternative ways of producing a behavioral outcome, and that is what is meant by understanding.

The point here is not that an approach guided by the improved control of behavior is wrong but that it certainly may be questioned. Specifically, it risks overlooking important though nonsalient aspects of the organism-environment relationship. On the other hand, one will not get far in seeking to understand all alternative ways of producing behavioral outcomes without explicit theoretical guidance, and that route, as Skinner has pointed out, is not immune to error either.

Finally, a word about the role of mediation in Skinner's system. Our treatment of the controversies arising from the opposition of the response-reinforcement and cognitive theories concluded that henceforth an S-R analysis would have to be critically concerned with processes mediating between observed stimuli and final behavioral output. It was also noted that Guthrie, despite his determined peripheralism, had recourse to intermediary processes. Skinner, too, makes reference to mediating events. We find, for example, that response-produced cues play a considerable role in accounting for behavior under schedules of reinforcement; that emotional reactions to nonreinforcement account for abrupt changes in response strength during extinction; that punishment suppresses behavior by means of competing emotional reactions; that much verbal behavior is controlled by other, covert verbal behavior; and that self-editing mediates overt verbal productions.

Recourse to mediation processes in accounts of learning is inevitable; indeed, learning by definition refers to behavior that cannot be accounted for by present stimuli only. What differentiates theories is not whether mediation is used, but how it is used. And true to his credo, Skinner's use of mediation is sparing and hedged with care to avoid properties other than those available to direct observation. Thus, the role of response-produced cues in schedules of reinforcement is experimentally tested by the "added clock" procedures (pp. 250–51), which mimic in fully objective terms the properties assigned to inter-

nal response-produced cues; emotional reactions to nonreinforcement
are patently indexed by such behaviors as attacking or biting the bar;
conditioned emotional reactions are operationally defined with great
care, and a paradigm is provided for objective analysis of their interaction with operant behavior; and covert verbal behaviors are held to
have no other properties than overt verbal behavior.

Most important, the emphasis in Skinner's treatment of mediated
behavior is different from that of other S-R theorists. Mediational
theories generally focus on processes within the organism that supplement the external stimulus in some fashion to account for the observed behavior. But Skinner is not willing to talk about events tran-
spiring within the organism; rather he usually seeks to account for
discrepancies between observed inputs and outputs solely by reference
to external events in the organism's *history*. For example, in discussing the concept of secondary reinforcement we noted (pp. 281–82)
that Miller's classic experiment on learned fear or anxiety assumed
that an internal state, anxiety, becomes conditioned to situational cues
and then functions as a drive to mediate the emission and learning
of new escape responses in the presence of those cues. Skinner's account of the same behavior was simply that the animal learns to escape the cues because it was shocked in their presence. Similarly,
self-editing of verbal behavior can be said to occur because the individual was punished for deficient verbal productions in the past; no
reference to intervening processes is necessary. In general, mediation
in Skinner's system is essentially *the organism's reinforcement history*.
While it is true that we never have complete knowledge of an organisms's reinforcement history and therefore that a reinforcement history is no more directly observed than is a covert mediating process,
it is also true that a reinforcement history is fully specifiable in terms
of observable environmental events and can be reconstructed in an
experimental subject in test of an explanatory role. By generally limiting his explanatory terms to present and past environmental stimuli,
Skinner was thus able to deal with mediated behavior but at the same
time keep his description at the level of observable events.

It must be concluded that Skinner has been remarkably successful
in building a comprehensive and fruitful interpretation of behavior
based quite strictly on observed stimulus-response correlations and "a
common sense exploration of the field." His reluctance to employ
central explanatory concepts was doubtless fueled in part by the general behaviorist reaction against mental psychologies. But it was primarily his own historical analysis of the reflex concept that convinced
him that explanations that go beyond observables invite meaningless

or false claims. Certainly, Skinner was the most rigid and successful of all S-R theorists in keeping to observed stimulus-response relationships. This is probably not unrelated to the fact that his system has been the most enduring.

7

Some Relations to Contemporary Theory: Current S-R Models and Information Processing Theory

Learning theory is a continuous and reactive matter and should be judged whole.

We have seen that global S-R theory was the product of several historical factors—the emergence of behaviorism as a reaction against the failures of mental psychology, the discovery of conditioning with its promise of truly generalizable principles of learning, and the influence of Darwinian evolutionary theory and the related accomplishments of the new science of animal psychology. These developments came together to provide the rationale for a general attack on the problem of learning. The hope was that a universal explanation of learning could be achieved through experimental stimulus-response analysis of the performance of several "representative" lower species. Present-day thinking would say that the sights of the global theorists were set too high and their methods inadequate. But when placed in historical context, their aims and methods were entirely reasonable.

It is true that global theory failed in the sense that an all-embracing explanation was not achieved; it is today an easy matter to single out phenomena that pose significant difficulties for each theoretical approach. But we have argued the need to read evidence in larger chunks, and from this perspective each of the approaches can lay claim to major achievements. To be a qualified success is the way of theory. Outright success is as fatal to a theory *as theory* as is outright failure. Theories persist if they are useful, and they are useful if they provide economical accounts of the known and guidance in dealing with the unknown. Global theories served these functions well for several decades. It must be remembered that the global theorists had the most difficult task facing a behavioral (objective) theory of learning—that of beginning. And the problem of beginning was that of bringing order to the field and rendering it amenable to scientific analysis. When Hull surveyed the field in 1935 he was dismayed to find literally a dozen different psychologies, most of questionable scientific standing, "earnestly defending themselves against a world of enemies" (1935a). We cannot solely credit Hull and the other global theorists with reducing this confusion, but surely their efforts were a major contributing factor. Granted, the principles and methods of global theory did not prove universal, but they were sufficiently generalizable and incisive to permit their fruitful extension to analyses of learning in other fields of psychology, such as child development, social psychology, clinical psychology, and educational psychology, extensions which in turn rendered immensely more difficult the attainment of a single theory of learning. And it was largely the research stimulated by global theory that defined the boundaries of phenomena requiring more specialized principles and techniques for their proper understanding. From these perspectives, it can be argued that the decline of interest in global theory was as much due to its successes as to its failures.

Our account of global theory has not covered a significant part of the "whole" necessary to its evaluation, and that part is its relationship to other theory and research in learning. As we have noted, interest in global theory construction and testing was replaced by interest in more limited theories designed to account for particular phenomena of learning, e.g., theories of such phenomena as discrimination learning, extinction, and the partial reinforcement effect. Several important "minitheories" coexisted with global theory, notably in the areas of discrimination learning and aversive learning, but the later period saw the active cultivation of minitheory. Many of these more restricted theoretical efforts were strongly influenced by the concepts of global theory and often reprised the conceptual oppositions of global theories. For example, the fundamental oppositions of cognitive and response-reinforcement theories were expressed in conflicting theories of discrimination learning, and opposing treatments of punishment proceeded from the response-reinforcement and Skinnerian formulations. The concepts of Guthrie's theory became the basis of several important quantitative models of learning. Minitheories based on elaboration of the r_g concept provided very fruitful analyses of such important phenomena as the action of incentives (expectations of rewards and punishments), the nature of secondary reinforcement, the partial reinforcement effect, punishment and avoidance learning, and discrimination learning. While a companion

volume would properly be required to detail the relation of global
theory to theories dealing with specific phenomena or subareas of
learning, we shall briefly consider several important theoretical efforts
that illustrate how the basic concepts of global theory have persisted
to guide the experimental analysis of learning.

SOME CONTEMPORARY ANALYSES

Frustration Theory

While it was early noted that withdrawal of reward may elicit emotional reactions (e.g., Skinner, 1938), global theory generally paid
little attention to the emotional by-products of nonreward. But Amsel's work (1958, 1962, 1967) has since made clear that behavior theory must make explicit provision for an active, emotional motivational component of nonreward conditions.

Amsel has argued that the failure to receive an anticipated reward
produces a primary emotional *frustration reaction* which, in interaction with the prevailing conditions, is an important determiner of the
organism's subsequent behavior in the face of nonreward. Note that
the frustration reaction is assumed to be elicited not by nonreward
per se but by failure to receive an *anticipated* reward, and to deal
with anticipated rewards Amsel chose the language of fractional response theory. He assumed that repeated rewards in a given situation
develop anticipatory reward responses (r_w-s_w's) in the manner of Hull's
r_g-s_g's, and it is the presence of r_w-s_w's that gives nonreward its frustrative effect. One consequence of the primary frustration reaction
(R_F) is immediate disruption of the learned behavior. For example,
during extinction of a continuously reinforced bar press response the
animal may be observed to become generally agitated and to periodically abandon bar pressing in favor of attacks upon the bar or food
dish. But the more interesting and significant consequences of the R_F
arise from the consideration that with repeated nonreward, the R_F
itself should become anticipatory. That is, the R_F is part of the total
goal reaction during nonrewards, and just as other parts of the goal
reaction become conditioned anticipatory responses, so fractions of
the frustration reaction (r_f-s_f's) should develop throughout the behavior sequence. By considering the action of the r_f-s_f at different stages
of conditioning, Amsel was able to account for a number of important effects associated with the withholding or reduction of reward.

The most important of these phenomena is the partial reinforcement effect (PRE)—the fact that intermittent reward leads to more persistent behavior than continuous reward. The PRE has evoked tremendous interest, doubtless because it is patently counterintuitive to find that rewarding a behavior on every occasion leads to less resistance to extinction than rewarding only some occurrences of the behavior. This paradox has proved as resistant to fully satisfactory resolution as it has proved provocative to theoretical analysis.

Amsel's explanation of the PRE involves a stage analysis of the behavior processes at work. While the analysis is applicable to any conditioned response, we shall develop it here with reference to partial versus continuous reinforcement of a simple approach response in a runway. Early in the training of partially rewarded subjects, rewarded trials strengthen approach behavior in the runway and begin the conditioning of anticipatory reward responses (r_r's). Since r_r's are poorly developed at this point, nonrewarded trials do not effect any significant amount of frustration. As training progresses, the r_r develops to provide the necessary condition for a frustration reaction on nonrewarded trials. However, the R_F and r_f will develop gradually since the r_r itself is only gradually strengthened through training. Hence we can expect a period during which the subject is rewarded for approaching in the presence of an initially weak but increasingly strong r_f. In this way, the s_f, the stimulation arising from the r_f, becomes an increasingly important part of the stimuli conditioned to evoke approach behavior in the runway. By the final stage of partial reward training, the subject has learned to approach the goal in the presence of strong anticipatory frustration reactions. When extinction of runway approach is then carried out, the sustained withdrawal of reward produces the very condition, frustration, that has become a cue for continued approach behavior; the subject therefore persists in approaching the nonrewarded goal.

In contrast, consider the subject who has acquired the runway approach behavior under an equal number of rewards but administered for each and every approach response. At the end of acquisition training, such a subject has strong anticipatory responses only for reward. The onset of extinction, then, elicits a large-magnitude frustration reaction for which the subject is not "prepared." As the frustration reaction becomes anticipatory with subsequent nonrewarded runs, the r_f can act only to compete with and weaken the behavior approach chain, thereby hastening extinction. In short, Amsel's analysis details how partial reinforcement literally trains the organism to persist in

Phases of experiment	(1) Preliminary learning	(2) Acquisition running response	(3) Extinction running response
Apparatus	(a) Short black wide box	(b) Long white narrow runway	(b) Long white narrow runway
Motivation	Hunger	Thirst	Thirst
Experimental conditions	*Running* Continuous (RC) Partial (RP) *Jumping* Continuous (JC) Partial (JP) *Climbing* Continuous (CC) Partial (CP)	*Running* Continuous reward	*Running* Continuous nonreward

Figure 7-1. Design of the Ross (1964) experiment.
(After Amsel, 1967, p. 18. Copyright by Academic Press, Inc. Reprinted by permission.)

the face of frustrative nonreward, whereas continuous reward simply sets the stage for a disruptive and competing frustration reaction.

An experiment by Ross (1964) provides a particularly compelling test of Amsel's hypothesized mechanism of persistence, $r_f\text{-}s_f \rightarrow R_{approach}$. In the first phase of a three-phase training procedure (see Figure 7-1), hungry rats learned to make one of three distinctly different approach responses in a short, black, wide box. One third of the rats simply ran across a short segment, another third jumped across a gap, and the remaining subjects climbed over a vertical obstacle to reach the goal. Half of each group received food on 50 percent of the acquisition trials with rewarded and nonrewarded trials intermixed throughout training, and the other half of each group was rewarded on every trial. In the second phase of training all subjects learned a continuously rewarded running approach response but under markedly different experimental conditions; their motivation was changed from hunger to thirst, and they ran in a long, white, narrow runway with high wire-mesh walls. The final phase involved extinction of the running response in the latter apparatus. The critical question was, what is the effect of phase 1 acquisition on phase 3 extinction?

Reasoning from Amsel's theory, we can answer as follows. Phase 2 continuous reward training should instate the anticipatory reward responses necessary to bring about a frustration reaction upon extinction of running in phase 3. The s_f will therefore be part of the stimulus complex during phase 3 extinction and should elicit whatever responses have been conditioned to it. Of course, subjects who received

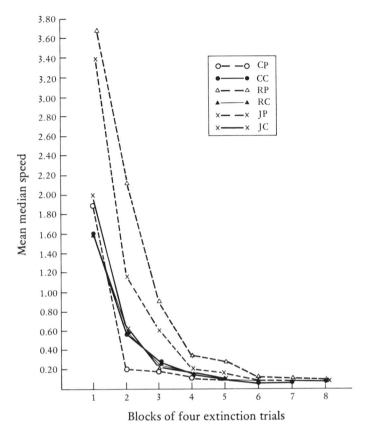

Figure 7-2. Running speeds during phase 3 extinction trials of the Ross experiment. (Adapted from Amsel, 1967, p. 20. Copyright by Academic Press, Inc. Reprinted by permission.)

100 percent reward in phase 1 have not acquired any specific response tendencies to the s_f, and hence for these subjects the frustration reaction in phase 3 will simply act to disrupt and retard approach behavior. But for subjects partially rewarded in phase 1, the s_f should have become conditioned to the specific responses trained in that phase. Since the running and jumping behaviors of phase 1 are fully compatible with the running called for in phase 3, transfer of these response tendencies via the s_f will serve to maintain running speed in phase 3, thus yielding a mediated partial reinforcement effect. That is, subjects partially rewarded for running and jumping in phase 1 should show greater resistance to extinction than those continuously rewarded. However, climbing a vertical obstacle is clearly incompatible with running, and hence the transfer of climbing responses to

phase 3 will depress approach behavior. Therefore, subjects partially rewarded for climbing in phase 1 should extinguish *more* rapidly than their continuously rewarded counterparts.

The phase 3 extinction results are shown in Figure 7-2. The figure plots running *speed* rather than traversal time, and thus higher scores indicate better maintenance of the running response. The data are in perfect accord with Amsel's theory. The three continuously rewarded groups (solid-line curves) are together in the middle of the figure and do not differ significantly from one another. Compared with these curves, the two upper dashed-line curves for the partially rewarded running and jumping groups show the PRE, as predicted. Both of these groups maintain significantly faster running than the continuously rewarded running and jumping groups. Again as predicted, the running speed of the partially rewarded climbing group (the lowest curve) is significantly lower than that of the continuously rewarded climbing group. Of considerable interest is the fact that animals in the partially rewarded climbing group were actually observed to repeatedly climb the high wire-mesh walls of the apparatus during phase 3.

It should be particularly noted that this test for the operation of the r_f-s_f mechanism was carried out under conditions that appear to preclude any other explanation of the data. Given the markedly different external environments of phases 1 and 3 as well as the different motivational states and overt behaviors involved, the only test condition common to phases 1 and 3 of the partially reinforced animals is nonreward. Therefore, it must be some common reaction to this condition that accounts for the transfer of phase 1 training to phase 3. And it seems reasonable to conclude that this common reaction is the frustration reaction, with its attendant links to phase 1 behaviors.

Amsel's theory has proved a particularly comprehensive explanation of the PRE. For one thing, it accounts for the fact that the PRE shows up even with long intertrial intervals during the rewarded period, a finding that has been troublesome for competing theories of the PRE based on the persistence of stimulus traces from nonreward trials to reward trials. Since in Amsel's theory the mechanism of persistence is based on within-trial contiguity of r_f-s_f→*approach*→*reward*, the PRE should be independent of absolute intertrial interval. The frustration analysis also predicts specific features of behavior during acquisition under partial reward. For example, since r_f-s_f builds up gradually, we would expect that comparison of the performance of partially and continuously rewarded subjects during acquisition would show an initial period of no difference followed by a period of greater behavior variability in partially rewarded subjects reflecting the early conditioning of competing anticipatory frustration responses, followed by the disappearance of that variability, reflecting the ultimate integration of r_f-s_f into the approach stimulus complex. These expectations, along with others stemming from motivational properties ascribed to the r_f, have been confirmed in a number of studies (Amsel, 1967).

Considerably more could be said about frustration theory, but our primary concern here is not with frustration theory itself but with frustration theory as an exemplar of the lineage of global theory. Enough has been said to make clear that Amsel's frustration theory is a particularly fruitful contemporary expression of Hull's fractional anticipatory response concept.

Expectancy Theory

Global theory implicitly assumed that innate instrumental response tendencies play a relatively trivial role in the modification of voluntary behavior. To be sure, the global theorists made provision for innate instrumental responses, but these were generally seen as what might be called *starter behaviors*, that is, responses which the organism might emit upon introduction to the learning situation but which quickly are replaced by the behavior favored by the strengthening operations of the theory. However, recent observations indicate that innate behaviors play a considerably more directive role in learning than was credited by global theory. These observations in turn have fostered conceptions which place learning in the context of adaptive biology as a whole. Included among these conceptions are theories which assume that organisms generally do not learn to make particular responses but rather learn expectancies about biologically significant events which then elicit species-characteristic behaviors appropriate to those expectancies. We shall briefly relate several of the phenomena that prompt such a view before considering an exemplar of expectancy theory itself.

Another Look at the "Superstition" Experiment. The reader will recall Skinner's experiment on superstition in the pigeon (pp. 236–38), in which he found that regularly presented but response-independent food reinforcement led to the acquisition of stable behavior patterns differing from bird to bird. Skinner interpreted this outcome as evidence of the strictly mechanical action of reinforcement, strengthening whatever behaviors happen to precede it. Staddon and

Simmelhag (1971) repeated Skinner's experiment but with more detailed observations of the behaviors involved. They defined 16 responses that exhaustively described the pigeon's activity in this situation, e.g., wing flapping, walking, turning in circles, preening, pecking at part of the chamber, and then kept a continuous record of these responses throughout the training. Like Skinner, Staddon and Simmelhag observed the development of behavior chains during the inter-reinforcement intervals, but their fine-grain analysis revealed patterns not expected from the assumption that reinforcement acts blindly on preceding behavior. Their records show that whatever behaviors immediately preceded the initial reinforcements, the birds eventually came to emit *pecking* responses, generally in the area of the magazine wall, just prior to receipt of reinforcement. The behavior chains that developed to fill the earlier portions of the inter-reinforcement intervals were found to be composed of motivationally relevant responses but exclusive of magazine-wall pecking. Staddon and Simmelhag argued that both the terminal pecking responses and the interim behavior chains are instances of species-characteristic behaviors emitted when the organism is in a situation where reinforcement has been received and is presently imminent (terminal response) or is presently not available (interim behaviors).

Instinctive Drift. In 1947 Keller and Marion Breland started a business concerned with the marketing of conditioned operant behaviors. Their aim was to apply learning principles to the traditional field of animal training to achieve the rapid production of "trick" performances that could be used to command audience interest for advertising or entertainment purposes. Several years later they reported the successful development of a number of animal acts for commercial application (Breland and Breland, 1951). For example, a number of chicken acts were developed for county fair booth exhibits. In one act, a hen played a tune on a toy piano, in another a hen performed a tap dance in costume, and in still another a hen "laid" a number of wooden eggs as called for by the audience. Sets of these acts were prepared and shipped all over the United States, the birds playing thousands of performances and showing to as many as 5,000 people in a day. These and other animal acts were shaped on the basis of the principles of reinforcement, successive approximations, and intermittent reinforcement.

A second account ten years later (Breland and Breland, 1961) reported the conditioning for various commercial purposes of some 39 species totaling over 6,000 individual animals, but this report also emphasized some disconcerting breakdowns of conditioned operant behaviors. For example, a raccoon had been readily conditioned to pick up a coin and drop it into a small metal coin bank, but when required to deposit two coins, the raccoon began to rub the coins together for prolonged periods even though this behavior only prevented reinforcement. In another instance, a chicken had been conditioned to hit a small baseball past miniature toy fielders to secure a food reward. But when the walls of the display cage were removed for photography, the bird became very excited and began to chase after the ball, repeatedly pecking at it, even though this behavior was never reinforced. Another example involved the conditioning of pigs to deposit four or five wooden coins in a piggy bank for a single food reinforcement. While the pigs conditioned successfully in this performance, the behavior began to deteriorate with continued practice; deposit of the coins became slower and slower and was progressively replaced by repeated rooting and tossing of the coins.

In these and similar cases described by the Brelands the natural food-getting behavior of the animal appeared to intrude upon the conditioned behavior. The raccoon was showing the sort of rubbing responses that might result in removal of the exoskeleton of a cray fish; the pig's rooting behavior is a characteristic food-seeking response in this species, and the hen's pecking response is, of course, its dominant food-getting behavior. As Breland and Breland put it, these are instances in which "learned behavior drifts toward instinctive behavior."

Auto-Shaping. In the standard operant procedure, the pigeon is trained to peck at a lighted plastic disk (key) through the reinforcement of successive approximations to the desired behavior. But in 1968 Brown and Jenkins reported that key pecking can be auto-shaped simply by unconditional presentation of food after the key is lighted momentarily. In their procedure the key was periodically illuminated for an eight-second period that was followed immediately by a four-second presentation of the food hopper regardless of the bird's behavior. The subjects had earlier been trained to eat out of the hopper. After a number of such lighted key–food pairings the birds began to make orienting movements toward the lighted key, and these movements were followed in later trials by pecking responses directed at the key itself. As Brown and Jenkins noted, the emergence of regular key pecking in these circumstances requires appeal to the species-specific tendency of the pigeon to peck at things it looks at since key pecking is not the initial behavior established by the reinforcement and since the pecking response is irrelevant to receipt of the reinforcement.

An important extension of the Brown and Jenkins experiment was contributed by Williams and Williams (1969). The latter investigators repeated the essential features of the auto-shaping experiment except that key pecking *prevented* reinforcement. That is, a response key was again illuminated several seconds before grain was presented, but pecks on the lighted key turned off the key and prevented presentation of the food hopper. Surprisingly, Williams and Williams observed the development and maintenance of key pecking under these conditions, even though key pecking actually reduced the reinforcement available to the subject! Again, the tendency of a hungry pigeon to peck at things it looks at is revealed as a powerful determiner of the experimental outcome.

Avoidance Learning. In the avoidance paradigm an animal must learn to make some experimenter-designated response within a specified time period in order to postpone or prevent the occurrence of a noxious stimulus. While successful avoidance behavior is often rapidly established, there also have been frequent reports of extreme difficulty or outright failure in avoidance conditioning. This inconsistency in outcome has been particularly troublesome for learning theory because successful and unsuccessful avoidance experiments cannot be distinguished on the basis of the stimulus-response relationships and reinforcement contingencies imposed. Rather, as Bolles's trenchant analysis (1970) has made clear, the distinguishing feature is the response employed. Thus, if a rat must learn to jump out of a compartment to avoid an impending shock, a single shock experience in the compartment may produce successful avoidance thereafter. It would be only slightly more difficult for a rat to learn to avoid shock by moving in timely fashion from an area where shocks are delivered to a consistently safe part of the apparatus. However, if the rat is required to shuttle back and forth on successive trials between areas that are alternately shocked and safe, successful avoidance may require hundreds of trials and may never be achieved by some animals. And if the rat must make a bar pressing or wheel turning response to avoid shock, successful avoidance is unlikely to be achieved.

We can make sense out of these discrepancies, argued Bolles, if we consider that strong species-specific defense reactions (SSDR's) are likely to be evoked by noxious stimulation. Avoidance learning will be rapid or slow to the degree that the conditioning situation favors an SSDR. If we make the reasonable assumption that the rat's SSDR's to traumatic stimulation consist of freezing, fleeing, or fighting, rapid acquisition of jump out and one-way running avoidance can be accounted for by the fact that these situations support the fleeing reaction, whereas the freezing and aggressive reactions only lead to more shock; hence fleeing the shock area rapidly becomes the dominant and successful response. While fleeing is also supported in the shuttle-type avoidance situation, the animal must also run to the place it has just left on the preceding trial (and where it may have been shocked in the past); hence fleeing is likely to compete with the tendency to freeze and the result is relatively slow avoidance learning. Finally, bar pressing and wheel turning are arbitrary avoidance responses invented by the experimenter, and the animal's SSDR's can only conflict with the response to be learned; consequently, avoidance will be achieved with great difficulty, if at all. In Bolles's analysis, then, there is little learning in avoidance learning. Essentially, animals successfully avoid if the designated avoidance response is a dominant species-specific defense reaction to the noxious stimulus and they fail to avoid if it is not.

The four phenomena listed above exhibit a common pattern—in each a behavior-reinforcement relationship or contingency is overridden by species-characteristic reactions associated with the reinforcing stimulus itself. In each case, the specific behaviors generated under the experimental procedures can be better understood by reference to unlearned behavior the subject brings to the situation than by reference to the behavior-reinforcement contingencies in effect. While we should be cautious in generalizing from these observations, which after all involve the quite special behavioral propensities of a few species, they nevertheless urge more careful consideration of the role of unlearned behaviors in the typical learning situation. More generally, these phenemona encourage conceptions that minimize the importance of specific response acquisition while emphasizing instead central and motivational determinants of performance.

Bolles (1972) has sketched the outlines of just such a general model of learning. Following Tolman's concept of expectancy, Bolles proposed that learning involves the acquisition of two kinds of expectancies: (1) S-S* expectancies—that certain stimulus events, S, predict certain other biologically important events, S*; and (2) R-S* expectancies—that certain of the animal's own behaviors, R, predict certain biologically important events, S*. Further, an animal can be expected to have both innate S-S* and R-S* expectancies, although the latter are the more common. Again as in Tolman's theory, expectancies are assumed to interact appropriately with motivational states to produce performance; in Bolles's model S-S* and R-S* expectancies are simply assumed to combine to yield performance.

These assumptions readily fit the outcome of the usual appetitive learning experiment. The animal quickly learns that certain spatial and/or temporal cues predict reinforcement (the S-S* expectancy) and that certain responses in the presence of these cues are likely to be followed by reinforcement (the R-S* expectancy). If the animal is hungry, these expectancies combine to yield appropriate performance. But Bolles's model also accounts for the appearance of non-reinforced behavior, primarily through the mechanism of innate (as well as previously learned) R-S* expectancies. Of course, innate R-S* expectancies may operate in any learning experiment, but their contribution should be greatest under conditions likely to prevent the expression of strong, experiment-based R-S* expectancies, as when an experimenter-defined R-S* contingency is difficult to detect or is only periodically effective. In the superstition experiment there is, of course, no R-S* contingency, and consequently as temporal cues come to predict S* this expectancy combines with the strong, innate "peck-food" expectancy to yield the terminal pecking behavior; in the presence of other temporal cues predicting delay or nonoccurrence of S*, other (nonpecking) behaviors come to dominate. Similarly, in auto-shaping the "lighted key–food delivery" contingency leads to rapid formation of an S-S* expectancy which combines with the dominant peck-food expectancy to yield pecking behavior, oriented toward a salient feature of the environment. The instinctive drift phenomena reported by Breland and Breland can also be interpreted in terms of innate R-S* expectancies which become dominant when the experimenter-defined R-S* contingency is altered (as when the raccoon was required to increase the ratio of coin deposits to food) or disrupted (as when a change was made in the learning environment of the baseball-playing hen). And we have already detailed how the wide differences in rate of avoidance learning fit the Bolles conception. The rat quickly learns that particular cues predict shock and already "knows" several ways of achieving safety. Avoidance is rapid if the situation is compatible with these ways and slow if it is not.

Bolles's model, then, brings the various observations of biological constraints on learning within the framework of traditional learning theory. It does so essentially by supplementing Tolman's classical S-S expectancy concept with the assumption of R-S* expectancies which may be innate or acquired. Bindra (1972) and Estes (1969) have espoused similar versions of expectancy theory which assume that learning consists of the formation of S-S representations while the production of behavior as such is accounted for by preexisting S-R connections and by the motivational consequences of hedonic stimuli. These contemporary expressions of expectancy theory attest to the viability of Tolman's views and underscore the continuity in basic conceptualizations of the learning process. Indeed, current expectancy theories are at heart virtually identical with Morgan's original statement of cognitive theory (pp. 37–39).

Mathematical Learning Theory

We earlier noted that Hull (pp. 93–94) had urged mathematical treatment of learning and had himself sought to interpose quantitative description and prediction between the postulate and test stages of his program whenever possible. Hull's efforts in this vein were an important stimulus to the development of quantitative models, and so it is somewhat ironic that it was not Hull's concepts that eventually proved most congenial to mathematical learning theory but rather those of Guthrie.

Estes's 1950 paper, "Toward a Statistical Theory of Learning," opened the way to effective mathematical treatment of learning phenomena. The key, as Estes showed, is that the learning situation must first be represented in *psychological* terms that are amenable to mathematical analysis. Estes accomplished this largely by employing the assumptions of Guthrie's theory. Specifically, Estes expressed these assumptions in the following way.

1. The response in the typical learning situation may be conceived of as a class, or population, of different but functionally equivalent movements (cf. Guthrie's movement-act distinction). For example, the bar press response may be broken down into a variety of movement patterns each of which has the end result of displacing the bar to some criterion extent.

2. Similarly, the stimulus in the to-be-learned S-R relation may be regarded as a population of independent environmental aspects, of which only a sample is effective at any given time. For example, the conditioned stimulus in a classical conditioning experiment may be treated as a combination of elements which differs somewhat from trial to trial. The elements constitute stimulus aspects inherent in presentation of the CS itself, e.g., the particular values of intensity, directionality, duration present on a given occasion, and those arising from less controlled sources, such as the subject's posture during CS

presentation. Under this view, to present the "same stimulus" repeatedly in a learning experiment is to present successive stimulus *samples* which differ somewhat with respect to element composition.

3. (The sampling assumption). All elements in the stimulus population are equally likely to be sampled on any trial or occasion.

4. (The conditioning assumption). Upon each occurrence of a response, i.e., of a specific movement, all elements in the stimulus sample present on that occasion become *fully* conditioned to the response and will evoke only that response on the occasion of their next presentation. The corollary of this assumption is that the conditioning of a stimulus element to one response automatically involves the breaking of any preexisting relations between that element and any other response.

5. (The reinforcement assumption). Reinforcement is any condition which ensures that successive occurrences of members of the response population will each be contiguous with a new sample of elements from the stimulus population.

The reader will readily see the similarity to Guthrie's conception that the learning of acts proceeds by the accumulation of one-trial conditionings of specific stimulus pattern–movement conjunctions, each conditioned on the basis of sheer contiguity of the stimulus-response elements. But Estes adopted the foregoing assumptions not solely or even primarily because he was persuaded that they best accommodate the general data of learning, but largely because they enable application of mathematical probability theory.

Mathematically, probability is defined as the expected frequency of occurrence of an event relative to its maximum possible frequency of occurrence. The probability of a response may then be defined as the expected frequency of occurrence of the response (i.e., of members of the response class) relative to the maximum possible frequency under a particular set of conditions. Since the model tells us that response occurrence will be a direct function of the prior occurrence of stimulus elements in conjunction with the response, the way is opened to describe changes in response probability in terms of the changing proportion of stimulus elements conditioned to the response. That is, the probability of the response will be the probability of occurrence of stimulus elements already conditioned to the response.

The focus of Estes's model, then, is on the probability that a sampled stimulus element is conditioned to the target response. This probability can be specified given estimates of θ, the probability that a stimulus element will be sampled on any given trial or occasion, and of p_1, the probability of occurrence of the target response at the outset of trial 1. Under the assumptions of the model, the latter probability will also be the probability of occurrence of stimulus elements already linked to the response. Both of these parameters can be derived from data generated under the experimental conditions at issue. The probability p_1 can be estimated simply by observing how a sample of subjects respond at the outset of learning. The probability θ can be calculated from samples of the changing proportion of target responses over successive trials and is essentially observed learning rate.

Given the basic assumptions of the model and estimates of p_1 and θ, and assuming reinforcement of the target response, then p_2, the probability that a stimulus element is conditioned to the target response at the beginning of the next trial is

$$p_2 = p_1 + \theta(1-p_1)$$

That is, with probability p_1 the element is already conditioned to the behavior and of course remains so conditioned when the reinforcement occurs, and with probability $1-p_1$ the element was previously unsampled in relation to the behavior but becomes conditioned on this trial when sampled with probability θ. If p_1 is zero, then p_2 becomes θ, i.e., becomes the proportion of stimulus elements sampled on trial 1. Tracking the changing probabilities of response over successive reinforced trials, p_3, p_4, p_5, etc., is a matter of repeatedly reapplying the above equation to the probability obtained from each succeeding trial. Thus, $p_3 = p_2 + \theta(1-p_2)$, and so on. The result of these repeated calculations is a predicted learning curve.

Figure 7-3 shows predicted curves when the probability of response on trial 1 is .20 and θ is variously estimated at .05, .10, or .20. It can be seen that θ has a marked effect on rate of learning. This follows from the consideration that when θ is low there will be much trial-to-trial variability in the elements sampled and so many samples will be required to bring a sizable number of elements into repeated conjunction with the response; when θ is high most elements will reappear in the successive samples within a relatively small number of trials and consequently the response rapidly becomes highly probable. Recall that Pavlov early noted the marked increase in speed of conditioning which attended the isolation of his subjects in a special chamber that reduced environmental variability.

It is interesting to note that stimulus sampling theory, as the foregoing approach has been termed, has been applied primarily to hu-

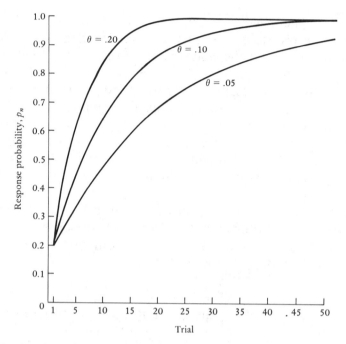

Figure 7-3. Graphs of the basic probability function when $p_1 = .20$ and θ is set at .05, .10, and .20.
(From E. R. Hilgard and G. H. Bower, *Theories of Learning*, Fourth Edition, p. 383. © 1975. Reprinted by permission of Prentice-Hall, Inc., Englewood Cliffs, N.J.)

man learning rather than to the simpler processes of animal behavior. This has been true of the development of mathematical learning theory in general, and in consequence a variety of relatively complex tasks have now been modeled at the human level with considerable success.

One such problem which received early and extensive analysis from stimulus sampling theory is *probability learning*. In this situation the subject is asked to guess on each trial which of two or more alternative events will occur, e.g., illumination of a red versus a green light. After the subject has stated his or her guess, either the red or the green light is briefly illuminated and the subject then makes a prediction of red or green for the next trial and so on through succeeding trials. Each light event occurs on some predetermined proportion of the trials, but the sequence of events is random and so there is no way the subject can arrive at perfect trial-to-trial prediction of the red/green outcomes. From the subject's point of view, the task is to try to "outguess" the experimenter on as many trials as possible. How will humans behave in such a probabilistic situation? Will their guesses eventually come to match the actual probabilities (proportions) of the events? Will they show a preponderance of predictions of the more frequent outcome in order to maximize their proportion of correct guesses? Will their guesses "track" immediately preceding outcomes in some systematic fashion?

Stimulus sampling theory provides unambiguous, quantitative predictions about human performance in this situation. With reference to the two-choice probability learning task illustrated above, let R_1 and R_2 refer to the subject's prediction of events E_1 and E_2, respectively. The theory assumes that E_1 and E_2 function as reinforcers that condition all elements sampled on the occasion of their occurrence to R_1 and R_2, respectively. If an element conditioned to R_1 is sampled on a trial when E_2 occurs, then this element switches its connection to R_2 in the all-or-nothing fashion demanded by the theory. Therefore, E_1 occurrences have the effect of increasing the probability that an element is connected to R_1 and decreasing the probability that an element is connected to R_2, whereas E_2 occurrences have the obverse effects. By applying separate equations as shown above for the probability of conditioning elements to R_1 and R_2 given the scheduled occurrences of E_1 and E_2, predicted learning curves and asymptotic response proportions can be generated. For example, the theory predicts that the proportions of R_1 and R_2 responses will, with extended testing, equal the actual probabilities of E_1 and E_2. This matching prediction has been confirmed for a variety of relatively neutral outcome events.[1] In addition, the theory has proved capable of predicting other detailed aspects of probability learning performance, such as individual learning curves generated under specific probabilities and scheduling (sequencing) of the outcome events.

Figure 7-4 illustrates a not uncommon degree of correspondence between predictions from stimulus sampling theory and actual performance in simple probability learning tasks. The data are from an experiment by Atkinson (1956), who had college students predict which of two lights (E_1, E_2) would occur by pressing one of two corresponding telegraph keys (R_1, R_2). Each subject made 144 predictions over a random sequence of E_1, E_2 outcomes. For different groups of subjects, E_1 occurred on a relatively high (Group A) or relatively low (Group B) proportion of the trials. Also, for both groups there were a number of interspersed trials on which neither E_1 nor E_2 occurred. The predicted values in Figure 7-4 were derived under the assumption (and in part to test the assumption) that nonoccurrence of the reinforcing event has no effect on probability learning,

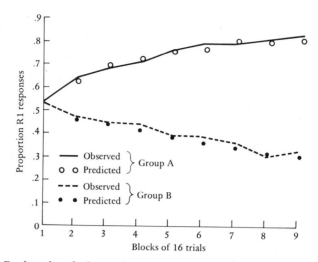

Figure 7-4. Predicted and observed curves representing mean proportion of R1 per 16-trial block.

(After Atkinson, 1956, p. 31. Copyright 1956 by the American Psychological Association. Reprinted by permission.)

i.e., that nonreinforcement does not change the proportion of elements connected to R_1 or R_2. Under this assumption, Atkinson could simply ignore nonreinforced trials in the calculation of response probabilities and, by applying the basic probability equation with appropriate parameter values for Groups A and B, was able to predict the proportion of R_1 for each group on successive trial blocks. As can be seen, there is very close agreement between predicted and observed values.

We shall not attempt to summarize here the many additional applications and achievements of stimulus sampling theory, nor shall we review subsequent adaptation of the theory to another class of mathematical models (Markov chain models) designed to accommodate observations of one-trial verbal and concept learning. Hilgard and Bower (1975) provide an excellent summary of this work. The point to register here is the resurfacing of the assumptions of Guthrie's theory as a central psychological rationale within this continuing stream of mathematical modeling efforts.

It is true as Bower notes (Hilgard and Bower, 1975) that mathematical models are neutral with respect to particular theories of learning. That is, mathematics is a descriptive mode available to any theoretical orientation. But it is clear that Guthrie's ideas have been an important heuristic stimulus to the development of quantitative models by learning psychologists. Moreover, the often impressive fit of stimulus sampling and Markov chain models to human performance cannot be considered neutral with respect to the principle of all-or-none transitions in learning.

Contemporary Operant Analyses

It is more difficult to point to unambiguous instances of Skinnerian influence in contemporary theoretical developments since Skinnerian psychology eschews particular theoretical concepts. One must look for developments that center on empirical principles while emphasizing the control or production of behavior within operant methodology. A good deal of current work exhibits these features, as indicated by the recent publication of major reference works on operant research (Honig and Staddon, 1977) and on applications of operant research (Leitenberg, 1976). From among many analyses that might be cited as sympathetic to a Skinnerian orientation, some contributions by Premack will be taken as exemplars here, primarily because they also concern two substantive themes in Skinner's work—the principle of reinforcement and the application of reinforcement principles in accounts of language development.

The Relativity of Reinforcement. The global theorists tended to think of reinforcement in absolute terms in the sense that stimuli were conceived of as being either reinforcing for all behaviors or not reinforcing for any behavior. Although agreement was never reached on the property that would make an event uniformly reinforcing for behavior, few learning psychologists took exception to Skinner's view that once a stimulus has been observed to strengthen some behavior, we can be confident that that stimulus will strengthen any other behavior of the same organism. While this assumption has never been systematically tested, it is certainly true that common reinforcers such as food can be used to strengthen a wide variety of responses. However, Premack (1959, 1965) has suggested a new principle of reinforcement which is fundamentally inconsistent with a simple, absolute conception of reinforcement.

Premack's principle is a strictly empirical principle induced from observation of familiar instances of reinforcement. Consider the prototypical case—a hungry rat presses a bar, a pellet of food is delivered, and bar pressing becomes more likely. On the one hand, we might hypothesize that the strengthening effect of the food reward is due to drive reduction, or sensory stimulation, or ingestive behavior per se, or still some other factor. But if we confine ourselves solely to what has happened, we may state that there has been a sequence of two activities the second of which is more probable than the first. That is, eating has followed bar pressing and is patently more probable than bar pressing since in the presence of both the cues for bar pressing (i.e., the bar) and the cues for eating, the animal chooses to eat. On the basis of such observation, Premack argued, first, that reinforcers are better conceived of as activities than as stimuli, and second, *that of any two activities the more probable one will reinforce the less probable one.*

This principle has several important implications. First, events cannot be simply classified as reinforcing or not reinforcing since whether or not a given activity will be reinforcing depends upon the relative probability of the activity with which it is paired. Consider four activities, A, B, C, D, and suppose we have found that the probabilities of these activities are such that A>B>C>D. By Premack's principle, A can be used to strengthen any of the remaining activities in the set, whereas D cannot be used as a reinforcer; B can be used to strengthen C and D but not A, and C will reinforce only D. A second implication follows from the consideration that it is possible to change response probability by operations that are independent of reinforcement, e.g., by deprivation operations; hence, observed reinforcement relations between activities should be reversible through the reversal of their relative probabilities.

Contrary to the traditional view, then, Premack's analysis indicates that reinforcers cannot be classified in yes/no fashion and are not necessarily trans-situational. An activity (or presentation of a stimulus that occasions a given activity) may be reinforcing in one situation and not at all reinforcing in another situation, depending upon changes in the probability of the activity itself and depending upon the relative probability of the activity with which it is paired.

Obviously, test of these propositions requires an independent measure of response probability. Premack (1965) argued that such a measure is given by response rate in a free operant (i.e., unconstrained) situation, or alternatively, by the subject's preferences for the activities when allowed free choice among them. An early experiment by Premack (1959) used the latter procedure to test the prediction from the response probability principle that it should be possible to use a traditionally *reinforced response* (manipulation) as a reinforcer for a traditionally *rewarding activity* (eating). In an initial 15-minute session, first-grade children were given access to a pinball machine and a nearby candy dispenser and allowed to both play the machine and eat as much candy as they chose. The candy consisted of constant-size chocolate bits delivered one at a time as the child consumed the previous piece. In this test, 61 percent of the children ("players") made more pinball machine responses than candy eating responses, and for 39 percent of the children ("eaters"), the reverse was true. (The mean duration of each playing response was about equal to the time required to ingest a single piece of candy.) In a second 15-minute session several days later, the children were again exposed to the pinball machine and the candy dispenser, but now one or the other device was inoperative until the other activity occurred. In one contingency relation, eating-playing, each operation of the pinball machine was contingent upon prior ingestion of a piece of candy. In the playing-eating contingency, delivery of each piece of candy was contingent upon prior operation of the pinball machine. A random half of the players and eaters were tested under each of the foregoing contingencies.

The results were clear-cut. When the second activity in the sequence was the higher-probability activity, there was a significant increase in eating by players and in playing by eaters; for both types of subjects the response increment was about five times greater than when the second activity in the sequence was the lower-probability response.

In another study Premack (1962) tested the predicted reversibility of the reinforcement relation. Rats were first allowed either continuous access to an activity wheel but access to water for only one hour per day, or were allowed free access to water but access to the activity wheel only one hour per day. These two deprivation conditions were found to make drinking more likely than running and running more likely than drinking, respectively, when the wheel and water were made simultaneously available. According to the model, it should be possible to reinforce either running with drinking or drinking with running, depending upon the deprivation conditions employed. The tests were conducted in an activity wheel equipped with a locking device and a retractable water tube. To make drinking contingent upon running, the tube was retracted, the wheel was freed, and availability of the tube was made contingent upon several revolutions of the wheel. Conversely, to make running contingent upon drinking, the wheel was locked, the tube was presented, and release of the wheel was made contingent upon a predetermined number of licks on the tube. As predicted, Premack found that *in the same subjects and depending upon which deprivation parameters were used,* running reinforced drinking and drinking reinforced running.

Still other studies with both monkeys (Premack, 1963) and children (Homme et al., 1963) have shown that the reinforcing power of various nonconsummatory activities is proportional to their free operant rates. All of these experiments testify to the validity of the principle that a "reward" is simply any response that is independently more probable than another response.

The idea that reinforcers are best conceived of as activities is particularly appealing in relation to everyday human behavior, which does not appear very dependent upon traditional biological reinforcers. Common activities or classes of activities, such as reading, writing, painting, competing, working, and conversing, clearly have reward value for many individuals and are sought as ends in themselves. A general recognition of the reinforcing function of activity sequences in human affairs is strongly implicit in the familiar admonition to "put business before pleasure." And frequent application of the response probability rule is seen in such practices as requiring a child to complete dinner before dessert or piano practice before baseball playing.

In addition to its applicability in the practical control of behavior, Premack's principle is basically inimical to formal theoretical analysis of reinforcement. Concepts such as purpose, drive reduction, arousal level, goals, and primary and secondary rewards become of questionable utility in the face of the need to secure independent measures of response probability—and in the face of the adequacy of those measures for the prediction of behavior. Moreover, the principle renders meaningless a search for properties that might define a "reinforcer" in any absolute terms, since it makes clear that reinforcement can be defined only relative to the immediate behavioral context.

In these pragmatic and atheoretical aspects, Premack's principle stands as a very significant extension of the empirical analysis of reinforcement.

Language Learning in the Chimpanzee. It has been credo among linguists that only human beings are capable of language and that the study of animal learning is thus largely irrelevant to the understanding of language acquisition. We find Chomsky saying, "Anyone concerned with the study of human nature and human capacities must somehow come to grips with the fact that all normal humans acquire language, whereas acquisition of even its barest rudiments is quite beyond the capacities of an otherwise intelligent ape" (1968, p. 59).

It is understandable, then, that considerable stir has been created by the recent claims of several investigators that it is possible to teach language to chimpanzees (Gardner and Gardner, 1969; Premack, 1970, 1971; Rumbaugh, Gill, and von Glasersfeld, 1973). These claims have been hotly contested, as they should be, since the question is fundamental not only to theories of language and language acquisition but to our conception of the nature of man.

Efforts to teach chimpanzees to *talk* have failed consistently, which is perhaps not surprising given the anatomically limited vocal apparatus of the chimp. But Premack reasoned that the defining feature of language is not vocalization per se but what language *does*, or *accomplishes*, in the organism-environment interaction. Therefore, if a set of responses can be made to exhibit the functional properties of language, that set may be said to constitute a language whether or not it shows the physical and structural features of human language. The similarity to Skinner's functional analysis of language is obvious.

The behavior Premack chose to work with involved selection among metal-backed plastic bits varying in shape, size, color, and texture, each plastic bit serving as a "word." By arranging the plastic bits, or words, on a magnetized slate, the chimpanzee could "talk" with the experimenter and vice versa.

The first step involved training the chimpanzee, Sarah, to use plastic bits to label the elements of a simple social transaction—giving fruit. An apple was placed just out of Sarah's reach and a blue triangle—the plastic word for apple—within her reach. By shaping and reinforcement procedures, Sarah was induced to "request" the apple by placing the triangle on the magnetized board. Each time Sarah placed the proffered triangle she was rewarded with a piece of apple. Sarah quickly learned this game, and new fruits were introduced, each with its own plastic word, to be requested in the same manner. After learning the various fruit–plastic word associations, Sarah had to pass tests that required her to choose from among simultaneously presented plastic words the appropriate word for a particular fruit. Next, the trainer giving the fruit was varied, each trainer being designated by a distinctive plastic bit. Now in order to secure food, Sarah had to pick up and correctly order on the board the words designating the trainer and fruit present (see Figures 7-5 and 7-6). When Mary was the trainer and the fruit was apple, Sarah had to write *Mary apple*. If Sarah wrote *apple Mary*, reward was withheld until Sarah placed the words in correct order. After mastering tests involving appropriate selection among trainer and fruit words, the sequence was extended to the recipient of the fruit and the action involved. For

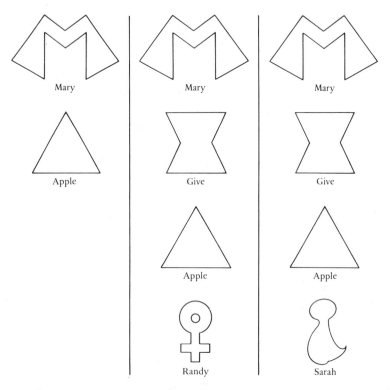

Figure 7-5. In examining Figures 7-5 to 7-10 it should be borne in mind that the plastic bits presented to Sarah varied in texture and in color as well as in shape and size. This circumstance made it relatively easy for Sarah to discriminate the plastic bits themselves.
(After Premack, 1970, p. 55. Reprinted from *Psychology Today Magazine*. Copyright ©
1970 Ziff-Davis Publishing Company.)

example, as seen in Figure 7-5, Sarah might be required to compose *Mary give apple Sarah* (securing the fruit for herself) or *Mary give apple Randy* (thus denying the fruit to herself but for a reward even more valued than apple). In this way, Sarah learned to map a series of whole social transactions.

Having trained Sarah to use words (plastic bits) to label elements of her experience, Premack proceeded to probe Sarah's capacity to use syntax, the organizational structure of word strings. Here again, as throughout Sarah's language learning, Premack relied heavily on reinforcement procedures.

Sarah's lessons in sentence structure began with training in use of the preposition *on*. After teaching Sarah plastic words for red, green, blue, and yellow, the trainer put the words *green on red* on the board,

Figure 7-6. This picture shows Elizabeth (one of Sarah's successors) giving an apple to trainer Amy after having created the sentence "Elizabeth give apple Amy." (Photo courtesy of Ellen Cohn-Arntz.)

placed a red card on the table, handed Sarah the green card and induced her to place it on the red one. Sarah quickly learned this placement response (see Figure 7-7). The sentence on the board was then reversed to *red on green*, and Sarah was given the red card as the one to be placed on top. After mastering these sequences, Sarah was tested in two ways for ability to apply the preposition to new sentences. First, the placement response was requested with novel color combinations. If the trainer wrote *yellow on blue* Sarah had to place the yellow card on the blue one, and so on. Second, the trainer gave Sarah the appropriate plastic words (including *on*) and required her to place them on the board so as to correspond to the *trainer's* placement of cards. If the trainer put the blue card on the green one, Sarah had to write *blue on green*. Sarah's performance was largely correct over series of each test. In similar fashion she was subsequently taught the use of *side* and *front of*. These performances suggest that Sarah had come to understand one contribution of syntax— word order—to sentence meaning.

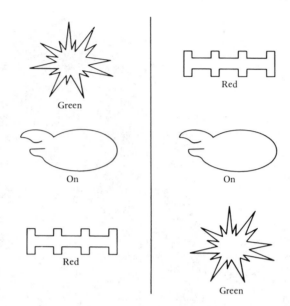

Figure 7-7.
(After Premack, 1970, p. 56. Reprinted from *Psychology Today Magazine.* Copyright ©
1970 Ziff-Davis Publishing Company.)

Next consider the sensitizing of Sarah to information in the hierarchical organization of a sentence. The aim was to get Sarah to
correctly respond to a sentence such as *Sarah insert banana pail apple dish* by putting the banana in the pail and the apple in the dish.
The *components* of the actions required are, of course, given by the
word items, but the desired *relation* among these components must
be understood from the sentence structure. That is, on the basis of
the component words alone it is not clear whether the apple and dish
should not also go in the pail or perhaps be left alone; it is only by
an inference from the hierarchical organization of the words that the
correct meaning is given.

Using the procedures previously described for the social transaction
training, Sarah was first taught to respond correctly to the four statements that can be derived from the target sentence—*Sarah insert banana pail, Sarah insert banana dish, Sarah insert apple pail, Sarah
insert apple dish.* Training then continued with pairs of the statements combined one above the other, as in column 1 of Figure 7-8.
When these combinations were gradually reduced to the target sentence as in columns 2 and 3 of the figure, Sarah's performance remained at 80 to 90 percent correct. Finally, Sarah's performance was
not disturbed when novel fruits were later substituted or when *take*

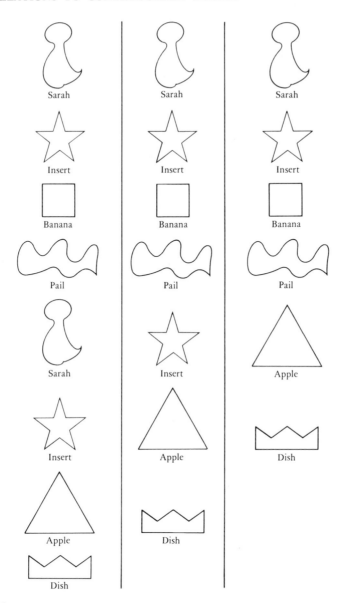

Figure 7-8.
(After Premack, 1970, p. 57. Reprinted from *Psychology Today Magazine*. Copyright ©
1970 Ziff-Davis Publishing Company.)

out was substituted for *insert*. Clearly Sarah could be shaped into
using the organizational structure of sentences.

Teaching of the interrogative function was based on Sarah's ability
to distinguish between pairs of identical objects and pairs of different

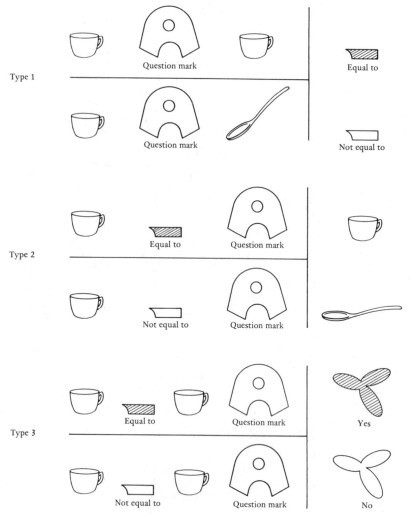

Figure 7-9.
(After Premack, 1970, p. 58. Reprinted from *Psychology Today Magazine*. Copyright ©
1970 Ziff-Davis Publishing Company.)

objects. The experimenter placed two cups before Sarah, gave her the
plastic word for *same*, and required her (by the food reinforcement
contingency) to place it between the cups. On other trials, a cup and
spoon would be presented and Sarah had to place the word *different*
between them. After this brief training, Sarah was found able to correctly select *same* or *different* and apply it appropriately to completely
new pairs of same and different items. Then, by adding a question
mark symbol to the procedure Sarah was taught three types of ques-

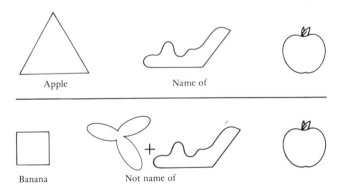

Apple   Name of

Banana   Not name of

Figure 7-10.
(After Premack, 1970, p. 58. Reprinted from *Psychology Today Magazine.* Copyright ©
1970 Ziff-Davis Publishing Company.)

tions, as illustrated in Figure 7-9. In question type 1, Sarah's task was
to replace the question mark with the right word, *same* (equal to) or
different (not equal to). In type 2, Sarah had to replace the question
mark with the object that was like or unlike the first item in the series.
In type 3, Sarah had to replace the question mark with *yes* or *no*.
Sarah was ultimately able to answer all three types of questions for a
great variety of items, including novel items as well as recombinations
of old items.

A simple extension of the foregoing procedure enabled teaching of
the concept *is the name of.* Sarah was given the plastic word *apple*
and a real apple and required to place the symbole for *name of* between them (see Figure 7-10). On other occasions she was given incorrect pairings of word and item, as in the lower portion of Figure
7-10, and had to put the symbol for *not name of* between the items.
Sarah quickly learned to put the correct symbol between many word-
object pairs and even proved able to correctly use in sentences the
names of items learned in this way. Here we have language being
used to teach language.

The language behaviors reviewed above hardly exhaust Sarah's accomplishments but should be sufficient to indicate the nature of Premack's approach. By similar procedures, Sarah was also taught plur-
alization; quantifiers (all, none, one, several); logical connectives (if
. . . then); the copula (the connective *is*); conjunctions; class concepts of color, size, and shape; and still other language functions.
More generally, Sarah supplied abundant evidence that she could use
the plastic words to accomplish functions operationally similar, if not

identical, to those accomplished by the words of human language. But does this mean Sarah has language? Not so, say a number of critics (e.g., Brown, 1973) who argue that however much Sarah's various performances resemble language they are nevertheless merely conditioned behaviors tied to the conditions of training or similar situations. But, of course, from a Skinnerian point of view that is precisely what language *is*, i.e., the development and maintenance of a class of operants through reinforcement and the generalization of those operants on the basis of contextual (situational and linguistic) similarity. And as for the charge that Sarah's language is of limited scope and generative power (Sarah attained a vocabulary of about 120 words and showed little production of novel word combinations), it can be countered that one should expect a rather childlike level of performance given the relatively limited extent of Sarah's training and the fact that this type of research is in its early stages.

The controversy over the precise significance of Sarah's achievements, and those of other "talking apes," is a complex matter going beyond our present concerns.[2] Not the least complication in this issue is that authorities on language are not in agreement on the criteria that constitute an appropriate test for the existence of a language. Our point in reviewing Premack's work is to show just how far one may progress in shaking credos by systematic application of a purely empirical, functional analysis.

Summary

This section has reviewed a few of the many significant links between global theory and current theoretical efforts. For purpose of illustration we have simplified somewhat the connection between global theory and the current work discussed. For example, Estes's stimulus sampling theory was presented as a formalization of Guthrie's theory, and that is generally acknowledged to be the case. But certainly Hull's pioneering efforts at quantification helped prepare the way for Estes's work. Also, any discussion of probability learning could point to an early analysis by Tolman and Brunswik (1935) which first discussed the probabilistic nature of most S-S relations in the real world and suggested the matching test. And Estes himself (1950) acknowledged not only Guthrie's contribution to the framework of stimulus sampling theory but also the model's indebtedness to Skinner's concept of the generic nature of the stimulus and response. In similar fashion, we could delineate additional influences, from within and without global theory, on the work of the remaining investigators discussed in this section. Nor would any of these investigators consider themselves to be Hullian, Tolmanian, Guthrian, or Skinnerian, certainly not in the sense of espousing the psychological world view of these major theorists. Rather, the contemporary theoretical mode is eclectic. Global theory and global theorists as such do not persist. But as the present review attests, the fundamental concepts of global theory persist, often in recombinant form, and it is impossible to properly evaluate current work on learning without knowledge of global theory.

INFORMATION PROCESSING AND ANIMAL LEARNING

In Chapter 1 it was noted that the fields of animal learning and human memory began within a few years of one another but followed largely independent courses of inquiry. It was also noted, however, that recent developments in the study of verbal learning and memory have exerted significant influence on animal learning theory. This section describes the general nature of the rapprochement between these historically separate areas. No effort is made to provide here a comprehensive coverage of contemporary research on animal learning. Rather, the aim is simply to convey the flavor of current research carried out within an information processing framework and to indicate its major continuities and discontinuities with earlier work.

Laboratory study of verbal learning and memory began with Ebbinghaus's pioneering experiments (1885). In his procedure subjects learned lists of nonsense syllables, such as *jid, wuh,* and *xat,* generally by repeated presentation of the syllables one at a time in fixed order until the subject could correctly anticipate each item. Then some time later the subject would be asked to recall as many of the items as possible. By using nonsense syllables, Ebbinghaus was able to minimize the contribution of prior verbal experience to his experimental measures of learning and memory. Nonsense syllable learning rapidly became a widely used paradigm for controlled study of factors affecting acquisition and retention of verbal material, factors such as amount and schedule of practice, list length or amount of material, and degree of similarity among the items to be learned. Of course, work also proceeded on the memorization of meaningful material as well (e.g., Bartlett, 1932), and this research generally centered on the role of organizational factors in learning and memory.

But in all of this work the focus was consistently on long-term retention, that is, on retention measured over a period of days or weeks since time of learning. It was not until 1959 that procedures were introduced that enabled delineation of the retention function over very brief periods, for example, periods as brief as those of the intertrial intervals of the typical learning experiment.

In what Melton (1963) has termed "a moment of supreme skepticism of laboratory dogma" (which had for decades dictated the study of long-term retention), Peterson and Peterson (1959) determined the recallability of single consonant trigrams, e.g., X-J-R, after intervals of 3, 6, 9, 12, and 18 seconds. The general procedure was for the experimenter to spell a consonant syllable and immediately speak a three-digit number. The subject then counted backward by threes or fours from this number until the onset of a light signaling the end of the retention interval. For example, the experimenter might say "X J R 3 0 9"; the subject then immediately began saying "309, 306, 303, 300," etc., continuing in this manner until onset of the light, at which point the subject attempted to recall the consonant syllable. Both syllables and numbers were spoken at the rate of one per second, paced by a clicking metronome. Each subject was tested with eight different consonant trigrams at each of the five recall intervals. The purpose of the counting backward requirement was to prevent rehearsal, that is, to keep the subjects from repeating the consonants to themselves during the retention interval.

Figure 7-11 shows the results of this experiment. The startling finding is that retention of this simple bit of material drops off steeply within seconds of input and is at a level of only 10 percent after 18 seconds! Murdock (1961) replicated the Petersons' finding and also showed that the same retention function prevails when three unrelated common words are used instead of three consonants (see Figure 7-11). The rapid decline in short-term retention, then, obtains for meaningful as well as for meaningless "chunks" of material.

The Petersons' experiment points to the existence of a short-lived storage process different from the long-term memory familiar to us from everyday experience. The data of Figure 7-11 suggest that information about an environmental event enters the short-term store and, if not acted upon in any way by the subject, persists for but a very brief period before being lost forever.

A number of experiments triggered by the Petersons' study (see Atkinson and Shiffrin, 1968; Melton, 1963) confirmed the utility of distinguishing between *short-term memory* (STM), a process of limited capacity and duration instigated by events just experienced, and

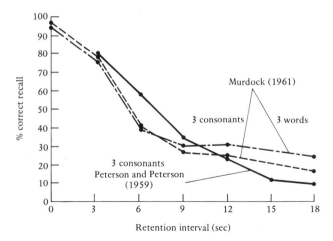

Figure 7-11. Percentage of three-consonant trigrams and three-word trigrams correctly recalled after varying retention intervals.
(Adapted from Melton, 1963, p. 9. Copyright by Academic Press, Inc. Reprinted by permission.)

long-term memory (LTM), a process of unlimited capacity and duration established by repetitive learning experiences. Some neurological evidence also supports this behavioral dichotomy. Milner (1959, 1967) has reported that human patients with bilateral lesions in the hippocampal brain area show no loss of preoperatively acquired memory but seem incapable of acquiring new information in any permanent way. At the same time, the patient has excellent short-term memory for new events but this memory can be made to persist only as long as the patient continues to think about the events. The patient behaves as if the lesions have interfered with a link between a short-term store and a long-term store, each of which remains intact.

If the STM/LTM categorization is correct, the critical question for the learning theorist becomes, how is information transferred from STM to LTM? What action can be taken by the organism to rescue the content of STM from its rapid trajectory into oblivion? The evidence (Atkinson and Shiffrin, 1968; Craik and Lockhart, 1972) suggests that we can distinguish several levels of action or "processing," each differentially effective in producing transfer of information to LTM. At the lowest level, the mere presence of an item in STM may result in some transfer of information about the item to LTM; however, at this minimal level of attention, the rate of transfer would be low and many repetitions of the environmental experience (i.e., many entries of the item into STM) might be necessary to build up a lasting

trace. A somewhat higher level of processing would be simple rehearsal of the item, i.e., sustained attention to or repetition of the
item, as when one continuously repeats a telephone number just
looked up so that it is not forgotten before dialing. Repetition of the
item both extends its STM life and enhances transfer to LTM. For
example, it seems reasonable to suppose that if subjects in the Peterson and Peterson experiment had been allowed to repeat the trigrams
to themselves during the retention interval there would have been 100
percent short-term recall and considerable long-term retention of the
trigrams as well.

Still higher levels of processing concern the ways in which the
stimulus input may be encoded by the subject. By encoding is meant
the alteration or elaboration of the short-term store by information
from long-term memory, as when a new input is seen as an instance
of a familiar category or is related to a set of similar items. The nature
and extent of such higher-level processing may greatly accelerate the
transfer of information to LTM, particularly when the encodings link
the new items to a rich network of preexisting material in LTM. A
striking instance of this type of rapid learning was encountered by the
author in the course of replicating the Peterson and Peterson experiment as a classroom project. Consonant trigrams were selected from
the same standard source (Hilgard, 1951) used by the Petersons and
their procedure repeated in all details, with pairs of students serving
as experimenter and subject for one another. The results beautifully
replicated the Petersons' data but for the glaring exception of memory
for one consonant syllable which for all subjects remained at 100
percent even after the longest retention interval. A check of the stimuli revealed the unwitting inclusion of the trigram L B J—the famous
initials of then-President Lyndon Baines Johnson. Mere presentation
of the trigram doubtless triggered a multitude of preexisting associations which immediately linked the trigram to long-term memory.

Once material has been transferred to long-term memory, by whatever means, recall of the material requires a *search* of the long-term
store, and the success of this search will be determined by a number
of variables, including the strength of the original trace, the number
of potentially interfering similar items of information, the availability
of cues or prompts to aid the search, and the strategies employed in
the search.

Within the framework of the STM/LTM categorization, then, the
question of how organisms acquire and display new behavior as a
function of experience becomes the question of how organisms acquire and recall long-term memory for experience. And the research

on verbal learning and memory suggests this is largely a matter of *how organisms process environmental input, or information*, particularly in the period immediately following its reception.

Examined from the perspective of our learning theory concerns, the information processing conception poses two questions: (1) can the events of the traditional learning experiment be validly cast in cognitive, information processing terms rather than in the usual S-R terms? and (2) can the facts of conditioning be accommodated to the conception of a transfer of information from STM to LTM?

Several lines of evidence suggest an affirmative answer to these questions.

Secondary Reinforcement as Information

In an experiment which antedated much of the information processing literature, Egger and Miller (1962) tested the hypothesis that a necessary condition for establishing a stimulus as a secondary reinforcer is that the stimulus provide information about the occurrence of primary reinforcement. Rats were first trained to bar press for food and then given a period of secondary reinforcement training with the bars removed. In the latter phase all animals received, at approximately one-minute intervals, 135 paired presentations of a flashing light and a tone, each light-tone combination terminating in the delivery of a food pellet. However, one of these stimuli (S_1; light for half the animals, tone for the other half) always came on one half second before the other (S_2; again split equally across subjects), both S_1 and S_2 terminating 1.5 seconds later with delivery of the food pellet (see upper half of Figure 7-12). Since both S_1 and S_2 occur close in time to receipt of a primary reinforcer, both should acquire secondary reinforcing power if, as global theory generally assumed, temporal contiguity to a primary reinforcer is the critical factor in establishment of a secondary reinforcer. But note that the second, shorter stimulus (S_2) provides no new information about the occurrence of primary reinforcement because the first stimulus (S_1) has already given reliable information that the food pellet is to come. Hence, under the information hypothesis, only S_1 should acquire secondary reinforcing power.

Half the rats (Group A) were trained as described above. The remaining animals (Group B) received the same S_1/S_2 presentations paired with food delivery but in addition received 110 interspersed presentations of S_1 alone, i.e., additional presentations of S_1 *not* fol-

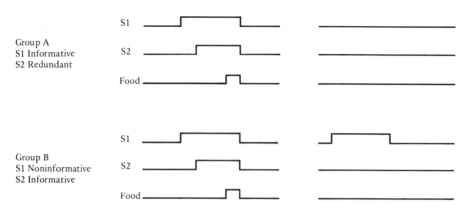

Figure 7-12. Schematic representation of the secondary reinforcement training phase of the Egger and Miller (1962) experiment. Departures from baselines show occurrences of S1, S2, and food-pellet delivery in their temporal relationships to one another in the two basic treatment conditions.

lowed by S2 or by food (see lower half of Figure 7-12). For these animals S2 should have information value because it always precedes food delivery whereas S1 precedes food and no food about equally often. Hence, the information hypothesis predicts that S2 should acquire reinforcing power for Group B animals but not for Group A.

To test the reinforcing power of S1 and S2, the bar was reinserted and presses were followed by one-second presentations of S1 or S2 on an FR 3. No food pellets were ever given during this test phase.

As predicted, S2 was a stronger secondary reinforcer (resulted in a significantly higher rate of bar pressing) for Group B than for Group A, and S1 was a significantly more effective reinforcer for Group A than for Group B. It should be noted that Groups A and B received exactly the same number of pairings of S1 and S2 with food and in exactly the same immediate stimulus context, so that any differences in the reinforcing power of S1 and S2 presumably lie in different information value.

In a second paper Egger and Miller (1963) used the same general procedure to further test the information hypothesis, but in this experiment the temporal distribution of food itself was used to render an initially neutral stimulus either redundant or informative. The conditions of the secondary reinforcement training phase are schematized in Figure 7-13. In the redundant condition (top diagram of the figure) delivery of a single pellet of food always predicted the delivery, two seconds later, of three additional pellets. Thus, an intervening stimulus (a flashing light or a tone) was completely redun-

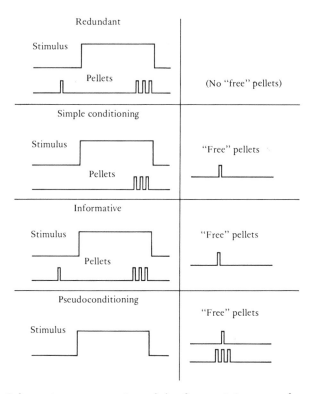

Figure 7-13. Schematic representation of the four training procedures in the secondary reinforcement phase of the Egger and Miller experiment.
(After Egger and Miller, 1963, p. 133. Copyright 1963 by the American Psychological Association. Reprinted by permission.)

dant with the single-pellet delivery. The second level of the diagram represents the traditional procedure for producing a secondary reinforcer; the stimulus is not redundant because there is no other way of predicting the delivery of the three pellets. In the third level of the diagram the redundancy of level 1 is removed simply by having a single pellet of food occur at other times without being followed by the stimulus or the additional pellets; therefore, the stimulus is a more reliable predictor of the three-pellet reward than is the single pellet which precedes the stimulus. Finally, the bottom diagram represents the control for pseudoconditioning, in which presentation of the stimulus and the pellets are never paired.

As in the preceding experiment, completion of the secondary reinforcement training phase was followed by several test sessions in which bar pressing was followed by brief presentation of the training stimulus on an FR 3. Figure 7-14 shows average cumulative response curves for the various groups during the first of these sessions. It can be seen that the informative stimulus proved to be comparable in reinforcing power to the stimulus in the traditional conditioning procedure (the two upper curves of Figure 7-14). But the redundant stimulus, even though it had the same immediate associations with food as the informative stimulus, was not a significantly stronger reinforcer than the pseudoconditioning stimulus, which had never been paired with food at all (the two lower curves).

The Egger and Miller studies provide clear evidence for the proposition that an informative stimulus is likely to be a reinforcing stimulus.

Classical Conditioning: Contiguity or Contingency?

It is always an exciting moment in science when a new perspective is suddenly imparted to a familiar, thoroughly researched basic phenomenon. This kind of discovery within the familiar is rare because it is so difficult to take a fresh perspective upon the known. But this is precisely what Robert Rescorla accomplished for Pavlovian conditioning and its proper control procedures.

At the time of Rescorla's analysis (1966, 1967) the procedures of Pavlovian conditioning had been applied for more than half a century and were universally accepted as appropriate to the investigation of behavior changes within that paradigm. But these procedures, as Rescorla pointed out, were based on an implicit assumption that Pavlovian conditioning works because of the temporal *pairings* of the conditioned stimulus (CS) and unconditioned stimulus (UCS). Given this view, the appropriate control to demonstrate true conditioning involves the testing of an independent group of subjects with the CS apart from any history of pairing with the UCS. Thus, traditional controls involve such procedures as (1) a single presentation of the CS alone to test its unconditioned effects, (2) testing response to the CS after the same number of CS presentations as received by the experimental subjects to test the contribution of familiarity with the CS, (3) testing response to the CS after repeated presentations of the UCS alone to check possible sensitization from UCS presentations, and (4) testing response to the CS after a number of presentations of both CS and UCS, but never close together in time. Note that all of these procedures involve the explicit *unpairing* of CS and UCS, which, of course, is appropriate to the view that pairing is the critical

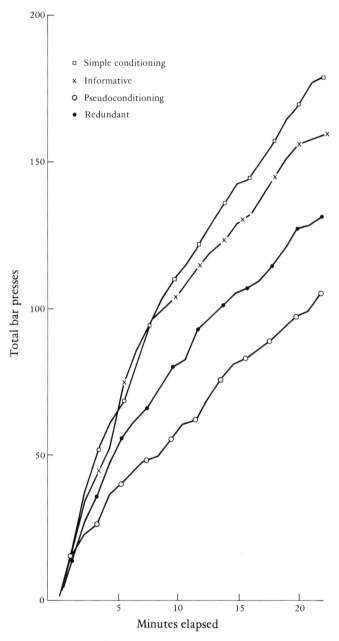

Figure 7-14. Average cumulative response curves during test for secondary reinforcement.
(Adapted from Egger and Miller, 1963, p. 136. Copyright 1963 by the American Psychological Association. Reprinted by permission.)

factor in conditioning and hence control procedures ought to retain features of the experimental procedure while excluding that factor.

But these control procedures are not appropriate if it is assumed that Pavlovian conditioning works because of the temporal *contingency* between CS and UCS. By contingency is meant a predictive relation; occurrence of the CS informs that occurrence of the UCS, and no other event, is imminent. If conditioning depends on prediction of UCS occurrence from CS occurrence, the traditional controls provide inappropriate baselines because their explicit unpairing of CS and UCS means that the CS must come to predict *nonoccurrence* of the UCS. The appropriate baseline condition under the contingency conception is one in which the subject is taught that the CS is *irrelevant* to the UCS, and this can be accomplished by a procedure in which both CS and UCS are presented randomly and independently with no contingency whatsoever between them. In such a procedure, the CS and UCS might occur in conjunction by chance interspersed with random occurrences of the CS alone and the UCS alone. The important feature is that over the entire series the CS not provide any information about UCS occurrence.

If the contingency analysis is correct, it can be expected that pairings of CS and UCS within a truly random schedule of CS and UCS occurrences will not result in any conditioning, even though the number of such pairings might be sizable. In addition, a truly random control baseline should show excitatory conditioning, i.e., higher probability of conditioned response (CR) to the CS, in an experimental group for which CS presentation predicts UCS occurrence, and should show inhibitory conditioning, i.e., lower probability of CR to the CS, in an experimental group for which CS occurrence predicts absence of the UCS.

Rescorla (1966) provided evidence in confirmation of these expectations. Dogs were first trained to make a shock-avoidance response in a two-compartment shuttlebox under a Sidman schedule (Sidman, 1953). In this procedure there is no external cue to the impending shock; the subject simply has a fixed time to make a given response in order to avoid shock, and each response restarts the fixed interval to shock. In Rescorla's experiment the dogs had 30 seconds to jump over a barrier from one compartment into the other if shock was to be avoided. If the subject failed to avoid, shock continued until the jumping response occurred, at which point the subject again had 30 seconds to jump the barrier before the next shock. Each jumping response restarted the 30-second response-shock interval, and thus the subject could continually avoid shock by jumping the barrier at a rate that exceeded once every 30 seconds.

After several days of this training, each subject was confined to one half of the shuttlebox and given Pavlovian fear conditioning. Dogs in Group R (random CS-UCS relation) were given 24 shocks (UCS) on a variable interval schedule with a mean interval of 2.5 minutes. Twenty-four 5-second tones (CS) were independently programmed throughout the session in such a way that tone onset was equiprobable at any time. Dogs in Group P (positive CS-UCS relation) were treated in identical fashion to Group R except they received only those shocks programmed to occur in the 30 seconds following each tone onset. Dogs in Group N (negative CS-UCS relation) were treated identically to Group R except they received only those shocks which were not programmed to occur within 30 seconds after tone onset. Thus for Group P, CS onset predicted imminent UCS occurrence; for Group N, CS onset predicted that UCS was not imminent; and for Group R, CS onset provided no information about UCS occurrence since shock was equally likely for these subjects throughout the session.

After several days of such Pavlovian conditioning and independent Sidman avoidance training, all subjects received a single test session to probe conditioning to the CS. In this session the Sidman avoidance schedule remained in effect, and 24 presentations of the five-second tone CS were superimposed on the avoidance behavior at variable intervals. The rationale of this procedure was that if the CS had acquired control of the animal's fear reaction during the Pavlovian conditioning phase, then presentation of the CS during avoidance behavior should alter the subject's general level of fearfulness and, consequently, the rate of avoidance responding.

Figure 7-15 shows the rate of barrier jumping in the test session during the periods immediately preceding and following CS presentation. It can be seen that in the 30 seconds preceding CS onset (the portion of the graph marked "pre"), all groups exhibited a moderate, stable rate of barrier jumping sufficient to avoid shock. Onset of the five-second CS resulted in a marked *increase* in barrier jumping for Group P but a marked *decrease* in response rate for Group N, with both groups returning to the base response rate by the end of the 30-second post-CS interval. For Group R, CS onset had little or no effect. The period of the graph marked "post" simply shows that avoidance responding remained at a moderate, stable rate in all groups following the 30-second post-CS interval.

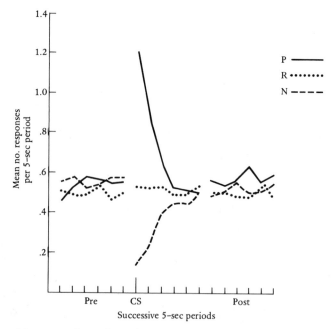

Figure 7-15. Mean number of avoidance responses per successive 5-second periods immediately preceding and following CS onset during the test session. (From Rescorla, 1966, p. 384. By permission.)

These differential effects follow nicely from the contingency analysis. The CS for Group P signaled increased probability of the UCS during the Pavlovian conditioning and consequently acquired the power to elicit fear, thus resulting in increased jumping rate in the test phase; the CS for Group N signaled decreased probability of the UCS during the Pavlovian conditioning and consequently acquired the power to inhibit fear, thus resulting in decreased avoidance response rate in the test phase. Since the CS had no predictive value for Group R animals it did not become a modulator of fear and thus had no effect on their test performance, *even though Group R animals had received as many contiguous occurrences of CS and UCS as subjects in Group P*. These effects are clearly consistent with the view that the critical factor in Pavlovian conditioning is not the pairing of CS and UCS but the degree to which the CS allows the subject to predict UCS occurrence.

The Egger and Miller and the Rescorla analyses indicate that the information value of a stimulus is an important determinant of its

discriminative and reinforcing effects. Other recent lines of investigation provide some insight into the specific mechanisms whereby
informative stimuli acquire control over behavior.

Blocking

Kamin (1968, 1969) has made a particularly fruitful application of
the conditioned emotional response (CER) paradigm (see pp. 257–
60) to study selective learning within Pavlovian conditioning. The
CER paradigm tests the conditioned status of a CS in terms of its
power to alter ongoing operant behavior. (The Rescorla experiment
above employed a variant of the basic CER procedure.) In Kamin's
application, rats were first trained to a stable rate of bar pressing for
food on a variable interval schedule. Then a series of CS-shock pairings were imposed on the ongoing VI food-reinforced behavior. The
CS, which was three minutes in duration, consisted of either white
noise or diffuse light or white noise and light together and was followed by a ½-second shock. Following CS-UCS pairings, subjects
received a single test presentation of the CS alone, again superimposed on the VI behavior, to assess the power of the CS to alter bar
pressing. This was expressed in terms of a suppression ratio, $B/(A+B)$,
where B represents the number of bar presses during the three-minute
CS and A the number of bar presses during the three minutes immediately preceding the CS. If the CS has no effect on bar pressing,
the ratio would approach .50, whereas strong suppressive power would
be indicated by ratios approaching .00. The suppression ratio, then,
indexes the degree to which the CS elicits a conditioned emotional
response as a function of its pairings with shock.

Within this general procedure, Kamin's basic discovery is given by
the experimental design and outcome outlined below.

| Group | Conditioning phase | | Test phase |
	I	II	
A	N (16)	LN (8)	L .45
B	—	LN (8)	L .05
C	—	N (24)	L .44

Prior to the conditioning phases listed in the table, the three groups of animals (A, B, C) had each acquired stable VI behavior. The CS's employed with each group during CER conditioning are noted in abbreviated form; L, N, and LN refer to the light, noise, and the compound of light + noise, respectively. The number of shock pairings with each CS is indicated in parentheses immediately following the CS notation. Finally, the CS employed during the test trial is listed (light for all groups) together with the median suppression ratio for the group on the test trial.

The finding of basic interest is given by comparison of Groups A and B. The test results for Group B show the amount of control acquired by the light as a result of eight pairings of light + noise with shock; as can be seen, test presentation of the light produced nearly complete suppression of bar pressing in Group B (ratio of .05). However, for Group A, which had the same compound conditioning of light + noise but preceded by conditioning trials with noise alone, test presentation of the light resulted in very little suppression of bar pressing (ratio of .45). These results were taken by Kamin to show that prior conditioning to a stimulus element "blocks" conditioning to a new element placed in compound with the first. That the prior conditioning to the noise in Group A produced essentially complete blocking of conditioning to the light is shown by the fact that Group A's suppression ratio is virtually identical to that of Group C, which had never been conditioned with the light.

Additional experiments demonstrate that the blocking effect is not specific to the particular sequence of stimuli employed. For example, when the foregoing experimental procedure is repeated but with the roles of the light and noise stimuli reversed, a total block of compound conditioning to the noise is produced by prior conditioning to the light.

The phenomenon of blocking is, of course, flatly contrary to the general view that any stimulus occurring in temporal contiguity with a UCS acquires control over the conditioned response. Compounded stimulus elements occurring with a UCS *may* acquire control over the response, as the data from Group B show, but *not* if they are compounded with a previously conditioned element, as shown by the data of Group A.

The reader has doubtless noticed that the blocking cue and the blocked cue have the same logical status as the informative cue and redundant cue, respectively, in the Egger and Miller experiments. That is, for Group A the noise stimulus in conditioning phase II already predicts the UCS as a consequence of its phase I pairings, whereas the light stimulus in phase II predicts nothing new. The light does not acquire control over the response, much as the redundant stimulus in the Egger and Miller studies failed to acquire secondary reinforcing power. This comparison suggests that redundancy is the basis of the blocking effect and that blocking will not occur if the added cue in conditioning phase II is made informative. Kamin tested this possibility by repeating the blocking procedure but with an increase in shock intensity in conditioning phase II as compared with the level employed during phase I. Specifically, the following design was used.

	Conditioning Phase		
Group	I	II	Test phase
A	N-1 ma (16)	LN-1 ma (8)	L .45
B	N-1 ma (16)	LN-4 ma (8)	L .14
C	N-4 ma (8)	LN-4 ma (8)	L .36

The notations 1 *ma* and 4 *ma* refer to shock intensity—1 and 4 milliamps. Group A is the standard blocking procedure and yields the full blocking effect (median suppression ratio of .45). Group B is treated identically except that the light is informative in conditioning phase II because it precedes a shock different from that used in phase I. Note that this change virtually eliminates the blocking effect (ratio of .14). Group C shows that the elimination of blocking is not a simple consequence of employing an intense shock during the compound conditioning trials since a clear blocking effect is manifested (ratio of .36) when the same intense UCS is employed throughout the experiment. Statistically, the suppression ratios of Groups A and C do not differ, but both are significantly higher than that of Group B. These data clearly support the view that blocking occurs because of the redundancy of the superimposed element.

But how does redundancy prevent a stimulus from acquiring control over behavior? Kamin suggests that conditioning occurs only when the UCS instigates some "mental work" on the part of the subject, and this mental work occurs only if the UCS is surprising, that is, unpredicted. Thus, in the early trials of the traditional conditioning experiment, the subject has no way of predicting occurrence of the UCS following the CS and hence a surprising UCS activates the covert response processes that result in learning the CS-UCS relation. In the blocking experiment, occurrence of the UCS following onset of the blocked stimulus element is essentially not surprising because UCS occurrence is predicted by the simultaneously presented preconditioned stimulus element; therefore, the subject does not engage in the processes necessary for learning the relationship between the blocked cue and UCS occurrence.

This leaves open the question of the nature of those critical covert processes. Kamin suggests they are concerned with reviewing (rehearsing) the memory store of recent stimulus input. More specifically, he speculates that in order for learning to take place the UCS must provoke the subject into a backward scanning of its memory for recent events, and this scanning is prompted only by an unpredicted UCS. This conception, of course, is very similar to the language of information processing theory, which maintains that transfer of input to long-term (permanent) storage is a function of the processing accorded that input in the short period immediately following its reception.

To test this line of analysis Kamin applied the blocking procedure but with an effort to surprise the animal very shortly after each pairing of the compound cue and UCS. Animals were first conditioned to the noise stimulus in the usual phase I conditioning procedure. Then in the conditioning of phase II, the subjects received the usual pairings of the light + noise compound and shock. However, on each compound conditioning trial, five seconds following delivery of the UCS, an extra (surprising) shock was administered. The second, surprising shock was of the same intensity and duration as the first. When these subjects were subsequently presented with the light CS in the test phase, they exhibited a median suppression ratio of only .08. Consistent with Kamin's analysis, the blocking effect was entirely eliminated by delivery of the unpredicted shock shortly following compound CS-UCS pairings.

Priming in STM

Building on the idea that surprising events are rehearsed (and thus learned) and expected events are not, Wagner (1976) has advanced a systematic integration of conditioning and information processing concepts. Central to Wagner's thinking is the idea that an expected event is one that is prerepresented (primed) in short-term memory (STM) at the moment of its occurrence, whereas a surprising event is not so primed. Priming of STM may come about in two ways. First, actual occurrence of the event will leave a short-lived representation in STM; this is termed *self-generated priming*. Second, as a consequence of prior learning, situational cues may elicit a representation of the event from long-term storage; this is termed *retrieval-generated priming*. To put the matter somewhat loosely, an event is expected if the subject is thinking about it, and the subject may be thinking about it either because the event has just occurred or because the situation reminds the subject of it.

When an event occurs which is primed in STM, i.e., which matches representations in STM, there is little if any processing of that input and it quickly decays from STM. But an event which is not primed is by definition a surprising event and rehearsal is engaged, with the result that the input remains longer in STM and thus is registered to some degree in LTM.

Within this general conception, Wagner and his colleagues have pursued a provocative experimental analysis of the implications of the information processing framework for our understanding of conditioning and other simple forms of behavior change. Several of these experiments will be described by way of illustrating the promise of this line of analysis.

It will be recalled that the information processing model assumes STM to have limited capacity; if an episode is to be maintained in STM by means of rehearsal it must be at the expense of attention to some other event. Therefore, if a surprising CS-UCS episode commands rehearsal, it follows that such an episode should not only be learned itself but should *interfere* with the learning of temporally adjacent episodes since it deprives them of rehearsal. In one of several tests of this prediction, Wagner, Rudy, and Whitlow (1973) arranged for surprising or unsurprising events to occur close in time to a newly instituted CS-UCS relationship.

The experimental preparation involved eyelid conditioning in the rabbit with a 100-millisecond shock to the area of the eye as the UCS and anticipatory lid deflection as the conditioned response. The CS was 1,100 milliseconds in duration and came on 1,000 milliseconds before the UCS. The relatively complex experimental procedure involved a number of CS's which, for the sake of simplicity, will not be specifically described other than to note that they were from different modalities and were, of course, counterbalanced over experimental treatments.

In an initial training phase, the rabbits received daily eyelid conditioning sessions with a CS_A reinforced by shock and a CS_B non-

reinforced until rate of eyelid deflection approached 100 percent and
0 percent, respectively, upon presentation of CS_A and CS_B. At the
termination of this phase, then, occurrence of the UCS following
presentation of CS_A (CS_A-UCS) and nonoccurrence of the UCS following presentation of CS_B (CS_B-$\overline{UCS}$) have the status of expected
episodes. On the other hand, CS_A-$\overline{UCS}$ and CS_B-UCS would be
incongruent with the training and constitute surprising episodes.

In the second phase of the experiment, all subjects were trained
with two new cues, CS_C and CS_D, both of which were shock-
reinforced on every presentation, with presentations of CS_C and CS_D
alternating at 15-minute intervals. During this training each CS_C-
UCS pairing was followed ten seconds later by a distinctive post-trial
event but each CS_D-UCS pairing was not. For one group of rabbits
(Group C) the post-trial event was congruent with the first phase of
training, involving occurrence of either CS_A-UCS or CS_B-$\overline{UCS}$. For
a second group (Group I) the post-trial event was always incongruent
with phase 1 training, involving either CS_A-$\overline{UCS}$ or CS_B-UCS. The
essential experimental steps can be summarized as follows.

	Phase 1	Phase 2
Group C	CS_A-UCS CS_B-$\overline{UCS}$	CS_C-UCS, followed by CS_A-UCS or by CS_B-$\overline{UCS}$ CS_D-UCS, no post-trial event
Group I	CS_A-UCS CS_B-$\overline{UCS}$	CS_C-UCS, followed by CS_A-$\overline{UCS}$ or by CS_B-UCS CS_D-UCS, no post-trial event

Thus, for Group C each CS_C-UCS acquisition trial was followed
shortly by a nonsurprising event, and for Group I each CS_C-UCS
acquisition trial was followed shortly by a surprising event. For both
groups, CS_D-UCS sequences were not followed by any distinctive
event.

The outcome of these treatments is depicted in Figure 7-16, which
shows acquisition of conditioned responding to CS_C and CS_D. For
Group C there was no difference in acquisition to CS_C and CS_D; a
congruent post-trial episode had an effect comparable to no post-trial
event at all. But for Group I rate of conditioning to CS_C was significantly lower than to CS_D. As predicted, a surprising event interfered
with the learning of a new episode, but only if it was a temporally
adjacent new episode.

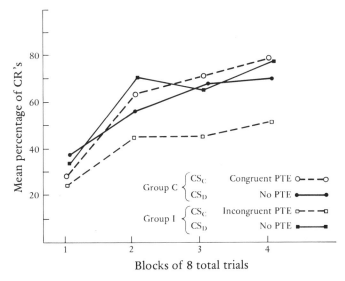

Figure 7-16. Acquisition of conditioned eyelid response as a function of congruent or incongruent post-trial events (PTE). Each CS_C–UCS pairing was followed by a congruent PTE in Group C and by an incongruent PTE in Group I. For both groups, CS_D–UCS pairings were not followed by any distinctive PTE.
(From Wagner, Rudy, and Whitlow, 1973, p. 418. Copyright 1973 by the American Psychological Association. Reprinted by permission.)

In related experiments Wagner et al. studied the effect of placing the incongruent post-trial event at different intervals following CS_C-UCS pairings. Figure 7-17 summarizes the major findings. Acquisition of conditioned responding to CS_C was increasingly depressed the shorter the interval between CS_C-UCS and the post-trial event, over the range from 3 seconds to 300 seconds. This is just what would be expected under the assumption that the incongruent post-trial event terminated a rehearsal process which itself decays with lengthening post-trial interval.

These data indicate that surprising UCS's have their effects by commanding a limited post-stimulus processing capacity. If this is true, and if we were somehow able to look into the subject's short-term memory store, we should find that representation of a surprising UCS is better retained than that of an expected UCS. A test of this proposition clearly requires a relatively direct probe of STM, and Terry and Wagner (1975) have provided just such a probe by an ingenious three-part experimental procedure centering, again, on eyelid conditioning in the rabbit.

In phase 1 of training, a CS (CS_R, the releasing stimulus) was

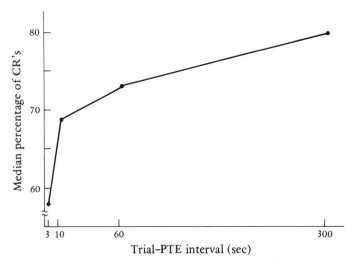

Figure 7-17. Median percentage of conditioned eyelid responses over ten acquisition trials as a function of the interval between CS_C–UCS pairing and an incongruent post-trial event (PTE).
(From Wagner, Rudy, and Whitlow, 1973, p. 421. Copyright 1973 by the American Psychological Association. Reprinted by permission.)

shock-reinforced or nonreinforced in the manner of the foregoing experiment, depending upon the occurrence or nonoccurrence of the same brief shock UCS (the preparatory stimulus) 2, 4, 8, or 16 seconds prior to CS_R. In Group 1, when the preparatory shock stimulus was presented prior to CS_R (designated as P episodes) CS_R was reinforced, and when the preparatory shock was not presented prior to CS_R (designated as $\bar{P}$ episodes) CS_R was not reinforced. In Group 2 just the opposite contingencies were arranged—given the preparatory shock, CS_R was not reinforced, but when the preparatory shock was not presented, CS_R was reinforced. Note that the range of intervals separating the preparatory stimulus and CS_R approximates STM span, and thus the subject could use any persisting information about presence/absence of the preparatory stimulus to predict occurrence/nonoccurrence of shock following CS_R onset.

The upper panels of Figure 7-18 show conditioned eyelid deflections to CS_R during the later stages of training under this procedure. It can be seen that Group 1 made more responses to CS_R on P than on $\bar{P}$ occasions, and the opposite was true for Group 2. These patterns, of course, are entirely consistent with the significance of the preparatory UCS for predicting reinforcement/nonreinforcement of CS_R. It can also be seen that the influence of the preparatory shock

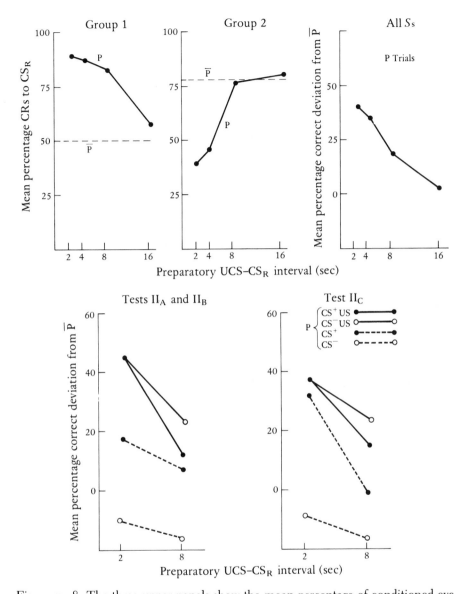

Figure 7-18. The three upper panels show the mean percentage of conditioned eye-blink responding to CS_R (the releasing stimulus) as a function of occurrence (P) or nonoccurrence ($\bar{P}$) of a preparatory shock UCS prior to CS_R. The two lower panels show the modulation of this conditioned behavior when the preparatory UCS was itself preceded by a CS^+ or a CS^-. See text for further explanation.
(Adapted from Terry and Wagner, 1975, pp. 127, 129. Copyright 1975 by the American Psychological Association. Reprinted by permission.)

in this regard declined quite sharply with increasing interval between the preparatory shock and CS_R. At the 16-second interval there was little difference between P and $\bar{P}$ episodes in either group, suggesting that this interval approached the limit of STM for the preparatory stimulus. The upper right-hand panel of the figure shows the combined performance of Groups 1 and 2 on P episodes; the panel presents the mean percentage of correct conditioned responses, i.e., responses appropriate to the signal function of the preparatory stimulus, expressed as deviation from the level of responding on $\bar{P}$ episodes.

These data permit the assumption that the subjects were differentially likely to respond to CS_R, depending upon the presence versus absence of a representation of the preparatory UCS in STM. To provide plausible grounds for such an assumed relation between a preparatory UCS, STM, and conditioned responding was the point of phase 1 of training, for it is then possible to ask whether varying the surprisingness of that preparatory UCS leads to its greater persistence in STM as measured by increased responding to CS_R.

In order to vary the surprisingness of the preparatory UCS it was first necessary to undertake another phase of eyelid conditioning, this time involving discrimination between two new 1,100-millisecond CS's (designated CS^+ and CS^-) from modalities other than that employed with CS_R. Each onset of CS^+ was followed 1,000 milliseconds later by a 100-millisecond shock, the same brief shock employed in the first phase of training; presentations of CS^- were never followed by shock. This discrimination training was continued until all subjects displayed consistent lid closure to CS^+ and neglible responding to CS^-.

It was now possible to apply the critical test. In this phase CS_R was again presented as in phase 1, sometimes preceded by the preparatory shock UCS and sometimes not, but now the preparatory UCS was itself preceded either by CS^+ or by CS^- with the same temporal relation that prevailed between CS^+ and this same UCS in discrimination training. As a consequence of the prior discrimination training, a preparatory UCS preceded by CS^+ would be a fully expected UCS, whereas a UCS preceded by CS^- would clearly be surprising. Therefore, if a surprising UCS is better retained in STM than an expected UCS, the sequence CS^--UCS should result in a greater number of correct (contingency appropriate) responses to CS_R than the sequence CS^+-UCS.

The lower panels of Figure 7-18 present results from several test sessions, the data combined over Groups 1 and 2 in the manner of the upper right-hand panel. The panel labeled II_A and II_B shows data from separate test sessions in which the interval between the preparatory UCS and CS_R was either two or eight seconds. Panel II_C shows the same measures but from later sessions, in which the two- and eight-second intervals were intermingled. The two upper (solid line) functions in each panel are the critical test results. These show the percentage of correct responses to CS_R as a function of whether CS^+ or CS^- preceded the preparatory UCS. It can be seen that these sequences did not result in differential responding at the two-second interval. But at the eight-second interval in both types of test session, subjects were significantly more likely to exhibit appropriate conditioned response to CS_R following a preparatory UCS that was surprising (i.e., that followed CS^-) than following a preparatory UCS that was expected (i.e. that followed CS^+). Since response to CS_R depended upon representation in STM of the preparatory UCS, these data clearly point to the conclusion that a surprising UCS is better retained in STM than is an expected UCS.

Finally, the two lower (dotted line) functions in the panels show responses to CS_R when preceded by CS^+ and CS^- *alone*, i.e., unaccompanied by the preparatory UCS, during the same test sessions. As may be seen, presentation of CS^+ alone elevated responding to CS_R above that seen on $\bar{P}$ episodes, and presentation of CS^- alone depressed response to CS_R below that on $\bar{P}$ episodes. These observations control for the possibility that the differential effects of the CS^+-UCS and CS^--UCS sequences were due simply to different contributions of the CS's per se. These observations also support the concept of retrieval-generated priming of STM, in that the level of responding to CS_R following CS^+ alone and CS^- alone suggests these cues acted to retrieve from LTM a representation of "UCS" and "no UCS," respectively.

These experiments by Wagner and his associates greatly strengthen an information processing analysis of conditioning. They add considerable operational specificity to the concepts of surprisingness/expectedness, short-term memory, and post-stimulus processing. In particular, they support the propositions that UCS's are surprising to the degree they are not represented (primed) in short-term store, that surprising UCS's direct a limited rehearsal capacity and thereby exert a selective effect on the learning of contemporaneous events, that surprising UCS's last longer in STM than expected events and thereby promote learning, and that well-established CS's act to prime STM by retrieving from LTM a representation of the UCS.[3]

In addition, there is no reason to suppose that the information processing conception holds only for CS-UCS relations. The organism may be particularly apt to attend to biologically significant stimuli and their accompanying signals, but once *any* stimulus has been attended to, it can be expected to leave some representation in STM, and that representation should be subject to the laws governing STM content. A phenomenon that particularly invites analysis from this perspective is *habituation*, the response decrement that results from mere repetition of a stimulus, as when an organism orients less and less to repeated presentations of a novel stimulus.

Wagner (1976) extended his information processing conception to habituation by assuming that the relationship postulated to hold for priming and post-stimulus processing also holds for priming and overt response elicitation. That is, he assumed that a stimulus is likely to evoke its unconditioned response to the degree that it is not prerepresented in STM.

Application of this proposition to habituation is straightforward—the decline in responsiveness with stimulus repetition reflects increasing priming of STM. Thus, the first presentation of a novel stimulus may evoke a large startle response since the stimulus is totally unexpected (nonprimed) in the given situation, but repetitions of the stimulus increase the probability of priming (self-generated or retrieval-generated), with consequent reduction in the startle reaction.

This conception immediately generates a number of expectations regarding the habituation process. For example, it would be expected that short intervals between stimulus repetitions would lead to relatively rapid, large response decrements since self-generated priming of STM, that is, priming via persistence of information from an immediately preceding occurrence of the stimulus, should increase with decreasing interstimulus intervals. But longer interstimulus intervals can be expected to lead to longer-term retention of habituation response decrement via the mechanism of retrieval-generated priming. The latter relationship follows because (1) longer interstimulus intervals should result in more post-stimulus rehearsal of the stimulus since there is lower likelihood of self-generated priming and the stimulus is therefore more persistently surprising over a series and (2) longer interstimulus intervals allow more time for post-stimulus rehearsal of the episode and hence better registration in LTM of the stimulus and its situational cues. Davis (1970) found just such relationships to hold for habituation of the startle response. Wagner (1976) provides a detailed discussion of the Davis experiment as well as of a number of additional habituation experiments that support extension of the information processing conception as a general model of behavior change.

Learning Theory and Information Processing: Continuities and New Directions

The experiments by Egger and Miller, Rescorla, Kamin, and Wagner and his associates reviewed in this section are meant to exemplify a general pattern of findings from many recent studies of animal conditioning. That pattern suggests that learning is fundamentally a matter of information acquisition, where information refers to representation of stimulus and response events that bear a predictive relation. Knowledge of predictive relations enables anticipatory representations (expectations) within environmental or environmental-behavioral sequences. The learning process itself is set in motion whenever the organism is confronted with input that is discrepant with expectations; discrepant input activates control processes (rehearsal and other mechanisms that are little understood at this time) which result in acquisition of information about contemporaneous stimulus and response relations.

At first blush these propositions seem a radical departure from global S-R theory. But to some degree the departure is a matter of form rather than of substance. It must be remembered that global theorists also saw the central problem of learning as that of accounting for the acquisition of knowledge about the environment, as our account of Hull's and Tolman's theories sought to make clear. And to say that learning results in anticipatory representations of environmental sequences is to state in cognitive terms what global theory said in response terms. Furthermore, we saw that the outcomes of the classic S-R theoretical controversies left S-R theorists in consensual agreement that learning might result from sequences of merely perceptual responses with minimal consequent reinforcement.

But there are significant departures in the information processing conception, particularly in the idea that discrepancy between present input and memory for past input instigates the processes that result in learning, and in the idea that the critical processes in learning are those traditionally linked to the study of memory. Certainly the data reviewed in this section make it difficult to maintain that contiguity is the critical condition for learning. And as the habituation analysis suggests, organisms may learn about any discrepant input, not just that involving hedonically toned stimuli. However, contiguity and hedonic consequences will doubtless retain central roles in any theory of learning since they act to limit the relationships organisms attend to and therefore learn about.

The information processing framework, then, suggests a new direction in learning theory. But it is very important to note that the information processing conception does not deal with the question of how information gets translated into action. This, of course, is the very problem that S-R theory sought to come to grips with from the outset—how can we account for what the organism will learn *to do* in any situation? A learning theory based on what the organism thinks about but which "leaves the organism buried in thought" (Guthrie's characterization of cognitive theory) is of no utility whatsoever unless and until it is coupled with rules that relate those cognitive processes to behavior, for only then do the postulated processes and changes become scientifically and pragmatically meaningful. Thus, the idea that organisms acquire information about predictive environmental relations to the degree that present input is discrepant with memory of past input is an important addition to our principles of learning, but any application of this proposition requires a knowledge both of the environmental conditions that define that discrepancy and of the conditions that direct expression of that information in behavior. In these regards, the full body of classical research and theory on animal learning, which have always sought an objective and integrated account of learning and performance, will remain central to our understanding.

These considerations point to a possible convergence of the separate pathways pursued by investigators of human memory and investigators of animal learning. We have throughout this text sought to establish the utility of conditioning principles to analysis of human learning and behavior. The utility of information processing principles in analysis of animal conditioning now seems equally well established. Of course, we should always be cautious in interpreting operational parallels between the behaviors of different subject populations, but it appears more reasonable than ever to accept the age-old assumption of the S-R psychologist that human and animal learning involve the same or highly similar processes. We appear to be at the beginning of a period of increased interaction and integration of the rich research traditions of human and animal learning. The prospect is exciting.

The coming years will doubtless see an intensive effort to understand the central processes that mediate learning, and thus there will be raised anew what is perhaps the most fundamental question in learning theory. That question concerns the *nature* of central mediating processes—should these processes be conceived of as following the same laws as observable stimulus-response relations or as following different laws? Present-day cognitive psychology has emphatically rejected the first alternative. It is the premise of present-day cognitive psychology that central processes transform stimulus input in ways that destroy direct correspondence between observable inputs and outputs. In this view, cognitive processes, or ways of knowing, become the key to understanding the higher organism. At the same time, observable stimuli and responses become most imperfect indexes to cognitive processes and at best mere guides for speculation about the nature and principles of cognition. The cognitive psychologist contends that S-R learning theory has failed to appreciate the critical transforming role of cognition as a consequence of having based its models of learning upon relatively simple organisms. But our review of global learning theory makes clear that the role of cognition was not ignored or unappreciated by most S-R theorists; instead, the problem of accounting for knowledge was a primary concern from the outset. The issue for the S-R theorists was not how to circumvent or negate cognitive processes but rather how to deal with them most meaningfully.

To argue that central (unobservable) processes follow different laws than overt behavior is an absolutely fundamental change. To take this step is certainly to slacken one's grip on the reins of theoretical inquiry. It means conceding, as the mainstream S-R theorist would never concede, that psychological processes are not knowable by direct observation—by the primary criterion of scientific knowledge.

We are dealing here with the issue that gave rise to behaviorism, and its proper consideration would, of course, require review of the evidence supporting the present-day cognitive psychologist's position. But at base the issue is a matter of how best to deal with phenomena that appear to challenge the dictum that behavior is a function of stimulation. In reviewing the classic controversies that stemmed from the opposition of response-reinforcement theory and cognitive theory (the sympathetic forerunner within the S-R fold of present-day cognitive theory), we concluded that while the cognitive theorist was consistently able to challenge a peripheral model, the response-reinforcement theorist, by using mediation models based on observable S-R relations, was usually able to account for the claims of the cognitive theorist and at the same time predict more specific or additional aspects of the phenomena. This is a lesson worth remembering.

Notes

Chapter 2

1. John Broadus Watson was a most remarkable person. Here is a man who accomplished a radical transformation of his field and then, at an age when most scientists are still trying to make their first mark, left the field forever! Watson was 42 when he was forced to resign his position at Johns Hopkins University as a result of a (for the time) scandalous episode involving divorce and remarriage to his assistant. Having no choice but to begin anew, Watson started at the lowest level of an advertising concern and found himself "studying the rubber boot market on each side of the Mississippi River . . . when I was [still] front-page news in Baltimore" (Watson, 1936, p. 279). A few months later he was selling Yuban coffee to retailers and wholesalers in Pittsburgh, Cleveland, and Erie. Displaying an amazing resilience, he quickly rose to prominence in the advertising world, noting that "I began to learn that it can be just as thrilling to watch the growth of a sales curve of a new product as to watch the learning curve of animals or men" (p. 280). One suspects he would have been a leader in virtually any field. He continued to write articles on behaviorism for the popular press, but the academic world had turned its back on him and he never rejoined its ranks. His life was a fascinating but in many ways tragic struggle with the intellectual and cultural forces of his time. Extended biographical study of Watson is long overdue. Cohen's recent biography (1979) is an excellent beginning.

2. What is not so often quoted is the remainder of Watson's statement in which he somewhat qualifies his confident prediction, as follows. "I am going beyond my facts and I admit it, but so have the advocates of the contrary and they have been doing it for many thousands of years. Please note that when this experiment is made I am to be allowed to specify the way the children are to be brought up and the type of world they have to live in" (1924b, p. 82).

Chapter 3

1. It is of some interest to establish the circumstances prompting the *first* use of white rats in psychological research. Zeaman tells the story that the reason psychologists studied rats is that they could not keep them out of the laboratory! "Studying other animals required food to be kept around. Rats came in uninvited, to help themselves, so psychologists decided (it shouldn't be a total loss) to study them" (Zeaman, 1976, p. 297). While we might prefer

Zeaman's humor, the actual circumstances are otherwise, although not without some humor of their own.

White rats were first used by Colin Stewart in the psychological laboratories of Clark University, Worcester, Massachusetts, which saw some of the earliest experimental studies of animal learning. In 1894 Stewart began a several-year study of activity levels employing wild gray rats as experimental subjects, but after a year abruptly switched to the use of albino rats. In responding to later inquiries about this first use of albino rats, Stewart said, "If anyone wants to know why I changed from wild gray rats to *white rats* in 1895, let him work with gray rats for a year" (In Miles, 1930, p. 334). The gray is a quite vicious animal and very difficult to handle, whereas the albino is placid and, as Stewart found, a generally superior experimental animal. Stewart also recalled that he purchased some of his albino subjects from a local "bird and animal store." Adolph Meyer, who was in the Clark laboratories at about the same time, states that he built up the albino colony at Clark and gave a portion of it to H. H. Donaldson at the University of Chicago, who used them in development of the well-known Wistar strain, which has been widely used by learning psychologists, including Waston, and continues in use to this day.

As we might have guessed, then, the initial selection of the albino as an experimental subject was not a matter of careful weighing of the scientific pros and cons of the matter. However, the circumstances do say something about the validity of the law of effect. And there is a pet shop in Worcester, Massachusetts, which has more scientific distinction than it knows.

Miles (1930) has published correspondence from the principals which documents these early developments.

2. In view of present-day criticisms that S-R psychology has presented its subjects with "artificial" learning tasks, Small's concern to choose a natural task for his subjects is particularly noteworthy. Quoting Small, "[T]he experiments must conform to the psycho-biological character of an animal if sane results are to be obtained. . . . Conforming with such considerations, appeal was made to the rat's propensity for winding passages. A recent magazine article upon the Kangaroo Rat, by Mr. Ernest Seton Thompson illustrates well the radical character of this rodent trait. Mr. Thompson gives a diagram of the Kangaroo Rat's home-burrow, the outline of which bears a striking resemblance to that of the apparatus used in these experiments. It suggests that the experiments were couched in a familiar language. Not only do they conform to the sensori-motor experiences of the animals, but they also fall in with their constructive instinct relative to home building" (Small, 1901, pp. 206–208). Thompson (1900) had indeed observed and reported a very mazelike nest of underground trails, complete with alternative routes to the goal (egress) and with blind passageways to fool would-be predators.

While such concern for the ecological validity of the experimental situation is commendable in light of present-day thinking, it must, unfortunately, be weighed against two other aspects of Small's approach. First, the kangaroo rat is a distinctly different species from the albino and one may well doubt the comparability of their behavior tendencies. Second, a history of the Hampton Court maze (Law, 1891) suggests that the maze was commissioned by King William III. (Small had patterned his maze after a drawing of the Hampton Court maze which appeared in the *Encyclopaedia Britannica*.) It is interesting to speculate that the "normal problem" which Small set for his rat subjects and, more important, the prototypical situation for the scientific study of learning, was a design favored by a king for the amusement of his guests. Incidentally, the Hampton Court maze still exists, and its half mile of hedge-lined passageways still amuses and frustrates visitors and tourists at Hampton Court Palace.

3. In retrospect, the fact that s_1r_1---s_2 cognitions were largely ignored in discussions of Tolman's theory is somewhat puzzling, given the explicit expressions of a behavioral component of cognition in early statements of the theory and given the common criticism that Tolman

"left out the response." Thus, we find Tolman saying in 1937, "A sign-gestalt expectation (hypothesis) is to be defined as an 'intervening variable'—a set—in the animal whereby the latter adjusts for the fact that such and such a spatially and temporally immediately present stimulus-object will lead (as the result of a particular type of behavior or lack of behavior on the part of the animal himself) to such and such a spatially and temporally more distant stimulus-object" (1937, p. 209). The selective attention of the field to cognition of the s_1---s_2 type appears to reflect a tendency to perceive the theory in terms of its most distinctive aspect. But in the final analysis, the overlooking of $s_1 r_1$---s_2-type cognition is not a matter of major importance, since the response in that type of cognition is really conceived of by Tolman as essentially another *stimulus*. That is, it is a matter of the organism *perceiving* a three-term sequence composed of s_1, its own behavior (r_1) in the presence of s_1, and s_2. Hence r_1 is as much a perceptual event as s_1 and s_2.

Chapter 4

1. Another seemingly trivial circumstance that may operate in the same fashion as prehandling was noted by Kanner (1954), who observed that the Blodgett effect has never been found when guillotine-type doors, rather than swinging doors, are used to prevent retracing in the maze. In contrast to swinging doors, guillotine doors evoke little if any fear in rats; therefore when guillotine doors are used, Experimental and Control groups are more likely to be equal in adaptation to the maze situation at time of first reward.

2. We refer to the seven positive studies listed in MacCorquodale and Meehl's excellent review (1954), plus Kendler's 1947 study. MacCorquodale and Meehl list both the Kendler experiment and a Maltzman (1950) experiment as showing negative outcomes, but in the writer's view neither of these experiments should be so classified. In Kendler's study two groups of rats, one strongly motivated for food and water (Group M), the other fully satiated (Group S), first were given equal experience running to both sides of a T-maze that contained food in one goal and water in the other. Group M subjects were allowed to consume the rewards in each goal during this phase, and Group S subjects, of course, showed no interest in the rewards. Subsequently, the two groups were tested under strong motivation either for food or for water. Not surprisingly, Kendler found that Group M performed better than Group S on the test trials. But while the performance of Group S subjects was not tested statistically, they appeared to show evidence of latent learning; for example, they attained 83 percent correct choice of the appropriate goal on one of the test days. In a footnote Kendler refers the reader to Spence, Bergman, and Lippitt for discussion of whether his Group S showed learning, and in discussing the Kendler (1947) study, Spence et al. (1950) confirm that Group S gave a significant number of appropriate choice responses in the test phase.

   In Maltzman's study, rats satiated for food and water were induced to run the maze by being isolated for a considerable period in confinement cages prior to each trial and then allowed access at the end of the run to a relatively large endbox containing a cagemate, the latter box being attached directly to the end of the choice arm. As Maltzman argued, these conditions appear to have imposed a strong motivational-reward condition, and hence it is questionable to include this experiment in the "irrelevant need absent" category.

3. Not all experiments employing separation of sequence and incentive learning yielded positive outcomes regarding latent learning, but the majority did. Our review has been confined to those experiments that best allow illustration of the theoretical exchange that took place here.

4. On the possibility that perceptual responses might become anticipatory and function much as r_g's, Hull wrote in 1934, "In addition to . . . obvious forms of fractional anticipatory behavior, it is probable that rich variety exists, of which we as yet know little or nothing. It

is conceivable, for example, that the mere visual stimulation by a light pattern may evoke hidden but characteristic reactions in the organism which could play the anticipatory role. At all events, it is not purely gratuitous to assume that both goal reactions and pregoal reactions may possess characteristic fractional components which will be drawn forward to the beginning of all action sequences originally leading to them . . ." (1934, p. 43).

5. The response-reinforcement analysis of sensory preconditioning described here has been presented in similar form by a number of authors, notably Bitterman, Reed, and Kubala (1953); Kimble (1961); and Osgood (1953).

Our application of response-reinforcement theory to sensory preconditioning assumes a drive reduction principle of reinforcement. While other mechanisms of reinforcement are possible that might be less strained here, the Hullian drive reduction conception is applied for the sake of consistency with the framework of the previous chapter. The basic points at issue are not seriously affected by the particular form of reinforcement invoked.

Chapter 5

1. Guthrie was somewhat inconsistent in his conception of what might constitute a conditioned stimulus pattern, at times appearing to intend only a particular combination of stimulus aspects that has occurred in contiguity with the response, and at other times appearing to mean any combination of stimulus aspects that have occurred at *some* time (i.e., not necessarily at the same time) in contiguity with the response. Under the latter conception, which most interpretors favor, response evocation will be an increasing function of the number of stimulus aspects, or elements, in the situation that have occurred in conjunction with the response.

2. Moore and Stuttard (1979) have pointed out that the reaction seen over trials 1 to 41 in Figure 5-1 may be a species-typical greeting response elicited by the sight of the experimenters. Using an experimental setup similar to that of Guthrie and Horton, Moore and Stuttard found that cats were likely to make such rubbing contacts with the vertical releae pole when a human observer was visible but not in the absence of a human observer. This observation led Moore and Stuttard to suggest that the stereotypy reported by Guthrie and Horton is not due to the operation of Guthrie's principle of learning but rather is simply species-typical behavior elicited inadvertently by the experimenters themselves. Similarly, the rolling response seen in Figure 5-2 may also be a form of greeting in the cat. But Moore and Stuttard's interpretation does not speak to the rapid stereotyping of other behavior in the Guthrie and Horton experiment, such as the behavior seen over trials 42 to 71 of Figure 5-1, nor does it speak to the extended chains of stereotyped movements reported by Guthrie and Horton. However, we are left with the question of how much of the stereotypy in the Guthrie and Horton experiment can be legitimately attributed to a principle of postremity as opposed to species-specific processes.

Chapter 6

1. It should be noted that since Skinner published his argument regarding Sherrington's concept of the reflex, advances in electron microscopy have enabled the direct observation of the structure of the synapse. These and other advances in neurophysiology make it more difficult to maintain the position that neurophysiological conceptions of stimulus-response relationships are necessarily derivative of behavioral observation. But advances that give physical status to hypothetical neurophysiological processes or that uncover new physiological functions do not really affect the core of Skinner's argument, which is that any explanation *of behavior* in neurophysiological terms must inevitably be evaluated with reference to known facts about environment-behavior relationships, and therefore will be supplementary to those observations. (And we should note, too, that while the synapse may now be observed, the

more critical aspect of the concept of the synapse, namely, synaptic functioning, still eludes observation.)

For the sake of completeness, it might be noted that Skinner did acknowledge one contribution of Sherrington's analysis to a behavioral definition of the reflex. Skinner noted that by introducing the requirement of a central interaction between sensory and motor neurons (at the synapse), Sherrington's physiological definition of the reflex excludes such relatively uninteresting instances of S-R correlations as movements of the organism solely under the influence of mechanical forces (e.g., the movement of the paw of a dog when it is shaken), the activity of an effector in response to direct stimulation, and those correlations between discrete activities which are mediated by nonnervous mechanisms, as, for example, by hormonal action.

2. While our treatment classifies Skinner as an S-R theorist, it is important to note that he himself objected to the traditional S-R formulation on several grounds, notably on the ground that the S-R formula does not give sufficient weight to the environmental consequences of the response (Skinner, 1969). (The later portions of this chapter will make clear the basis of this objection.) However, in the general sense of attempting to understand behavior (R) as a function of stimulation (S), Skinner is certainly an S-R theorist.

3. An analysis of Skinner's superstitious behavior experiment by Staddon and Simmelhag (1971) suggests an important alternative interpretation, which is discussed in Chapter 7. The point we wish to stress here is the striking congruence Skinner found between his observations of superstitious behavior and the view that increased frequency of response is an inevitable consequence of the temporal contiguity of a response and a reinforcer.

4. The fact that it is possible to condition rate itself poses a problem for the assumption that response rate indexes response probability. One may, of course, argue that a schedule of reinforcement makes more likely a certain *organization, or pattern, of behavior* that meets the criterion of reinforcement. However, this solution raises the question of how to measure the "likelihood of occurrence" of that behavioral organization itself. One means of attacking this question is suggested by recent experiments which afford the organism a choice between two concurrently operating schedules of reinforcement. For example, a pigeon might be given access to two response keys, one of which rewards pecking on a VI 2 and the other of which rewards pecking on a VR 60, and allowed to spend as much time as it wishes on either key. In such situations it is found that the relative frequency of response on a given alternative matches the relative frequency of reinforcement for responses on that alternative. That is,

$$\frac{\text{Responses on A}}{\text{Responses on A} + \text{responses on B}} = \frac{\text{Reinforcements on A}}{\text{Reinforcements on A} + \text{reinforcements on B}}$$

This relation is called the *matching law* (Herrnstein, 1970), and there is evidence that animals achieve matching primarily in terms of differential time on the alternatives (Baum and Rachlin, 1969; Catania, 1966). Thus, time spent on equally available behavioral alternatives may provide an index of response probability that is independent of rate of responding itself.

5. Skinner early (1938) studied the case of zero correlation between bar pressing behavior and reinforcement delivery. At the same time, he considered the likely outcomes of various scheduling operations in relation to respondent conditioning, including schedules in which presentations of the unconditioned stimulus (S^1) and an initially neutral stimulus (S^0) are either negatively correlated or wholly uncorrelated. It is worth quoting Skinner in this connection. "The negative correlation of S^0 and S^1 in a respondent would have the form of the presentation of a reinforcing stimulus in the absence of S^0 and the absence of presentation in its presence. A different result is here to be expected, at least in the extreme case in which

S^0 is presented more or less continuously for short periods of time. A positive response should then appear upon cessation of S^0 (Pavlov). The case of no correlation whatsoever would have the form of random presentation of both S^0 and S^1, any coincidence being accidental. Again unlike the case of the operant, a conditioned reflex would presumably arise even if the procedure were instituted at the beginning, because S^0 would become part of the experimental situation to which conditioned value always becomes attached in conditioning of this type" (Skinner, 1938, p. 166). This analysis is particularly interesting because it anticipates the strong contemporary concern to explore the range of possible CS-UCS correlations in classical conditioning. The analysis also illustrates the thoroughness with which Skinner considered the logical end points of the basic behavior-reinforcement relation.

6. More recent studies (Azrin and Holz, 1966) have shown that under certain conditions punishment may produce long-term suppression of appetitive behavior and may also be an effective means of changing behavior when combined with reward for alternative responses.

7. It is worth noting that each of the global theorists shows the same pattern of development, namely, early expression of the central ideas followed by strong adherence to those ideas and persistent elaboration of the theoretical framework over several decades of continuing research or application. If there is a lesson here it is perhaps that establishment of a psychological system, even in days fertile to the development of general theory, requires early, major, and sustained effort.

8. Early evidence was available to Skinner that even such mild forms of social approval as an experimenter's "mm-hmm" uttered occasionally as a subject spoke was sufficient to increase the frequency of words it followed (Greenspoon, 1955).

9. The interested reader may wish to see Chomsky's review (1959) of *Verbal Behavior* and MacCorquodale's rejoinder to that review (1970).

10. For this analogy I am indebted to Dr. George L. Geis, who has developed the analogy in a number of presented papers.

11. In his later writings Skinner gave increasing recognition to the role of genetic determinants of human behavior, but as he saw little immediate possibility that behavior could be altered by genetic manipulation, he continued to stress control of environmental determinants as the means of changing human behavior.

12. It is interesting to note that in a sense every Skinner box experiment involves a latent learning procedure. This is so because in the typical training procedure the hungry animal is first exposed to the Skinner box with the bar in place, but the food magazine inoperative. This exposure is given in order to determine the preconditioned strength of the subject's bar press behavior (so-called operant level determination). During subsequent bar press acquisition, conditions are identical to the operant level determination except that bar presses produce food pellets. These two training phases are directly analogous to the nonrewarded exposure and subsequent rewarded runs, respectively, of the classic Blodgett latent learning experiment. However, comparability to the latent learning procedure is obscured by the fact that in the Skinner box procedure a period of magazine training intervenes between the simple exposure and bar press training phases. Nevertheless, it is intriguing to speculate that a "latent learning" (or latent inhibition?) effect operates to some unknown degree in Skinner box conditioning.

Chapter 7

1. The matching principle holds for outcome events which are informative in nature rather than rewarding or punishing in the traditional sense. If subjects were predicting the receipt

of different amounts of money, and were allowed to keep correctly guessed payoffs, then maximization would be the more likely finding, i.e., the subjects would be likely to consistently guess the higher payoff. Of course, this differential effect of reward magnitude can be treated within stimulus sampling theory by adding appropriate assumptions and parameters to compose a somewhat different model specific to probability learning situations with motivationally significant outcome events. Such flexibility of mathematical models is at once a strength and a weakness. On the one hand, models can be developed that do quite well in describing behavior in a specific situation; on the other hand, models tend to proliferate in response to the need to accommodate additionally discovered determinants of the phenomenon at issue.

2. Critical discussions of chimpanzee language behavior may be found in Brown (1973) and Sebeok and Umiker-Sebeok (1980). Briefer reviews of the major arguments are given in Benderly (1980), Limber (1977), and Wade (1980).

3. Our discussion of post-stimulation processing at the animal level has made reference only to the likelihood or duration of *rehearsal* of STM content. This is because rehearsal, or perseveration, being perhaps the simplest form of processing, seems appropriate to postulate in relation to animal learning. However, it is not intended to rule out the possibility of higher-order forms of post-stimulation processing at the animal level.

References

Amsel, A. The role of frustrative nonreward in noncontinuous reward situations. *Psychological Bulletin*, 1958, 55, 102–119.

Amsel, A. Frustrative nonreward in partial reinforcement and discrimination learning: Some recent history and a theoretical extension. *Psychological Review*, 1962, 69, 306–328.

Amsel, A. Partial reinforcement effects on vigor and persistence. In K.W. Spence & J.T. Spence (Eds.), *The psychology of learning and motivation: Advances in research and theory* (Vol. 1). New York: Academic Press, 1967.

Anrep, G.V. The irradiation of conditioned reflexes. *Proceedings of the Royal Society of London*, 1923, 94, Series B, 404–425.

Atkinson, R.C. An analysis of the effect of nonreinforced trials in terms of statistical learning theory. *Journal of Experimental Psychology*, 1956, 52, 28–32.

Atkinson, R.C., & Shiffrin, R.M. Human memory: A proposed system and its control processes. In K.W. Spence & J.T. Spence (Eds.), *The psychology of learning and motivation: Advances in research and theory* (Vol. 2). New York: Academic Press, 1968.

Azrin, N.H., & Holz, W.C. Punishment. In W.K. Honig (Ed.), *Operant behavior: Areas of research and application*. New York: Appleton-Century-Crofts, 1966.

Bartlett, F.C. *Remembering: A study in experimental and social psychology*. Cambridge: Cambridge University Press, 1932.

Bass, M.J., & Hull, C.L. The irradiation of a tactile conditioned reflex in man. *Journal of Comparative Psychology*, 1934, 17, 47–65.

Baum, W.H., & Rachlin, H.C. Choice as time allocation. *Journal of the Experimental Analysis of Behavior*, 1969, 12, 861–874.

Benderly, B.L. The great ape debate. *Science 80*, 1980, 1, 60–65.

Bindra, D. A unified account of classical conditioning and operant training. In A.H. Black & W.F. Prokasy (Eds.), *Classical conditioning*. New York: Appleton-Century-Crofts, 1972.

Bitterman, M.E., Reed, P.C., & Kubala, A.L. The strength of sensory preconditioning. *Journal of Experimental Psychology*, 1953, 46, 178–182.

Blodgett, H.C. The effect of the introduction of reward upon the maze performance of rats. *University of California Publications in Psychology*, 1929, 4, 113–134.

Blodgett, H.C., & McCutchan, K. Place versus response learning in the simple T-maze. *Journal of Experimental Psychology*, 1947, 37, 412–422.

Bolles, R.C. Species-specific defense reactions and avoidance learning. *Psychological Review*, 1970, 77, 32–48.

Bolles, R.C. Reinforcement, expectancy, and learning. *Psychological Review*, 1972, 79, 394–409.

Breland, K., & Breland, M. A field of applied animal psychology. *The American Psychologist*, 1951, 6, 202–204.

Breland, K., & Breland, M. The misbehavior of organisms. *The American Psychologist*, 1961, 16, 681–684.

Brogden, W.J. Sensory pre-conditioning. *Journal of Experimental Psychology*, 1939, 25, 323–332.

Brogden, W.J., Lipman, E.A., & Culler, E. The role of incentive in conditioning. *American Journal of Psychology*, 1938, 51, 109–117.

Brown, P.L., & Jenkins, H.M. Auto-shaping of the pigeon's key-peck. *Journal of the Experimental Analysis of Behavior*, 1968, 11, 1–8.

Brown, R. *A first language: The early stages*. Cambridge, Mass.: Harvard University Press, 1973.

Bugelski, R. Extinction with and without sub-goal reinforcement. *Journal of Comparative Psychology*, 1938, 26, 121–134.

Bunch, M.E. The effect of electric shock as punishment for errors in human maze learning. *Journal of Comparative Psychology*, 1928, 8, 343–359.

Bunch, M.E. Certain effects of electric shock in learning a stylus maze. *Journal of Comparative Psychology*, 1935, 20, 211–242.

Catania, A.C. Concurrent operants. In W.K. Honig (Ed.), *Operant behavior: Areas of research and application*. New York: Appleton-Century-Crofts, 1966.

Chomsky, N. Review of Skinner's *Verbal Behavior*. *Language*, 1959, 35, 26–58.

Chomsky, N. *Language and mind*. New York: Harcourt, Brace & World, 1968.

Cohen, D. *J.B. Watson: The founder of behaviorism*. London: Routledge & Kegan Paul, 1979.

Cooper, L.A., & Shepard, R.N. Chronometric studies of the rotation of mental images. In W.G. Chase (Ed.), *Visual information processing*. New York: Academic Press, 1973.

Coppock, W.J. Pre-extinction in sensory preconditioning. *Journal of Experimental Psychology*, 1958, 55, 213–219.

Cowles, J.T. Food tokens as incentives for learning by chimpanzees. *Comparative Psychological Monographs*, 1937, 14, 1–96.

Craik, F.I., & Lockhart, R.S. Levels of processing: A framework for memory research. *Journal of Verbal Learning and Verbal Behavior*, 1972, 11, 671–684.

Crowder, N.A., & Martin, G. *Trigonometry*. New York: Doubleday, 1961.

Darwin, C. *The descent of man, and selection in relation to sex*. New York: D. Appleton, 1871.

Davis, M. Effects of interstimulus interval length and variability on startle-response habituation in the rat. *Journal of Comparative and Physiological Psychology*, 1970, 72, 177–192.

Dennis, W. A comparison of the rat's first and second explorations of a maze unit. *American Journal of Psychology*, 1935, 47, 488–490.

Dennis, W., & Sollenberger, R.T. Negative adaption in the maze exploration of albino rats. *Journal of Comparative and Physiological Psychology*, 1934, 18, 197–206.

Dodson, J.D. The relative values of satisfying and annoying situations as motives in the learning process. *Journal of Comparative Psychology*, 1932, 14, 147–164.

Ebbinghaus, H. *Memory*, 1885. (Translated by H.A. Ruger & C.E. Bussenius.) New York: Teachers College, 1913. Paperback ed., New York: Dover, 1964.

Egger, M.D., & Miller, N.E. Secondary reinforcement in rats as a function of information value and reliability of the stimulus. *Journal of Experimental Psychology*, 1962, 64, 97–104.

Egger, M.D., & Miller, N.E. When is a reward reinforcing? An experimental study of the information hypothesis. *Journal of Comparative and Physiological Psychology*, 1963, 56, 132–137.

Elliott, M.H. The effect of change of reward on the maze performance of rats. *University of California Publications in Psychology*, 1928, 4, 19–30.

Estes, W.K. An experimental study of punishment. *Psychological Monographs*, 1944, 57, Whole No. 263.

Estes, W.K. Toward a statistical theory of learning. *Psychological Review*, 1950, 57, 94–107.

Estes, W.K. New perspectives on some old issues in association theory. In N. J. Mackintosh & W. K. Honig (Eds.), *Fundamental issues in associative learning*. Halifax: Dalhousie University Press, 1969.

Estes, W.K., & Skinner, B.F. Some quantitative properties of anxiety. *Journal of Experimental Psychology*, 1941, 29, 390–400.

Evans, J.T., & Hunt, J. McV. The "emotionality" of rats. *American Journal of Psychology*, 1942, 55, 528–545.

Ferster, C.B., & Skinner, B.F. *Schedules of reinforcement*. New York: Appleton-Century-Crofts, 1957.

Gardner, R.A., & Gardner, B.T. Teaching sign language to a chimpanzee. *Science*, 1969, 165, 664–672.

Gentry, G., Brown, W.L., & Lee, H. Spatial location in the learning of a multiple T-maze. *Journal of Comparative and Physiological Psychology*, 1948, 41, 312–318.

Gilbert, R.W. The effect of non-informative shock upon maze learning and retention with human subjects. *Journal of Experimental Psychology*, 1936, 19, 456–466.

Gilbert, R.W. A further study of the effect of non-informative shock upon learning. *Journal of Experimental Psychology*, 1937, 20, 396–407.

Gonzalez, R.C., & Diamond, L. A test of Spence's theory of incentive-motivation. *American Journal of Psychology*, 1960, 73, 396–403.

Greenspoon, J. The reinforcing effect of two spoken sounds on the frequency of two responses. *American Journal of Psychology*, 1955, 68, 409–416.

Guthrie, E.R. Conditioning as a principle of learning. *Psychological Review*, 1930, 37, 412–428.

Guthrie, E.R. Association as a function of time interval. *Psychological Review*, 1933, 40, 355–367.

Guthrie, E.R. *The psychology of learning*. New York: Harper & Brothers, 1935.

Guthrie, E.R. *The psychology of human conflict*. New York: Harper & Brothers, 1938.

Guthrie, E.R. Association and the law of effect. *Psychological Review*, 1940, 47, 127–148.

Guthrie, E.R. Conditioning: A theory of learning in terms of stimulus, response and association. In N.B. Henry (Ed.), *National Society for the Study of Education*. Forty-First Yearbook, Part II. Chicago: University of Chicago Press, 1942.

Guthrie, E.R. *The psychology of learning* (rev. ed.) New York: Harper & Brotheres, 1952.

Guthrie, E.R., & Horton, G.P. *Cats in a puzzle box*. New York: Rinehart, 1946.

Hall, C.S. Emotional behavior in the rat. *Journal of Comparative Psychology*, 1934, 18, 385–403.

Hefferline, R.F. An experimental study of avoidance. *Genetic Psychology Monographs*, 1950, 42, 231–334.

Heidbreder, E. *Seven psychologies*. New York: Century, 1933.

Herrnstein, R.J. On the law of effect. *Journal of the Experimental Analysis of Behavior*, 1970, 13, 243–266.

Hilgard, E.R. Methods and procedures in the study of learning. In S.S. Stevens (Ed.), *Handbook of experimental psychology*. New York: Wiley, 1951.

Hilgard, E.R., & Bower, G.H. *Theories of learning* (4th ed.). Englewood Cliffs, N.J.: Prentice-Hall, 1975.

Holland, J.G. Techniques for behavioral analysis of human observing. *Science*, 1957, 125, 348–350.

Holland, J.G., & Skinner, B.F. *The analysis of behavior: A program for self-instruction*. New York: McGraw-Hill, 1961.

Homme, L.E., deBaca, P.C., Devine, J.V., Steinhorst, R., & Rickert, E.J. Use of the

Premack principle in controlling the behavior of nursery school children. *Journal of the Experimental Analysis of Behavior,* 1963, 6, 544.

Honig, W.K., & Staddon, J.E.R. (Eds.). *Handbook of operant behavior.* Englewood Cliffs, N.J.: Prentice-Hall, 1977.

Honzik, C.H. The sensory basis of maze learning in rats. *Comparative Psychological Monographs,* 1936, 13, 113–123.

Hull, C.L. A functional interpretation of the conditioned reflex. *Psychological Review,* 1929, 36, 498–511.

Hull, C.L. Knowledge and purpose as habit mechanisms. *Psychological Review,* 1930, 37, 511–525. (a)

Hull, C.L. Simple trial-and-error learning: A study in psychological theory. *Psychological Review,* 1930, 37, 241–256. (b)

Hull, C.L. Goal attraction and directing ideas conceived as habit phenomena. *Psychological Review,* 1931, 38, 487–506.

Hull, C.L. The goal gradient hypothesis and maze learning. *Psychological Review,* 1932, 39, 25–43.

Hull, C.L. Differential habituation to internal stimuli in the albino rat. *Journal of Comparative Psychology,* 1933, 16, 255–273.

Hull, C.L. The concept of the habit-family hierarchy and maze learning: Part 1. *Psychological Review,* 1934, 41, 33–54.

Hull, C.L. The conflicting psychologies of learning—a way out. *Psychological Review,* 1935, 42, 491–516. (a)

Hull, C.L. The mechanism of the assembly of behavior segments in novel combinations suitable for problem solution. *Psychological Review,* 1935, 42, 219–245. (b)

Hull, C.L. Mind, mechanism, and adaptive behavior. *Psychological Review,* 1937, 44, 1–32.

Hull, C.L. Simple trial-and-error learning—An empirical investigation. *Journal of Comparative Psychology,* 1939, 27, 233–258.

Hull, C.L. Conditioning: Outline of a systematic theory of learning. In N.B. Henry (Ed.), *National Society for the Study of Education: Forty-First Yearbook,* Part II. Chicago: University of Chicago Press, 1942.

Hull, C.L. *Principles of behavior. An introduction to behavior theory.* New York: Appleton-Century-Crofts, 1943.

Hull, C.L. *A behavior system: An introduction to behavior theory concerning the individual organism.* New Haven, Conn.: Yale University Press, 1952.

Johnson, E.E. The role of motivational strength in latent learning. *Journal of Comparative and Physiological Psychology,* 1952, 45, 526–530.

Kamin, L.J. "Attention-like" processes in classical conditioning. In M.R. Jones (Ed.), *Miami symposium on the prediction of behavior, 1967: Aversive stimulation.* Coral Gables, Fla.: University of Miami Press, 1968.

Kamin, L.J. Predictability, surprise, attention, and conditioning. In B.A. Campbell & R.M. Church (Eds.), *Punishment and aversive behavior.* New York: Appleton-Century-Crofts, 1969.

Kanner, J.H. A test of whether the "nonrewarded" animals learned as much as the "rewarded" animals in the California latent learning study. *Journal of Experimental Psychology,* 1954, 48, 175–183.

Keller, F.S., & Schoenfeld, W.N. *Principles of psychology: A systematic text in the science of behavior.* New York: Appleton-Century-Crofts, 1950.

Kendler, H.H. A comparison of learning under motivated and satiated conditions in the white rat. *Journal of Experimental Psychology,* 1947, 37, 545–549.

Kendler, H.H., & Gasser, W.P. Variables in spatial learning: I. Number of reinforcements during training. *Journal of Comparative and Physiological Psychology,* 1948, 41, 178–187.

Kimble, G.A. *Hilgard and Marquis' conditioning and learning.* New York: Appleton-Century-Crofts, 1961.

Lashley, K.S., & Ball, J. Spinal conduction and kinesthetic sensitivity in the maze habit. *Journal of Comparative Psychology,* 1929, 9, 71–105.

Lashley, K.S., & McCarthy, D.A. The survival of the maze habit after cerebellar injuries. *Journal of Comparative Psychology,* 1926, 6, 423–433.

Law, E. *History of the Hampton Court Palace.* London: G. Bell, 1891.

Leeper, R. The role of motivation in learning: A study of the phenomenon of differential motivational control of the utilization of habits. *Journal of Genetic Psychology,* 1935, 46, 3–40.

Leitenberg, H. (Ed.), *Handbook of behavior modification and behavior therapy.* Englewood Cliffs, N.J.: Prentice-Hall, 1976.

Limber, J. Language in child and chimp? *American Psychologist,* 1977, 32, 280–295.

Lindley, S.B. The maze-learning ability of anosmic and blind anosmic rats. *Journal of Genetic Psychology,* 1930, 37, 245–267.

Loucks, R.B. An appraisal of Pavlov's systematization of behavior from the experimental standpoint. *Journal of Comparative Psychology,* 1933, 15, 1–45.

MacCorquodale, K. On Chomsky's review of Skinner's *Verbal Behavior. Journal of the Experimental Analysis of Behavior,* 1970, 13, 83–99.

MacCorquodale, K., & Meehl, P. On the elimination of cul entries without obvious reinforcement. *Journal of Comparative and Physiological Psychology,* 1951, 44, 367–371.

MacCorquodale, K., & Meehl, P.E. Edward C. Tolman. In W.K. Estes, S. Koch, K. MacCorquodale, P.E. Meehl, C.G. Mueller, W.N. Schoenfeld, & W.S. Verplanck, *Modern learning theory.* New York: Appleton-Century-Crofts, 1954.

MacFarlane, D.A. The role of kinesthesis in maze learning. *University of California Publications in Psychology,* 1930, 4, 277–305.

Maltzman, I.M. An experimental study of learning under an irrelevant need. *Journal of Experimental Psychology,* 1950, 40, 788–793.

Meehl, P., & MacCorquodale, K. A further study of latent learning in the T-maze. *Journal of Comparative and Physiological Psychology,* 1948, 41, 372–396.

Meehl, P., & MacCorquodale, K. A failure to find the Blodgett effect, and some secondary observations on drive conditioning. *Journal of Comparative and Physiological Psychology,* 1951, 44, 178–183.

Melton, A.W. Implications of short-term memory for a general theory of memory. *Journal of Verbal Learning and Verbal Behavior,* 1963, 2, 1–21.

Miles, W.R. On the history of research with rats and mazes: A collection of notes. *Journal of General Psychology,* 1930, 3, 324–336.

Miller, N.E. A reply to "sign-Gestalt or conditioned reflex?" *Psychological Review,* 1935, 42, 280–292.

Miller, N.E. Studies of fear as an acquirable drive: I: Fear as motivation and fear reduction as reinforcement in the learning of new responses. *Journal of Experimental Psychology,* 1948, 38, 89–101

Milner, B. The memory defect in bilateral hippocampal lesions. *Psychiatric Research Reports,* 1959, 11, 43–58.

Milner, B. Amnesia following operation on the temporal lobes. In O.L. Zangwill & C.W.M. Whitty (Eds.), *Amnesia.* London and Washington, D.C.: Butterworth, 1967.

Moltz, H. Latent extinction and the fractional anticipatory response mechanism. *Psychological Review,* 1957, 64, 229–241.

Moltz, H., & Maddi, S. Reduction of secondary reward value as a function of drive strength during latent extinction. *Journal of Experimental Psychology,* 1956, 52, 71–76.

Moore, B.R., & Stuttard, S. Dr. Guthrie and *Felis domesticus* or: tripping over the cat. *Science,* 1979, 205, 1031–1033.

Morgan, C.L. *Introduction to comparative psychology.* London: Walter Scott, 1894.

Murdock, B.B., Jr. The retention of individual items. *Journal of Experimental Psychology*, 1961, 62, 618–625.

Osgood, C.E. *Method and theory in experimental psychology*. New York: Oxford University Press, 1953.

Pereboom, A.C. A note on the Crespi effect. *Psychological Review*, 1957, 64, 263–264.

Peterson, L.R., & Peterson, M.J. Short-term retention of individual verbal items. *Journal of Experimental Psychology*, 1959, 58, 193–198.

Pierrel, R. A generalization gradient for auditory intensity in the rat. *Journal of the Experimental Analysis of Behavior*, 1958, 1, 303–313.

Pierrel, R., & Sherman, J.G. Barnabus, the rat with college training. In *Brown Alumni Monthly*, February 1963.

Premack, D. Toward empirical behavior laws: I. Positive reinforcement. *Psychological Review*, 1959, 66, 219–233.

Premack, D. Reversibility of the reinforcement relation. *Science*, 1962, 136, 255–257.

Premack, D. Rate differential reinforcement in monkey manipulation. *Journal of the Experimental Analysis of Behavior*, 1963, 6, 81–89.

Premack, D. Reinforcement theory. In M.R. Jones (Ed.), *Nebraska Symposium on Motivation*. Lincoln: University of Nebraska Press, 1965.

Premack, D. The education of S*A*R*A*H: A chimp learns the language. *Psychology Today*, September 1970, pp. 54–58.

Premack, D. Language in chimpanzee? *Science*, 1971, 172, 808–816.

Pressey, S.L. A simple apparatus which gives tests and scores—and teaches. *School and Society*, 1926, 23, 373–376.

Rescorla, R.A. Predictability and number of pairings in Pavlovian fear conditioning. *Psychonomic Science*, 1966, 4, 383–384.

Rescorla, R.A. Pavlovian conditioning and its proper control procedures. *Psychological Review*, 1967, 74, 71–80.

Restle, F. A theory of discrimination learning. *Psychological Review*, 1955, 62, 11–19.

Restle, F. Discrimination of cues in mazes: A resolution of the "place-vs.-response" question. *Psychological Review*, 1957, 64, 217–228.

Reynolds, B. A repetition of the Blodgett experiment on "latent learning." *Journal of Experimental Psychology*, 1945, 35, 504–516.

Romanes, G.J. *Animal intelligence*. New York: D. Appleton, 1881.

Romanes, G.J. *Mental evolution in man, origin of human faculty*. New York: D. Appleton, 1893.

Ross, R.R. Positive and negative partial-reinforcement extinction effects carried through continuous reinforcement, changed motivation, and changed response. *Journal of Experimental Psychology*, 1964, 68, 492–502.

Rumbaugh, D.M., Gill, T.V., & von Glasersfeld, E.C. Reading and sentence completion by a chimpanzee (Pan). *Science*, 1973, 182, 731–733.

Schwartz, B. *Psychology of learning and behavior*. New York: W.W. Norton, 1978.

Sebeok, T.A., & Umiker-Sebeok, D.J. (Eds.). *Speaking of apes: A critical anthology of two-way communication with man*. New York: Plenum Press, 1980.

Seidel, R.J. A review of sensory preconditioning. *Psychological Bulletin*, 1959, 56, 58–73.

Seward, J.P. An experimental study of Guthrie's theory of reinforcement. *Journal of Experimental Psychology*, 1942, 30, 247–256.

Seward, J.P. An experimental analysis of latent learning. *Journal of Experimental Psychology*, 1949, 39, 177–186.

Seward, J.P., Jones, R.B., & Summers, S. A further test of "reasoning" in rats. *American Journal of Psychology*, 1960, 73, 290–293.

Seward, J.P., & Levy, N. Sign learning as a factor in extinction. *Journal of Experimental Psychology*, 1949, 39, 660–668.

Sheffield, F.D. Avoidance training and the contiguity principle. *Journal of Comparative and Physiological Psychology*, 1948, 41, 165–177.

Sheffield, F.D., & Roby, T.B. Reward value of a non-nutritive sweet taste. *Journal of Comparative and Physiological Psychology*, 1950, 43, 471–481.

Sherrington, C.S. *The integrative action of the nervous system.* New Haven, Conn.: Yale University Press, 1906.

Shipley, W.C. An apparent transfer of conditioning. *Psychological Bulletin*, 1933, 30, 541.

Sidman, M. Avoidance conditioning with brief shock and no exteroceptive warning signal. *Science*, 1953, 118, 157–158.

Silver, C.A., & Meyer, D.R. Temporal factors in sensory preconditioning. *Journal of Comparative and Physiological Psychology*, 1954, 47, 57–59.

Skinner, B.F. On the conditions of elicitation of certain eating reflexes. *Proceedings of the National Academy of Science*, 1930, 16, 433–438.

Skinner, B.F. The concept of the reflex in the description of behavior. *Journal of General Psychology*, 1931, 5, 427–458.

Skinner, B.F. The generic nature of the concepts of stimulus and response. *Journal of General Psychology*, 1935, 12, 40–65.

Skinner, B.F. *The behavior of organisms. An experimental analysis.* New York: Appleton-Century-Crofts, 1938.

Skinner, B.F. "Superstition" in the pigeon. *Journal of Experimental Psychology*, 1948, 38, 168–172. (a)

Skinner, B.F. *Walden two.* New York: Macmillan, 1948. (b)

Skinner, B.F. Are theories of learning necessary? *Psychological Review*, 1950, 57, 193–216.

Skinner, B.F. *Science and human behavior.* New York: Macmillan, 1953.

Skinner, B.F. Critique of psychoanalytic concepts and theories. *Science Monthly*, 1954, 79, 300–305.

Skinner, B.F. A case history in scientific method. *American Psychologist*, 1956, 11, 221–233.

Skinner, B.F. The experimental analysis of behavior. *American Scientist*, 1957, 45, 343–371. (a)

Skinner, B.F. *Verbal behavior.* New York: Appleton-Century-Crofts, 1957. (b)

Skinner, B.F. Reinforcement today. *American Psychologist*, 1958, 13, 94–99. (a)

Skinner, B.F. Teaching machines. *Science*, 1958, 128, 969–977. (b)

Skinner, B.F. What is psychotic behavior? In B.F. Skinner, *Cumulative record*, New York: Appleton-Century-Crofts, 1959.

Skinner, B.F. *The technology of teaching.* New York: Appleton-Century-Crofts, 1968.

Skinner, B.F. *Contingencies of reinforcement: A theoretical analysis.* New York: Appleton-Century-Crofts, 1969.

Skinner, B.F. *Beyond freedom and dignity.* New York: Alfred A. Knopf, 1971.

Small, W. Experimental study of the mental processes of the rat. II. *American Journal of Psychology*, 1901, 12, 206–239.

Smith, S., & Guthrie, E.R. *General psychology in terms of behavior.* New York: Appleton, 1921.

Spence, K.W., Bergmann, G., & Lippitt, R. A study of simple learning under irrelevant motivational reward conditions. *Journal of Experimental Psychology*, 1950, 40, 439–551.

Spence, K.W., & Lippitt, R.O. "Latent" learning of a simple maze problem with relevant needs satiated. *Psychological Bulletin*, 1940, 37, 429 (abstract).

Spence, K.W., & Lippitt, R.O. An experimental test of the sign-Gestalt theory of trial-and-error learning. *Journal of Experimental Psychology*, 1946, 36, 491–502.

Spence, K.W., & Shipley, W.C. The factors determining the difficulty of blind alleys in maze learning by the white rat. *Journal of Comparative Psychology*, 1934, 17, 423–436.

Staddon, J.E.R., & Simmelhag, V.L. The "superstition" experiment: A reexamination of its implications for the principles of adaptive behavior. *Psychological Review*, 1971, 78, 3–43.

Stewart, C. Stewart on the early use of the rat. In W.R. Miles, On the history of research with rats and mazes: A collection of notes. *Journal of General Psychology*, 1930, 3, 324–336.

Strain, E.R. Establishment of an avoidance gradient under latent-learning conditions. *Journal of Experimental Psychology*, 1953, 46, 391–399.

Symonds, P.M. Laws of learning. *Journal of Education Psychology*, 1927, 18, 405–413.

Terry, W.S., & Wagner, A.R. Short-term memory for "surprising" vs. "expected" unconditioned stimuli in Pavlovian conditioning. *Journal of Experimental Psychology: Animal Behavior Processes*, 1975, 104, 122–133.

Thistlethwaite, D. An experimental test of a reinforcement interpretation of latent learning. *Journal of Comparative and Physiological Psychology*, 1951, 44, 431–441.

Thompson, E.S. The kangaroo rat. *Scribner's Magazine*, 1900, 27, 418–427.

Thompson, R.F. Sensory preconditioning. In R.F. Thompson & J.F. Voss (Eds.), *Topics in learning and performance*. New York: Academic Press, 1972.

Thorndike, E.L. Animal intelligence: An experimental study of the associative processes in animals. *Psychological Review*, Monograph Supplement No. 8, 1898.

Thorndike, E.L. *Animal intelligence: Experimental studies*. New York: Macmillan, 1911.

Tighe, T.J. Concept formation and art: Further evidence on the applicability of Walk's technique. *Psychonomic Science*, 1968, 12, 363–364.

Tinklepaugh, O. An experimental study of representative factors in monkeys. *Journal of Comparative Psychology*, 1928, 8, 197–236.

Tolman, E.C. Purpose and cognition: The determiners of animal learning. *Psychological Review*, 1925, 32, 285–297.

Tolman, E.C. A behavioristic theory of ideas. *Psychological Review*, 1926, 33, 352–369.

Tolman, E.C. A behaviorist's definition of consciousness. *Psychological Review*, 1927, 34, 433–439.

Tolman, E.C. *Purposive behavior in animals and man*. New York: Century, 1932.

Tolman, E.C. The acquisition of string-pulling by rats—conditioned response or sign-Gestalt? *Psychological Review*, 1937, 44, 195–211.

Tolman, E.C. The determiners of behavior at a choice point. *Psychological Review*, 1938, 45, 1–44.

Tolman, E.C. A stimulus-expectancy need-cathexis psychology. *Science*, 1945, 101, 160–166.

Tolman, E.C. Principles of purposive behavior. In S. Koch (Ed.), *Psychology: The study of a science* (Vol. 2). New York: McGraw-Hill, 1959.

Tolman, E.C., & Brunswik, E. The organism and the causal texture of the environment. *Psychological Review*, 1935, 42, 43–77.

Tolman, E.C., & Gleitman, H. Studies in learning and motivation: I. Equal reinforcement in both end-boxes followed by shock in one end-box. *Journal of Experimental Psychology*, 1949, 39, 810–819.

Tolman, E.C., & Honzik, C.H. Introduction and removal of reward and maze performance in rats. *University of California Publications in Psychology*, 1930, 4, 257–275.

Tolman, E.C., Ritchie, B.F., & Kalish, D. Studies in spatial learning: I. Orientation and the short-cut. *Journal of Experimental Psychology*, 1946, 36, 13–24. (a)

Tolman, E.C., Ritchie, B.F., & Kalish, D. Studies in spatial learning: II. Place learning versus response learning. *Journal of Experimental Psychology*, 1946, 36, 221–229. (b)

Tolman, E.C., Ritchie, B.F., & Kalish, D. Studies in spatial learning: V. Response learning versus place learning by the non-correction method. *Journal of Experimental Psychology*, 1947, 37, 285–292.

Voeks, V.W. Postremity, recency, and frequency as basis for prediction in the maze situation. *Journal of Experimental Psychology*, 1948, 38, 495–510.

Wade, N. Does man alone have language? Apes reply in riddles, and a horse says neigh. *Science*, 1980, 208, 1349–1351.

Wagner, A.R. Priming in STM: An information-processing mechanism for self-generated or retrieval-generated depression in performance. In T.J. Tighe & R.N. Leaton (Eds.), *Habituation: Perspectives from child development, animal behavior, and neurophysiology*. Hillsdale, N.J.: Erlbaum, 1976.

Wagner, A.R., Rudy, J.W., & Whitlow, J.W. Rehearsal in animal conditioning. *Journal of Experimental Psychology Monograph*, 1973, 97, 407–426.

Warden, C.J., & Aylesworth, M. The relative value of reward and punishment in the formation of a visual discrimination habit in the white rat. *Journal of Comparative Psychology*, 1927, 7, 117–128.

Washburn, M.F. *The animal mind* (3rd ed.). New York: Macmillan, 1926.

Watson, J.B. Kinaesthetic and organic sensations: Their role in the reactions of the white rat to the maze. *Psychological Review*, Monograph Supplement, No. 33, 1907.

Watson, J.B. *Behavior: An introduction to comparative psychology*. New York: Holt, 1914.

Watson, J.B. *Psychology from the standpoint of a behaviorist*. Philadelphia: Lippincott, 1919.

Watson, J.B. *Psychology from the standpoint of a behaviorist*, 2nd ed. Philadelphia: Lippincott, 1924.(a)

Watson, J.B. *Behaviorism*. New York: W.W. Norton, 1924.(b)

Watson, J.B. John B. Watson. In C. Murchison (Ed.), *A history of psychology in autobiography*. Worcester, Mass.: Clark University Press, 1936.

Watson, J.B., & Rayner, R. Conditioned emotional reactions. *Journal of Experimental Psychology*, 1920, 3, 1–14.

Wickens, D.D., & Briggs, G.E. Mediated stimulus generalization as a factor in sensory preconditioning. *Journal of Experimental Psychology*, 1951, 42, 197–200.

Williams, D.R., & Williams, H. Auto-maintenance in the pigeon: Sustained pecking despite contingent non-reinforcement. *Journal of the Experimental Analysis of Behavior*, 1969, 12, 511–520.

Wolfe, J.B. Effectiveness of token-rewards for chimpanzees. *Comparative Psychological Monographs*, 1936, 12, 1–72.

Zeaman, D. The ubiquity of novelty—familiarity (habituation?) effects. In T.J. Tighe & R.N. Leaton (Eds.), *Habituation: Perspectives from child development, animal behavior and neurophysiology*. Hillsdale, N.J.: Erlbaum, 1976.

Zeaman, D., & Radner, L. A test of the mechanisms of learning proposed by Hull and Guthrie. *Journal of Experimental Psychology*, 1953, 45, 239–244.

Zener, K. The significance of behavior accompanying conditioned salivary secretion for theories of the conditioned response. *American Journal of Psychology*, 1937, 50, 384–403.

Author Index

Subject Index